DIFFERENTIAL EQUATIONS: THEORY AND APPLICATIONS

DIFFERENTIAL EQUATIONS: THEORY AND APPLICATIONS

Maria Catherine Borres

ARCLER
PRESS

www.arclerpress.com

Differential Equations: Theory and Applications

Maria Catherine Borres

Arcler Press
2010 Winston Park Drive,
2nd Floor
Oakville, ON L6H 5R7
Canada
www.arclerpress.com
Tel: 001-289-291-7705
001-905-616-2116
Fax: 001-289-291-7601
Email: orders@arclereducation.com

ISBN: 978-1-77361-403-8 (Hardcover)

ABOUT THE AUTHOR

Catherine is currently taking up Master of Arts in Education Major in Mathematics in Philippine Normal University – Manila. She is currently working as a Content Developer for Mathematics at the Affordable Private Education Center (APEC Schools).

TABLE OF CONTENTS

PREFACE

In many respects this book is perfect book for beginners. It combines a light-hearted approach, well rounded explanations and plenty of practice opportunities. It makes an ideal companion for those students who are commencing a course in differential equation at undergraduate or at graduate level.

The study of differential equations is a wide region in pure and applied mathematics, physics, and engineering. All of these disciplines are anxious with the properties of differential cquations of different types. Pure mathematics emphasis on the life and incomparability of solutions, while applied number emphasizes the rigorous occasion of the way for approximating solutions. Differential equations is an important legislature in modeling virtually every physical, technical, or biological process, from celestial motion, to bridge design, to interactions between neurons. Differential equations such as those used to solve real-life problems may not necessarily be directly solvable, i.e. do not have closed example solutions. Instead, solution tins be approximated using numerical methods.

Many fundamental traditions of physics and courtesy tins be formulated as differential equations. In biology and economics, differential equations are used to patterns the allotment of complex systems. The mathematical opinion of differential equations first developed together with the adeptness where the equations had originated and where the results found application. However, diverse problems, sometimes originating in quite distinct scientific fields, may give rise to identical differential equations. Whenever this happens, mathematical opinion seat the equations tins be viewed as a unifying outlook sitting diverse phenomena. As an example, consider assembly of enlightenment and sound in the atmosphere, and of shaft on the surface of a pond.

CHAPTER 1

BASIC CONCEPTS OF DIFFERENTIAL EQUATIONS

CONTENTS

1.1. INTRODUCTION

Mathematical models for real world phenomena often take the form of equations relating various quantities and their rates of change (derivatives), For example, the motion of a particle involves the distances covered in time t and velocity v or acceleration a. Now the rate of change $\frac{ds}{dt}$ of s with respect to t is the velocity v and rate of change $\frac{dv}{dt}$ of velocity with respect to t is the acceleration a. A particle moving in a straight line has an equation of motion as $s = f(t)$, where t is in seconds and s is in meters. Its velocity satisfies the equation

$$v=\frac{ds}{dt}=4t^2 + 5t-3$$

This leads us to the definition of a differential equation (D.E).

A differential equation is a mathematical equation that relates some function with its derivatives. In applications, the functions typically represent physical quantities, the by – product represent their rates of change, and the equation defines an association between the two. Because such relations are exceptionally common, differential equations amusement a prominent role in many disciplines including engineering, physics, economics, and biology.

In pure mathematics, differential equations are studied from several different perspectives, mostly concerned with their solutions-the set of functions that satisfy the equation. Only the simplest differential equations are solvable by explicit formulas; however, some ownership of solutions of a given differential equation may be determined without discovery their exact form.

1.1.1. Differential Equation and their Classification

An equation involving one dependent variable and its derivatives with respect to one or more independent variables is called differential equation. For example

$$\frac{dy}{dx}+ y\cos x= \sin x$$

$$\frac{d^2y}{dx^2}+xy\left(\frac{dy}{dx}\right)^2=0$$

$$\left[1+\left(\frac{dy}{dx}\right)^2\right]^{\frac{3}{2}}=\frac{d^2y}{dx^2}$$

1.1.2. Ordinary Differential Equation (O. D. E)

A differential equation, in which ordinary derivatives of the dependent variable with respect to a single independent variable occur, is called an ordinary differential equation (O.D.E). Above equations (i) , (ii) and (iii) are examples of Ordinary Differential Equation.

Partial Differential Equation

A differential equation involving partial derivatives of the dependent variable with respect to more than one independent variable is called a partial differential equation.

$$x\frac{\partial z}{\partial x}+y\frac{\partial z}{\partial y}=nx$$

$$\frac{\partial^2 u}{\partial x^2}+\frac{\partial^2 u}{\partial y^2}+\frac{\partial^2 u}{\partial z^2}=0$$

Order of Differential Equation

The order of a differential is the order of the highest derivative that occurs in the equation.

Degree of a Differential Equation

The degree of a differential equation is the greatest exponent of the highest order derivative that appears in the equation. (The dependent variable and its derivatives should be expressed in a form free from radicals and fractions).

- $\frac{dy}{dx}+y\cos x=\sin x$ (order 1, degree 1)
- $\frac{d^2y}{dx^2}+xy\left(\frac{dy}{dx}\right)^2=0$ (order 2, degree 1)

- $\left[1+\left(\frac{dy}{dx}\right)^2\right]^{\frac{3}{2}}=\frac{d^2y}{dx^2}$ (order 2, degree 2)

exponent of $\frac{d^2y}{dx^2}$ is 2 after removing the radical by squaring both sides of the equation

- $x\frac{\partial z}{\partial x}+y\frac{\partial z}{\partial y}=nx$ (order 1, degree 1)
- $\frac{\partial^2 u}{\partial x^2}+\frac{\partial^2 u}{\partial y^2}+\frac{\partial^2 u}{\partial z^2}=0$ (order 2, degree 1)

In this chapter, we shall study ordinary differential equation only.

Recall that a function $T:U\rightarrow V$, where U,V, are vector spaces over the same field is called linear if, for $\alpha,\beta\in F, x,y\in U$, $T(\alpha x+\beta y)=\alpha\,T(x)+\beta T(y)$.

1.1.3. Linear Differential Equation

An ordinary differential equation $F\left(x\,,\,y\,,\frac{dy}{dx},\frac{d^2y}{dx^2},\ldots,\frac{d^ny}{dx^n}\right)=0$ is said to be linear if F is a linear function of the variables $x\,,\frac{dy}{dx},\frac{d^2y}{dx^2},\ldots,\frac{d^ny}{dx^n}$. A similar definition applies to partial differential equations. Thus, the general linear ordinary differential equation of order n is

$$a_0(x)\frac{d^ny}{dx^n}+a_1(x)\frac{d^{n-1}y}{dx^{n-1}}+\ldots+a_{n-1}(x)\frac{dy}{dx}+a_n(x)\,y=F(x)$$

Where $a_0(x)$ is not identically zero.

The equation $\frac{d^3y}{dx^3}+2e^x\frac{d^2y}{dx^2}+y\frac{dy}{dx}=x^3$ is not linear because of the term $y\frac{dy}{dx}$.

It should be carefully noted that in a linear ordinary differential equation

- The dependent variable y and its derivative are all of degree one.
- No product of y or any of its derivative appear.
- No transcendental function of y and / or its derivative occur.

A differential equation that is not linear is called a nonlinear differential equation. Differential equations occur in the mathematical formulation of many problems in science and engineering. Some such problems are:

- Determining the motion of projectile, rocket, satellite or planet.
- Finding the charge or current in an electric circuit.
- Study of chemical reactions.
- Determination of curves with given geometrical properties.

Solution of a Differential Equation

A solution (or integral) of a differential equation is a relation between the variables, not containing derivatives, such that this relation and the derivatives obtained from it satisfy the given differential equation identically.

For example, the equation $\frac{dy}{dx} = -\mu y$ has a solution $y = c.e^{-\mu x}$, where c is an arbitrary constant.

The differential equation $\frac{d^2y}{dx^2} + y = 0$ has solution $y = A\cos x,\ y = B\sin x$ and $y = A\cos x + B\sin x$, where $A\ and\ B$ are arbitrary constants.

A solution of a differential equation which contains as many arbitrary constants as the order of the equation is called a general solution (or integral) of the differential equation. A solution obtained from the general solution by giving particular values to the constants is called a particular solution. The graph of a particular integral is called an integral curve of the differential equation.

Formation of a Differential Equation

Given a relation $f(x, y, c_1, c_2, \ldots, c_n) = 0$ (1)

between variables x, y and containing n constants $c_1, c_2, \ldots, c_n$ it is always possible to form a differential equation of order n such that the given relation (1) is the general solution of the equation. This is done by differentiating (1) n times thereby obtaining equations and then eliminating the n constants from the original relation and n derived equations. The method is illustrated by means of examples.

Example: The equation $y = x + a$ (1)

Represents a family of parallel straight lines for different values of a. Elimination of one constant 'a' requires two equations. The second equation is obtained by differentiating (1).

Thus $\frac{dy}{dx}=1$ is the differential equation of the relation (1), with a eliminated.

Example: From the differential equation by eliminating the two constants A and B from the relation $y = A\sin x + B\cos x.$ (1)

Solution: It is clear that three equations are required to eliminate two unknowns A and B. We obtain two other needed equations by successive differentiation of (1). Thus from (1), we have

$$\frac{dy}{dx} = A\cos x - B\sin x \quad (2)$$

$$\frac{d^2y}{dx^2} = -A\sin x - B\cos x = -y \text{, using (1)} \quad (3)$$

So, $\frac{d^2y}{dx^2} + y = 0$ (4)

is the required differential equation and (1) is its general solution.

Example: Find the differential equation of all parabolas whose axes are parallel to the $y-axis.$

Solution: General equation of a parabola whose axis is parallel to the $y-axis$ is

$$y = ax^2 + bx + c. \quad (1)$$

In order to obtain its differential equation, we have to eliminate a,b,c from (1). For that, we need three more equations. Differentiating (1) successively, we have

$$\frac{dy}{dx} = 2ax + b$$

$$\frac{d^2y}{dx^2} = 2a,$$

$$\frac{d^3y}{dx^3}=0.$$

The last equation does not contain any of the constants a,b and c. Thus, $\frac{d^3y}{dx^3}=0$ is the differential equation of all parabolas whose axes are parallel to the $y-axis$.

Initial and Boundary Condition

We have observed that general solution of a differential equation contains the same number of arbitrary constants, as is the order of the equation. Sometimes we need to find the solutions of differential equations subject to supplementary conditions. Two types of conditions will be often encountered.

Initial Condition

It is often required to find the solution of a differential equation subject to certain conditions. If the conditions relate to one value of the independent variable such as $y=y_0$ at $x=x_0$ (written as $y(x_0)=y_0$) and $\frac{dy}{dx}=y'(x_0)$ at $x=x_0$, where x_0 belongs to some interval $]\alpha\ ,\beta[$ then they are called initial conditions (or one- point boundary conditions) and x_0 is called the initial point. An initial value problem consists of a differential equation (of any order) together with a collection of initial conditions that must be satisfied by the solution of the differential equation and derivatives at the initial point.

1.1.4. Boundary Conditions

The problem of finding the solution of a differential equation such that all the associated constraints relate to two different values of the independent variable is called a two-point boundary value problem (or simply a boundary value problem). The associated supplementary boundary conditions are called two-point boundary conditions.

Example: Solve $\frac{dy}{dx}=2x$ (1)

Such that $y(1)=4.$

Solution: This is an initial value problem. We note that $y=x^2+c,$ c being arbitrary constant, is the general solution of (1). Since $y(1)=4,$ we have

$y(1)=1^2+c.$

Therefore, $4=1+c$ $\quad or \quad c=3.$

Thus, $y=x^2+3$ is the solution of the initial value problem (1).

Note that the general solution represents a family of parabolas for different values of $c.$ The solution $y=x^2+3$ is a particular member of the family that passes through $(1,4).$

Example: Solve $\frac{d^2y}{dx^2}+y=0$ (1)

Subject to the conditions $y\left(\frac{\pi}{4}\right)=\sqrt{2},\ y'\left(\frac{\pi}{4}\right)=\frac{1}{\sqrt{2}}.$

Solution: Since both the conditions relate to one value of $x,$ namely $x=\frac{\pi}{4},$ this is an initial value problem. We have already noted in previous example that

$y=A\sin x+B\cos x$ (2)

is the general solution of (1). Differentiating (2) w. r. t $x,$ we get

$y'=A\cos x-B\sin x.$ (3)

By the given conditions, we have respectively from (2) and (3)

$$y\left(\frac{\pi}{4}\right)=\sqrt{2}=A\sin\frac{\pi}{4}+B\cos\frac{\pi}{4}=\frac{A}{\sqrt{2}}+\frac{B}{\sqrt{2}}$$

$$y'\left(\frac{\pi}{4}\right)=\frac{1}{\sqrt{2}}=A\cos\frac{\pi}{4}-B\sin\frac{\pi}{4}=\frac{A}{\sqrt{2}}-\frac{B}{\sqrt{2}}.$$

Hence $A+B=2$ and $A-B=1$

Solving for $A\,and\,B,$ we find that

$$A=\frac{3}{2},B=\frac{1}{2}$$

With these values of $A\,and\,B,$ the particular solution of (1) is

$$y=\frac{3}{2}\sin x+\frac{1}{2}\cos x$$

Example: Verify that $y=c_1\cos x$ and $y=c_2\sin x$ are solutions of $\frac{d^2y}{dx^2}+y=0.$ Find a particular solution of the equation satisfying the boundary conditions

$$y(0)=1\,,\ y\left(\frac{\pi}{2}\right)=2.$$

Solution: We have $y=c_1\cos x.$

Differentiating twice, we get

$$\frac{dy}{dx}=-c_1\sin x$$

$$\frac{d^2y}{dx^2}=c_1\cos x=-y \quad \text{or} \quad \frac{d^2y}{dx^2}+y=0.$$

Thus, $y=c_1\cos x$ is a solution of $\frac{d^2y}{dx^2}+y=0$ (1)

Similarly, it can be checked that $y=c_2\sin x$ is also a solution of (1)

General solution of (1) is

$$y=c_1\cos x+c_2\sin x \tag{2}$$

Applying the boundary conditions, we obtain from (2)

$y(0)=1=c_1+0$ and

$$y\left(\frac{\pi}{2}\right)=2=0+c_2.$$

Thus $c_1 = 1$ and $c_2 = 2.$ Hence the particular solution of (1) satisfying the given conditions is

$$y = \cos x + 2\sin x$$

Example: Solve the boundary value problem $\frac{d^2y}{dx^2} + y = 0,\ y(0) = 1,\ y(\pi) = 5.$

Solution: Applying the boundary conditions to the general solution $y = A\sin x + B\cos x$

of the given equation, we have

$$y(0) = 1 = B$$

$$y(\pi) = 5 = -B$$

Thus, we obtain two values of B and we are unable to determine any definite value of $A.$ Hence, the boundary value problem has no solution. It follows that a boundary value problem need not always have a solution.

In above section we classified the differential equations and saw how a differential equation can be formed by eliminating constants from a given functional equation. The problem of finding general solution (integral) of a given differential may or may not exist. Even if the integral of a given equation exists, it may not be easy to find. We shall only discuss methods of solutions of special types of differential equations.

Separable Equations

Definition. A differential equation of the type

$$F(x)G(x)dx + f(x)g(y)dy = 0 \tag{1}$$

Is called an equation with separable variables or simply a separable equation. Equation (1) may be written as

$$\frac{F(x)}{f(x)}dx + \frac{g(y)}{G(y)}dy = 0$$

which can be easily integrated.

Example: Solve $\frac{dy}{dx}=\frac{x^2}{y}$ (1)

Solution: Equation (1) can be written as

$y\,dy=x^2dx.$

Integrating both the sides, we get

$\frac{y^2}{2}=\frac{x^3}{3}+c_1$ or

$3y^2=2x^2+c$

which is the required solution of (1).

Example: Solve $\frac{dy}{dx}=\frac{1}{x\tan y}$.

Solution: The given equation is separable and can be written as

$\frac{dx}{x}=\tan y\,dy$

Integrating, we have

$\ln|x|=-\ln|\cos y|+c$ or

$\ln|x|+\ln|\cos y|=c$

i.e., $\ln|x\cos y|=c$ or

$|x\cos y|-e^c=a(say)$

$x\cos y-a$ because $a>0$.

Example: Solve $x\sin y\,dx+\left(x^2+1\right)\cos y\,dy=0$ (1)

Solution: Dividing (1) by $\left(x^2+1\right)\sin y$, we have

$\frac{x}{x^2+1}dx+\cot y\,dy=0$

Which is a separable equation. Therefore,

$$\int \frac{x\,dx}{x^2+1} + \int \cot y\,dy = c_1 = \ln|c_2|$$

i.e., $\frac{1}{2}\ln\left(x^2+1\right) + \ln|\sin y| = \ln|c_2|$

or $\ln\left(x^2+1\right) + 2\ln|\sin y| = 2\ln|c_2|.$

Now, we can write $2\ln|\sin y| = \ln|\sin y|^2 = \ln(\sin y)^2 = \ln\sin^2 y$ and $2\ln|c_2| = \ln|c_2|^2 = \ln c_2^2 = \ln c,\ \textit{where}\ c_2^2 = c > 0$

Hence $\ln(x^2+1)\sin^2 y = c$ is the required solution.

Example: Solve the initial value problem

$$\frac{dy}{dx} = \frac{2x}{y + x^2 y},\quad y(0) = -2.$$

Solution: We have

$$\frac{dy}{dx} = \frac{2x}{y\left(1+x^2\right)}$$ or

$$y\,dy = \frac{2x}{1+x^2}dx$$

integrating, we obtain

$$\frac{y^2}{2} = \ln(1+x^2) + \ln c,$$ c being a constant.

$$\frac{y^2}{2} = \ln\left(c\left(1+x^2\right)\right)$$

$$y^2 = \ln(c^2\left(1+x^2\right)^2) \qquad \text{(i)}$$

Now setting $y(0) = -2$ into (1), we have

$4 = \ln c^2\ or\ c^2 = e^4.$

So (i) becomes

$$y^2 = \ln\left(e^4 \left(1+x^2\right)^2\right)$$

which is the required solution.

1.1.4. Homogenous Equations

Definition. A function $f(x,y)$ is called homogeneous of degree n if $f(tx,ty)=t^n f(x,y)$,

where t is a nonzero real number. Thus $\sqrt{xy}$, $\dfrac{x^{10}+y^{10}}{x^2+y^2}$ *and* $\sin\left(\dfrac{x}{y}\right)$ are homogeneous function of degree 1,8 and 0 respectively.

A first order differential equation

$$\frac{dy}{dx} = f(x,y) \qquad (1)$$

is said to be homogeneous function of any degree. If (1) is written in the form $M(x,y)dx+N(x,y)dy=0$. Then it is called homogeneous functions of the same degree.

The equation $\dfrac{dy}{dx} = \ln x - \ln y + \dfrac{x+y}{x-y} = \ln\dfrac{1}{\frac{y}{x}} + \dfrac{1+\frac{y}{x}}{1-\frac{y}{x}}$ is homogenous, but $\dfrac{dy}{dx} = \dfrac{y^3+2xy}{x^2} = y\left(\dfrac{y}{x}\right)^2 + 2\left(\dfrac{y}{x}\right)$ is not homogeneous.

Theorem. A homogeneous equation $\dfrac{dy}{dx} = g\left(\dfrac{y}{x}\right)$ (1) can be transformed into a separable equation (in the variables v and x) by the substitution $y = vx$.

Proof: Put $y = vx$ into (1). Then $\dfrac{dy}{dx} = v + x\dfrac{dv}{dx}$ and (1) becomes

$$v + \frac{xdv}{dx} = g(v)$$

or

$$v - g(v) + x\frac{dv}{dx} = 0$$

or $\left[v-g(v)\right]dx+x\,dv=0.$

This equation is separable and can be solved as in the previous section.

Example: Solve $\dfrac{dy}{dx}=\dfrac{x^3+y^3}{x^2y+xy^2}$

Solution: We have $\dfrac{dy}{dx}=\dfrac{1+\left(\frac{y}{x}\right)^3}{\left(\frac{y}{x}\right)+\left(\frac{y}{x}\right)^2}$

So that the equation is homogeneous. Setting $y=vx$ into (1), we obtain

$$v+x\frac{dv}{dx}=\frac{1+v^3}{v+v^2}=\frac{1+v^3}{v(1+v)}=\frac{v^2-v+1}{v}$$

$$x\frac{dv}{dx}=\frac{v^2-v+1}{v}-v=\frac{1-v}{v}$$

or $\dfrac{v}{1-v}dv=\dfrac{dx}{x}$

or $\left(-1+\dfrac{1}{1-v}\right)dv=\dfrac{dx}{x}$

integrating, we get

$$-v-\ln|1-v|=\ln|x|+\ln|c|$$

or $ln\left|cx(1-v)\right|+v=0$

Replacing $v\,by\,\dfrac{y}{x}$ in the above equation, we have

$$\ln\left|c(x-y)\right|+\frac{y}{x}=0$$

which is the required solution.

Example: Solve the initial value problem $\left(x^2+3y^2\right)dx-2xy\,dy=0$, $y(2)=6.$

Solution: Here

$$\frac{dy}{dx}=\frac{x^3+3y^2}{2xy}=\frac{3\left(\frac{y}{x}\right)^2+1}{2\left(\frac{y}{x}\right)} \qquad (1)$$

Which shows that the equation is homogeneous. Putting $y=vx$ into (1), we have

$$v+x\frac{dv}{dx}=\frac{3v^2+1}{2v}$$

$$x\,.\frac{dv}{dx}=\frac{3v^2+1}{2v}-v=\frac{v^2+1}{2v}$$

or $\frac{2v}{1+v^2}\,dv=\frac{dx}{x}$

on integrating, we get

$$\ln\left(1+v^2\right)=\ln|x|+\ln|c|$$

or $1+v^2=|cx|$

replacing $v\,by\,\frac{y}{x}$, we obtain

$$1+\frac{y^2}{x^2}=|cx|$$

or $x^2+y^2=|cx|\,x^2$

Now if $x\geq 0$, we can write this as

$$x^2+y^2=cx^3$$

or $y^2=cx^3-x^2$

or $y=\pm\sqrt{cx^3-x^2}$

Applying the initial condition, we get

$$y(2)=6=\pm\sqrt{8c-4}$$

or $8c-4=36$ *i.e.*, $c=5$

Hence $y=\sqrt{5x^3-x^2}$ is the required solution. We take the plus sign in the radical since $y(2)$ is positive.

1.1.5. Differential Equation Reducible to Homogeneous Form

Some differential equations are not homogeneous. But they can be reduced to homogeneous form.

The differential equation $(a_1x+b_1y+c_1)dx+(a_2x+b_2y+c_2)dy=0$ is not homogeneous. But it can be reduced to a homogeneous form as illustrated below:

Case 1. If $\frac{a_1}{a_2}\neq\frac{b_1}{b_2}$, then make the transformations $x=X+h, y=Y+k.$

The given equation becomes

$$(a_1X+b_1Y+a_1h+b_1k+c_1)dX+(a_2X+b_2Y+a_2h+b_2k+c_2)dY=0 \quad (1)$$

Let $h \, and \, k$ be the solution of the system of equations

$$\left.\begin{aligned} a_1h+b_1k+c_1&=0\\ a_2h+b_2k+c_2&=0\end{aligned}\right\}$$

Then for these values of $h \, and \, k,$ (1) reduces to the homogenous equation

$$(a_1X+b_1Y)dX+(a_2X+b_2Y)dY=0$$

in the variables $X \, and \, Y.$

Case II. If $\frac{a_1}{a_2}=\frac{b_1}{b_2},$ then put $z=a_1x+b_1y$ and the given equation will reduce to a separable equation in the variable $x \, and \, z.$

Example: Solve $\frac{dy}{dx}=\frac{2y-x+5}{2x-y-4}$ (1)

Solution: Putting $x = X + h,\ y = Y + k$

So the given equation becomes

$$\frac{dY}{dX} = \frac{2Y - X + 2K - h + 5}{2X - Y + 2hX\,k - 4} \qquad (2)$$

The solution of the system of equations

$$\left.\begin{aligned} -h + 2k + 5 &= 0 \\ 2h - \quad k - 4 &= 0 \end{aligned}\right\} \text{ is } h = 1,\ k = -2.$$

For these values of $h\, and\, k$, (2) reduce to

$$\frac{dY}{dX} = \frac{2Y - X}{2X - Y} \qquad (3)$$

which is homogeneous.

Putting $Y = vX$ into (3), we have

$v + X\dfrac{dv}{dX} = \dfrac{2v - 1}{2 - v}$ which is separable.

Therefore, $\left(\dfrac{2 - v}{v^2 - 1}\right) dv = \dfrac{dX}{X}.$

Integrating, we obtain

$$\ln\left|\frac{v-1}{v+1}\right| - \frac{1}{2}\ln\left|v^2 - 1\right| = \ln|X| + \ln|c_0|$$

or $$\ln\left|\frac{v - 1}{v + 1}\right|^2 = \ln|X|^2 + \ln|c_0|^2 + \ln\left|v^2 - 1\right|$$

$$\ln\left|\frac{v-1}{v+1}\right|^2 = \ln\left\{|c_0\, X|^2 .\ \left|v^2 - 1\right|\right\}$$

$$\left|\frac{v-1}{v+1}\right|^2 - c|X|^2 .\ |v-1|\ \ |v+1|\ ,\ where\, c = c_0^2$$

or $|v - 1| = c|X|^2 .\ |v+1|^3$

Replacing $v \, by \frac{Y}{X}$, this becomes

$$\left|\frac{Y-X}{X}\right| = c|X|^2 . \left|\frac{Y+X}{X}\right|^3$$

or $|Y-X| = c|Y+X|^3$ (4)

But $X = x-1$, $Y = y+2$. Hence (4) takes the form

$$|y-x+3| = c\,|y+x+1|^3$$

which is the required solution.

Example: Solve $(2x+y+1)dx+(4x+2y-1)dy=0$ (1)

Solution: Since $\frac{a_1}{a_2} = \frac{2}{4} = \frac{b_1}{b_2} = \frac{1}{2}$, we put $2x+y=z$. then (1) becomes

$$(z+1)dx+(2z-1)(dz-2dx)=0$$

or $(z+1-4z+2)dx+(2z-1)dz=0$

i.e., $3(1-z)dx+(2z-1)dz=0$ which is separable.

Dividing by $1-z$, we have

$$3\,dx+\frac{2z-1}{1-z}dz=0$$

or $$3.dx+\left(-2+\frac{1}{1-z}\right)dz=0$$

integrating, we obtain

$$3x-2z-\ln|1-z|=c_0$$

Replacing z by $2x+y$, it reduces to

$$3x-2(2x+y)-\ln|1-2x-y|=c_0$$

or $-x-2y-\ln|2x+y-1|=c_0$

or $\ln|2x+y-1|+x+2y=c, \ \ where\, c=-c_0$
is the required solution.

Exact Equation

Definition. The expression $M(x,y)dx+N(x,y)dy$ (1)
is called an exact differential if there exists a continuously differentiable function $f(x,y)$ of two real variables x and y such that the expression equals the total differential df. We know from calculus that

$$df=\frac{\partial f}{\partial x}dx+\frac{\partial f}{\partial y}dy.$$

Thus, if (1) is exact then

$$M(x,y)=\frac{\partial f}{\partial x}=f_x \ \ and \ \ N(x,y)=\frac{\partial f}{\partial y}=f_y.$$

If (1) is an exact differential then the differential equation

$$M(x,y)dx+N(x,y)dy=0$$

is called an exact equation.

Theorem. The differential equation $M(x,y)dx+N(x,y)dy=0$ (1)
is an exact equation8.U&hK□§ if and only if $\frac{\partial M}{\partial y}=\frac{\partial N}{\partial x}$, where the functions $M(x,y)$ and $N(x,y)$ have continuous first order partial derivatives.
Proof. Suppose that the equation (1) is exact so that $M\,dx+Mdy$ is an exact differential. By definition, there exists a function $f(x,y)$ such that

$$M(x,y)=\frac{\partial f}{\partial x}=f_x \ and \ N(x,y)=\frac{\partial f}{\partial y}=f_y$$

Then, $M_y=\frac{\partial M}{\partial y}=f_{xy}-\frac{\partial^2 f}{\partial y\,\partial x}$

$$N_x=\frac{\partial N}{\partial x}=f_{yx}=\frac{\partial^2 f}{\partial x\,\partial y}$$

Since $M\ and\ N$ possess continuous first order partial derivatives, we have $f_{xy}=f_{yx}$ and, therefore $\frac{\partial M}{\partial y}=\frac{\partial N}{\partial x}$ as desired.

Solution of an Exact Equation

If $M(x,y)dx+N(x,y)dy=0$ (1)

is an exact equation, then there exists a function $f(x,y)$ such that

$$df=\frac{\partial f}{\partial x}dx+\frac{\partial f}{\partial y}dy=M(x,y)dx+N(x,y)dy$$

Therefore $\frac{\partial f}{\partial x}=M\ and\ \frac{\partial f}{\partial y}=N$.

Integrating $\frac{\partial f}{\partial x}=M$ with respect to x, we have

$$f(x,y)=\int M\,dx+h(y). \tag{2}$$

The constant of integration $h(y)$ is an arbitrary function of y since it must vanish under differentiation w.r.t x.

Differentiating (2) $w.r.t.\,y$, we get

$$\frac{\partial f}{\partial y}=\frac{\partial}{\partial y}\left(\int Mdx\right)+h'(y)$$

i.e., $\frac{\partial f}{\partial y}=N=\frac{\partial}{\partial y}\left(\int Mdx\right)+h'(y)$

so $h'(y)=N-\frac{\partial}{\partial y}\left(\int Mdx\right)$

integrating the above equation $w.r.t\ y$, we obtain $h\ and$ hence $f(x,y)=c$ is the required solution of (1).

Example: Solve $(3x^2y+2)dx+(x^3+y)dy=0$ (1)

Solution: Here $M=3x^2y+2$ and $N=x^3+y$

$$\frac{\partial M}{\partial y}=3x^2, \quad \frac{\partial N}{\partial x}=3x^2$$

Thus $\frac{\partial M}{\partial y}=\frac{\partial N}{\partial x}$ and so the equation is exact.

To find the solution of (1), we note that the left hand side of the equation is an exact differential. Therefore, there exists a function $f(x,y)$ such that

$$\frac{\partial f}{\partial x}=3x^2y+2 \qquad (2)$$

And $\frac{\partial f}{\partial y}=x^3+y \qquad (3)$

Integrating (2) $w.r.t\ x$, we have

$$f(x,y)=x^3y+2x+h(y),$$

where $h(y)$ is the constant of integration. Differentiating the above equation $w.r.t\ y$ and using (3), we obtain

$$\frac{\partial f}{\partial y}=x^3+h'(y)=x^3+y$$

or $h'(y)=y.$

Integrating, we have

$$h(y)=\frac{y^2}{2}$$

Thus, $f(x,y)=x^3y+2x+\frac{y^2}{2}$

Hence the general solution of (1) is

$$x^3y+2x+\frac{y^2}{2}-c$$

Alternative Method

Integrating (2) and (3) $w.r.t\, x\, and\, y$ respectively, we have

$f(x,y)=x^3y+2y+h(y)$

and $f(x,y)=x^3y+\frac{y^2}{2}+g(x).$

Thus, $h(y)=\frac{y^2}{2}$ and $g(x)=2x$.

So, the general solution is

$x^3y+2x+\frac{y^2}{2}=c.$

Example: Solve the initial value problem

$\left(2y\sin x\cos x+y^2\sin x\right)dx+\left(\sin^2 x-2y\cos x\right)dy=0, y(0)=3$

Solution: Here $M=2y\sin x\cos x+y^2\sin x$

and $N=\sin^2 x-2y\cos x$

$\frac{\partial M}{\partial y}=2\sin x\cos x+2y\sin x$

$\frac{\partial N}{\partial x}=2\sin x\cos x+2y\sin x$

Thus $\frac{\partial M}{\partial y}=\frac{\partial N}{\partial x}$ showing that the given equation is exact.

Hence there exists a function $f(x,y)$ such that

$$\frac{\partial f}{\partial x}=2y\sin x\cos x+y^2\sin x \quad (1)$$

$$\frac{\partial f}{\partial y}=\sin^2 x-2y\cos x \quad (2)$$

Integrating (1) $w.r.t.\ x$., we have

erentiating this equation $w.r.t\, y$ and using (2), we get

$$\sin^2 x - 2y\cos x + h'(y) = \sin^2 x - 2y\cos x$$

i.e., $h'(y) = 0$ and so $h(y) = c_1$.

Hence the general solution of the given equation is

$$f(x, y) = c_2$$

i.e., $y\sin^2 x - y^2\cos x + c_1 = c_2$

$$y.\sin^2 x - y^2.\cos x = c_2 - c_1 = c$$

Applying the initial condition that when $x = 0,\ y = 3,$ we have $-9 = c$.

Hence $y^2\cos x - y\sin^2 x = 9$ is the required solution.

1.1.6. Integrating Factors

Definition. If the differential equation $M(x, y)dx + N(x, y)dy = 0$ (1)

is not exact but when it is multiplied by a function $\mu(x, y)$ and the resulting equation $\mu(x, y)M(x, y)dx + \mu(x, y)N(x, y)dy = 0$ is exact, then $\mu(x, y)$ is called an integrating factor (I.F) of the differential equation (1). The number of integrating factors of an equation may be infinite.

We list below (without proofs) some rules to find the integrating factors of equations of special types.

Rule 1. If $M(x, y)dx + N(x, y)dy = 0$ (1)

is not exact and $\dfrac{M_y - N_x}{N} = P,$ where P is a function of x only then (1) has an integrating factor $\mu(x)$ which also depends on x. $\mu(x)$ is solution of the differential equation

$$\frac{d\mu}{dx} = P\mu$$

i.e., $(x) = \exp\int P\,dx$

Note that $M_y = \frac{\partial M}{\partial y}$, $N_x = \frac{\partial N}{\partial x}$.

Rule II. If $\frac{N_x - M_y}{M} = Q$, where Q is a function of y only, then the differential equation $Mdx + Ndy = 0$ has an integrating factor $\mu(y) = \exp \int Q\, dy$.

Rule III. If $M\, dx + Ndy = 0$ (1)

is homogeneous and $xM + yN \neq 0$, then $\frac{1}{xM + yN}$ is an I.F of (1).

Rule IV. If $M\, dx + Ndy = 0$ is of the form $y\, f(x, y)\, dx + x\, g(x, y)\, dy = 0$

and $xM - yN \neq 0$, then $\frac{1}{xM - yN}$ is an I.F of (1).

The following differential formulas are useful in the calculation of certain exact equations:

$$d\left(\frac{y}{x}\right) = \frac{x\, dy - y\, dx}{x^2}$$

$$d\left(\frac{x}{y}\right) = \frac{y\, dx - x\, dy}{y^2}$$

$$d(xy) = x\, dy + y\, dx$$

$$d\left(x^2 + y^2\right) = 2\left(x\, dx + y\, dy\right)$$

$$d\left(\ln \frac{x}{y}\right) = \frac{y\, dx - x\, dy}{xy}$$

$$d\left(arc \tan \frac{x}{y}\right) = \frac{y\, dx - x\, dy}{x^2 + y^2}$$

Example: Solve $y\, dx + \left(x^2 y - x\right) dy = 0$

Solution. The equation is not exact. Rearranging the equation, we have

$$y\,dx - x\,dy + x^2 y\,dy = 0$$

or $$\frac{y\,dx - x\,dy}{x^2} + y\,dy = 0$$

Now it is an exact equation and may be written as $-d\left(\frac{y}{x}\right) + y\,dy = 0$. Integrating, we have

$$-\frac{y}{x} + \frac{y^2}{2} = c$$

or $xy^2 - 2y = cx$ is the general solution.

Example: Solve $\left(x^2 - 2x + 2y^2\right)dx + 2xy\,dy = 0$ (1)

Solution. Here $\dfrac{M_y - N_x}{N} = \dfrac{4y - 2y}{2xy} = \dfrac{1}{x}$

Therefore, I.F $\mu(x, y)$ is the solution of $\dfrac{d\mu}{dx} = \dfrac{\mu}{x}$ or $\mu = x$ is an I.F.

Multiplying the equation by x, we have

$$\left(x^3 - 2x^2 + 2xy^2\right)dx + 2x^2 y = 0.$$

This equation is exact. So we can easily find that its solution is $\dfrac{x^4}{4} - \dfrac{2x^3}{3} + x^2 y^2 = c_0$ or $3x^4 - 8x^3 + 12x^2 y^2 = c$ is the solution of (1).

Example: Solve $dx + \left(\dfrac{x}{y} - \sin y\right)dy = 0$.

Solution. Here, by rule II,

$$\frac{Nx - M_y}{M} = \frac{\frac{1}{y} - 0}{1} = \frac{1}{y}$$

is a function of y only. Therefore, $\mu(y) = \exp\int \dfrac{dy}{y} = e^{\ln y} = y$ is an I.F. Multiplying the equation by y, we have

$$y\,dx + \left(x - y\sin y\right)dy = 0$$

or $y\,dx + x\,dy - y\sin y\,dy = 0$

or $d\left(xy\right) - y\sin y\,dy = 0.$

Integrating, we get $xy + y\cos y - \sin y = c$ which is the required solution.

Example: Solve $\left(x^2 y - 2xy^2\right)dx - \left(x^3 - 3x^2 y\right)dy = 0$ (1)

Solution: The equation is homogeneous but not exact. We have

$$xM + yN = x^3 y - 2x^2 y^2 - x^3 y + 3x^2 y^2 = x^2 y^2 \neq 0$$

So, using Rule III, $\frac{1}{xM + yN} = \frac{1}{x^2 y^2}$ is an I.F. Therefore, multiplying (1) by $\frac{1}{x^2 y^2}$, we obtain

$$\left(\frac{1}{y} - \frac{2}{x}\right)dx - \left(\frac{x}{y^2} - \frac{3}{y}\right)dy = 0$$

This equation is exact. Integrating, we get

$$\frac{x}{y} - 2\ln|x| + 3\ln|y| = c$$

is the required solution.

Note that, here the term $\frac{x}{y}$, involving both $x\,and\,y$, has been taken only once.

Example: Solve $y\left(x\,.y + 2x^2 y^2\right) + x\left(xy - x^2 y^2\right)dy = 0$ (1)

Solution: The equation is of the form $y\,f\left(xy\right)dx + x\,g\left(xy\right)dy = 0$

Now, $xM - yN = x^2 y^2 + 2x^3 y^3 - x^2 y^2 + x^3 y^3 = 3x^3 y^3 \neq 0.$

Therefore, $\frac{1}{x^3 y^3}$ is an I.F.

Multiplying (1) by $\frac{1}{x^3 y^3}$, we have

$$\left(\frac{1}{x^2 y}+\frac{2}{x}\right)dx+\left(\frac{1}{xy^2}-\frac{1}{y}\right)dy=0.$$

This is an exact equation. Integrating, we get

$$-\frac{1}{xy}+2\ln|x|-\ln|y|=c$$

as the required solution.

Linear Equations

A first order ordinary differential equation (ODE) is linear in the dependent variable y and the independent variable x if it is or can be written in the form

$$\frac{dy}{dx}+P(x)y=Q(x), \qquad (1)$$

where $P \, and \, Q$ are functions of x.

Solution of a Linear Equation.

The linear equation $\frac{dy}{dx}+P(x)=Q(x)$ may be written as

$$\left[P(x)y-Q(x)\right]dx+dy=0 \qquad (2)$$

which is of the form $M\,dx+N\,dy=0$, where $M=P(x)y-Q(x) and \, N=1.$

Now $\frac{\partial M}{\partial y}=P(x) and \frac{\partial N}{\partial x}=0$

Thus (2) is not exact unless $P(x)=0$ in which case (1) is separable. However, an integrating factor (depending only on x) of (2) may be easily found. Let $\mu(x)$ be an I.F. of (2). Then multiplying (2) by $\mu(x)$, we get

$$\left[\mu(x)P(x)y-\mu(x)Q(x)\right]dx+\mu(x)dy=0 \qquad (3)$$

Now (3) is an exact equation if and only if

$$\frac{\partial}{\partial y}\left[\mu(x)P(x)y-\mu(x)Q(x)\right]=\frac{\partial}{\partial x}\left[\mu(x)\right]$$

This condition reduces to $\mu(x)P(x)=\frac{d}{dx}\left[\mu(x)\right]$

i.e., $\mu P(x)=\frac{d\mu}{dx}$

or $\frac{d\mu}{\mu}=P(x)dx,$

Integrating, we obtain

$$\ln|\mu|=\int P(x)dx$$

or $\mu=\exp[\int P(x)dx\,]>0.$

Thus $e^{\int P\,dx}$ *or* $[\exp\int P(x)dx\,]$ is an I.F. of the linear equation (1).

Multiplying (1) by $e^{\int P\,dx}$, we have $e^{\int P\,dx}\frac{dy}{dx}+e^{\int P\,dx}P(x)y=Q(x)e^{\int P\,dx}$

or $\frac{d}{dx}\left[ye^{\int P\,dx}dx\right]=Q(x)e^{\int P\,dx}$

Integrating this we obtain the solution of (1) in the form

$$ye^{\int P\,dx}=\int[Q(x)e^{\int P\,dx}]dx+c$$

Example: Solve $(x-1)^3\frac{dy}{dx}+4(x-1)^2y=x+1 \qquad (1)$

Solution: We write the equation in the standard form

$$\frac{dy}{dx}+\frac{4}{x-1}y=\frac{x+1}{(x-1)^3}$$

Here $P(x)=\frac{4}{x-1}$. Therefore, an I.F. of (1) is

$$\exp[\int \frac{4\,dx}{x-1}] = \exp\left[\ln(x-1)^4\right] = (x-1)^4$$

Multiplying (1) by this I.F., we get

$$(x-1)^4 \frac{dy}{dx} + 4(x-1)^3 y = x^2 - 1$$

or $\frac{d}{dx}\left[\pm(\aleph 1)^4 \right] \quad {}^2 \quad 1$

integrating, we obtain

$$y(x-1)^4 = \frac{x^3}{3} - x + c$$

which is the required solution.

Example: Solve $(x+2y^3)\frac{dy}{dx} = y$

Solution: We have $\frac{dy}{dx} = \frac{y}{x+2y^3}$

This equation is clearly not linear in y. But in the first order differential equation, the roles of x and y are interchangeable in the sense that either variable may be regarded as dependent variable. Let us regard x as dependent variable and y as independent variable. The equation may be written as

$$\frac{dx}{\pm} - \frac{1}{}x = 2y^3 \qquad (1)$$

which is linear in x with $I.F = \exp\left[\int\left(-\frac{1}{y}\right)dy\right] = \exp\left[\ln\frac{1}{y}\right] = \frac{1}{y}.$

Multiplying (1) by $\frac{1}{y}$, we get

$$\frac{1}{y}\frac{dx}{dy} - \frac{1}{y^2}x = 2y$$

or $\frac{d}{dx}\left(\frac{x}{y}\right) = 2y$

Integrating, we have $\frac{x}{y} = y^2 + c$ or $x = y\left(y^2 + c\right)$ is the required solution.

Example: Solve the initial value problem

$$\frac{dr}{d\theta} + r\tan\theta = \cos^2\theta, \qquad r\left(\frac{\pi}{4}\right) = 1$$

Solution. The equation is linear in r with

$$I.F. = \exp\left[\int \tan\theta\, d\theta\right] = \exp\left[\ln\sec\theta\right] = \sec\theta,$$

Multiplying the given equation by $\sec\theta,$ we have

$$\sec\theta\frac{dr}{d\theta} + r\sec\theta\tan\theta = \cos\theta$$

$$\frac{d}{d\theta}\left[r\sec\theta\right] = \cos\theta$$

Integrating, we obtain

$$r\sec\theta = \sin\theta + c$$

Applying the initial condition, we have

$$1.\sec\left(\frac{\pi}{4}\right) = \sin\left(\frac{\pi}{4}\right) + c$$

or $c = \sqrt{2} - \frac{1}{\sqrt{2}} = \frac{1}{\sqrt{2}},$

Therefore, $r\sec\theta = \sin\theta + \frac{1}{\sqrt{2}}$

$$r = \sin\theta\cos\theta + \frac{\cos\theta}{\sqrt{2}}$$

or $2r = \sin 2\theta + \sqrt{2}\cos\theta$ is the required solution.

1.2. THE BERNOULLI EQUATION

Definition. An equation of the form $\frac{dy}{dx}+P(x)y=Q(x)y^n$ (1)
is called the Bernoulli differential equation. This equation is linear if $n=0 or 1$. If n is not zero or 1, then (1) is reducible to a linear equation.

Dividing by y^n, (1) becomes

$$y^{-n}\frac{dy}{dx}+P(x)y^{1-n}=Q(x) \quad (2)$$

In (2), put $v=y^{1-n}$ then it reduces to

$$\frac{dv}{dx}+(1-n)P(x)v=(1-n)Q(x)$$

which is linear in v.

Note. Consider the equation

$$f'(y)\frac{dy}{dx}+P(x)f(y)=Q(x)$$

Letting $v=f(y)$, this equation becomes $\frac{dv}{dx}+P(x)v=Q(x)$ which is linear in v.

Example: Solve $\frac{dy}{dx}+\frac{xy}{1-y^2}=xy^{\frac{1}{2}}$ (1)

Solution: Dividing by $y^{\frac{1}{2}}$, (1) becomes

$$y^{-\frac{1}{2}}\frac{dy}{dx}+\frac{x}{1-x^2}\ y^{\frac{1}{2}}-x \quad (2)$$

Put $y^{\frac{1}{2}}=v$

or $\frac{1}{2}y^{-\frac{1}{2}}\frac{dy}{dx}=\frac{dv}{dx}$

then (2) reduces to

$$\frac{dv}{dx}+\frac{x}{2(1-v^2)}.\ v=\frac{x}{2} \quad (3)$$

This is linear in v.

$$I.F. = \exp\left[\int \frac{x}{2\left(1-x^2\right)}\,dx\right] = \exp\left[-\frac{1}{4}\ln\left(1-x^2\right)\right] = \left(1-x^2\right)^{-\frac{1}{4}}$$

Multiplying (3) by $\left(1-x^2\right)^{-\frac{1}{4}}$, we get

$$\left(1-x^2\right)^{-\frac{1}{4}}\frac{dv}{dx} + \frac{x}{2\left(1-x^2\right)^{\frac{5}{4}}}.\, v = \frac{x}{2\left(1-x^2\right)^{\frac{1}{4}}}$$

or $$\frac{d}{dx}\left[\left(1-x^2\right)^{-\frac{1}{4}} v\right] = -\frac{1}{4}\left[-2x\left(1-x^2\right)^{-\frac{1}{4}}\right]$$

Integrating, we have

$$v\left(1-x^2\right)^{-\frac{1}{4}} = -\frac{1}{4}\frac{\left(1-x^2\right)^{\frac{3}{4}}}{\frac{3}{4}} + c$$

or $$v = c\left(1-x^2\right)^{\frac{1}{4}} - \frac{1-x^2}{3}$$

or $$y^{\frac{1}{2}} = c\left(1-x^2\right)^{-\frac{1}{4}} - \frac{1-x^2}{3}$$

is the required solution of (1).

1.2.1. Equation Solvable for p

Example: Solve $x^2p^2 + xp - y^2 - y = 0.$ (1)

Solution: we factorize the left hand side of (1) to obtain

$$\left(x^2p^2 - y^2\right) + \left(xp - y\right) = \left(xp - y\right)\left(xp + y + 1\right) = 0$$

Therefore, either

$xp - y = 0$ (2)

or $xp + y + 1 = 0$ (3)

(2) gives $x\frac{dy}{dx} = y$ or $\frac{dy}{dx} = \frac{dx}{x}$

or $y = cx$. (4)

From (3), we have $xp = -(y+1)$

or $x\frac{dy}{dx} = -(y+1)$ or $\frac{dy}{y+1} = -\frac{dx}{x}$

which yields $x(y+1) = c.$ (5)

Combining (4) and (5), the required solution of (1) is $(y - cx)(xy + x - c) = 0.$

Example: Solve $xp^3 - (x^2 + x + y)p^2 + (x^2 + xy + y)p - xy = 0$ (1)

Solution: By inspection, we find that $p - 1$ is a factor of left hand side of (1). Thus, the given equation is

$$(p-1)\left[xp^2 - (x^2 + y)p + xy\right] = 0$$

or $(p-1)(xp - y)(p - x) = 0.$

Therefore, either

$p - 1 = 0$ (2)

or $xp - y = 0$ (3)

or $p - x = 0$ (4)

From (2), we have

$\frac{dy}{dx} = 1$ or $y = x + c$ (5)

From (3), we get

$x\frac{dy}{dx} = y$ or $y = cx$ (6)

From (4), we obtain

$\frac{dy}{dx} = x$

or $y=\frac{x^2}{2}+c \quad or \;\; x^2+2(y-c)=0.$ (7)

The general solution of (1) is obtained by combining (5), (6) and (7).

Thus $(y-x-c)(y-cx)(x^2-2y+2c)=0$ is the general solution of (1).

1.2.2. Equations solvable for y

Example: Solve $y=p^2x+p$ (1)

Solution. Differentiating (1) $w.r.t.x,$ we have

$$\frac{dy}{dx}=p=p^2+2xp\frac{dp}{dx}+\frac{dp}{dx}$$

or $\frac{dp}{dx}(2px+1)+p^2-p=0$

$$\frac{dp}{dx}=\frac{-p(p-1)}{2px+1}$$

$$\frac{dx}{dp}=\frac{-2px-1}{p(p-1)}$$

i.e., $\frac{dx}{dp}+\frac{2x}{p-1}=-\frac{1}{p(p-1)}$ (2)

which is linear in x.

$$I.F.=\exp\left[\int\frac{2}{p-1}dp\right]=\exp\left[\ln(p-1)^2\right]=(p-1)^2.$$

Multiplying (2) by $(p-1)^2$ and integrating, we get

$$x(p-1)^2=c-p+\ln p \tag{3}$$

$$x=\frac{c-p+\ln p}{(p-1)^2}.$$

Substituting this value of x into (1), we have

$$y = p^2.\frac{c - p + \ln p}{(p-1)^2} + p$$

$$y(p-1)^2 = p^2(c - p + \ln p) + p \qquad (4)$$

Thus, (3) and (4) constitute the required solution of (1) with p as a parameter.

Example: Solve $y + px = p^2x^4$ (1)

Solution. From (1), we get

$$y = p^2x^4 - px$$

Differentiating the above equation $w.r.t..x,$ we have

$$\frac{dy}{dx} = p = 4x^3p^2 + 2px^4\frac{dp}{dx} - p - x\frac{dp}{dx}$$

$$2p - 4x^3p^2 - 2x^4p\frac{dp}{dx} + x\frac{dp}{dx} = 0$$

$$2p(1 - 2px^3) + x(1 - 2px^3)\frac{dp}{dx} = 0$$

$$(1 - 2px^2)\left(2p + x\frac{dp}{dx}\right) = 0$$

Hence, either $1 - 2px^3 = 0$

or $2p + x\dfrac{dp}{dx} = 0$ (2)

The equation (2) gives

$$\frac{dp}{p} + 2\frac{dx}{x} = 0$$

or $\ln p + 2\ln x = \ln c$

or $px^2 = c$ or $p = \dfrac{c}{x^2}$

Substituting this value of p into (1), we obtain $y = c^2 - \frac{c}{x}$ or $xy - c^2 x + c = 0$

which is the required solution.

We have yet to deal with relation (2). If we eliminate p from (1) and (2), we get

$$y = -\frac{1}{4x^2} \qquad (3)$$

It is easy to check that (3) is also a solution of (1) and this solution does not involve any constant.

Equation Solvable for x

Example: Solve $xp = 1 + p^3$ (1)

Solution: We have $x = \frac{1}{p} + p$ (2)

Differentiating (1) $w.r.t.y,$ we have

$$\frac{dx}{dy} = \frac{1}{p} = -\frac{1}{p^2}\frac{dp}{dy} + \frac{dp}{dx}$$

or $\frac{1}{p} = \left(1 - \frac{1}{p^2}\right)\frac{dp}{dy}$

or $1 = \left(p - \frac{1}{p}\right)\frac{dp}{dy}$

Which is a separable equation. Therefore,

$$dy = \left(p - \frac{1}{p}\right)dp$$

Integrating, we get $c + y = \frac{p^2}{2} - \ln p$

i.e., $2y = p^2 - 2\ln p - 2c$ (3)

Thus, (2) and (3) constitute the solution of (1).

1.2.3. Clairaut's Equation

Definition. The equation $y = xp + f(p),$ (1)

is known as Clairaut's equation. Differentiating both sides of (1), $w.r.t.x,$ we get

$$p = p + xp' + f'(p).p'$$

Cancelling like terms, we have

$$xp' + p'f'(p) = 0$$

or

$$p'\left[x + f'(p)\right] = 0$$

Since one of the factors must be zero, two different solution arise.

- If $p' = 0,$ then $p = c$ and substitution of this value into (1) yields the general solution

$$y = cx + f(c)$$

- If $x + f'(p) = 0,$ then $x = -f'(p)$ and (1) can be rewritten as $y = -pf'(p) + f(p)$

Thus $x \, and \, y$ are both expressed as functions of p and we obtain the parametric equations

$$\left.\begin{aligned} x &= -f'(p) \\ y &= f(p) - p\,f'(p) \end{aligned}\right\} \qquad (2)$$

of a curve, representing a solution of (1). The solution (2) is called the singular solution. This solution is not deducible from the general solution p may be eliminated between the two equations in (2) to get a relation in $x \, and \, y$ involving no constant.

Example: Find the general solution and singular solution of

$$y = xp + \frac{1}{4}p^4 \qquad (1)$$

Solution. The general solution of the equation is $y = cx + \frac{1}{4}c^4.$ Differentiating (1) $w.r.t.x,$ we have

$$p = p + xp' + p^3 p'$$

or $x = -p^3$ (2)

We eliminate p from (1) and (2) and get

$$y = x(-x)^{\frac{1}{3}} + \frac{1}{4}(-x)^{\frac{4}{3}} = -\frac{3}{4}x(-x)^{\frac{1}{3}}$$

or $64y^3 + 27x^4 = 0$ is the singular solution.

Example: Solve $x^2(y - px) = yp^2$ (1)

Solution: This is not Clairaut's equation. Let us write

$x^2 = u$ and $y^2 = v$

Then, $2x\,dx = du$ and $2y\,dy = dv$.

Hence $\frac{y}{x}\frac{dy}{dx} = \frac{dv}{du}$

$$p = \frac{dy}{dx} = \frac{x}{y}\frac{dv}{du}$$

Substituting this value of p into (1), we get

$$x^2\left(y - \frac{x^2}{y}\frac{dv}{du}\right) = \frac{x^2}{y}\left(\frac{dv}{du}\right)^2$$

or $$y^2 - x^2\frac{dv}{du} = \left(\frac{dv}{du}\right)^2$$

or $$v - u\frac{dv}{du} = \left(\frac{dv}{du}\right)^2$$

or $$v = u\frac{dv}{du} + \left(\frac{dv}{du}\right)^2 \quad (2)$$

This is Clairaut's equation. Hence its general solution is

$$v = cu + c^2$$

or $y^2 = cx^2 + c$ is the general solution of (1). Differentiating (2) $w.r.t\, u$, we get

$$\frac{dv}{du} = u\frac{d^2v}{du^2} + \frac{dv}{du} + 2\frac{dv}{du}\frac{d^2v}{du^2}$$

or $\frac{d^2v}{du^2}\left(u + 2\frac{dv}{du}\right) = 0$, then $v = cu + c^2$, which is the general solution. If $u + 2\frac{dv}{du} = 0,$ then

$$u = -2\frac{dv}{du} \tag{3}$$

And $$v = u\frac{dv}{du} + \left(\frac{dv}{du}\right)^2 \tag{4}$$

give the singular solution.

Eliminating $\frac{dv}{du}$ from (3) and (4), we obtain

$$v = -\frac{u^2}{4}$$

or $u^2 + 4v = 0$ is the singular solution.

Replacing $u, v\, by\, x^2 and\, y^2$ respectively, we have

- $y^2 - cx^2 + c^2$ as the general solution.
- $x^4 + 4y^2 = 0$ as the general solution.

1.2.4. Envelope

Let $f(x, y, c) = 0$ be a one-parameter family of curves. Suppose all members of the family of curves are drawn for various values of the parameter c arranged in order of magnitude. The two curves that correspond to two consecutive values of c will be designated as neighboring curves. The locus of the ultimate points of intersection of neighboring curves is called the envelope of the family $f(x, y, c) = 0.$

Let A, B, C represents three neighboring intersecting members of the family. Let P be the point of intersection of $A \, and \, B \, and \, Q$ be the point of intersection of $B \, and \, C$. By definition $P \, and \, Q$ are points on the envelope. Thus the curve B and the envelope have two contiguous points common, and therefore, have ultimately a common tangent. Hence B and the envelope touch each other. In the same way, we may show that the envelope touches any other member of the family. Thus, a curve E which, at each of its points, touches someone of the curves of the family is the envelope of the family.

For the family of circles with centre on the $x-axis$ and radius 1, that is, the family of circles with equation $(x-1)^2 + y^2 = 1,$

The pair of lines $y = \pm 1$ is the envelope. It is easy to see that the lines $y = \pm 1$ touch each member of the family of the circles.

Equation of the Envelope

To find an equation of the envelope of the family $f(x, y, c) = 0,$ consider two neighboring curves and

$$\left. \begin{aligned} f(x, y, c) &= 0 \\ f(x, y, c+h) &= 0 \end{aligned} \right\} \tag{1}$$

Find the intersection of these two curves and let $h \to 0.$ The point of intersection then must approach the point of contact of the curve $f(x, y, c) = 0$ with the envelope. At the point of intersection of the equation

$$\frac{f(x, y, c+h) - f(x, y, c)}{h} = 0 \tag{2}$$

is true as well as (1). Letting $h \to 0,$ we have from (2)

$$f_c(x, y, c) = 0 \tag{3}$$

and from (1)

$$f(x, y, c) = 0 \tag{4}$$

If we eliminate c from (3) and (4), we obtain an equation of the envelope. The eliminant is called the c - discriminant of the family $f(x,y,c)=0$.

The c-discriminant may contain loci other than the envelope.

Example: The family of parabolas $(x-c)^2-2y=0$ has the x-axis as envelope.

Singular Solution

Let

$$f(x,y,p)=0 \tag{1}$$

be a nonlinear first order differential equation in which the left-hand side is a polynomial in p. The general solution of this differential equation will be a one parameter family

$$f(x,y,c)=0 \tag{2}$$

The envelope E of (2) is a curve which, at each of its points, touches some one member of the family (2). At a point of contact P of the envelope and a member of (2), the values of x,y,p are the same. But the values of x,y,p for the curve at P satisfy (1). Thus, the envelope of family (2) is a solution of the differential equation (1).

Example: Consider $p^2-xp+y=0.$

This is Clairaut's equation and its general solution (replacing simply p by c) is

$$y=cx+c^2 \tag{1}$$

Now we find the envelope of the family (1)

$$f(x,y,c)\equiv y-cx+c^2=0 \tag{2}$$

$$f_c\equiv\frac{\partial f}{\partial c}=-x+2c=0 \tag{3}$$

Substituting the value of c from (3) into (2), we have

$$y-x\left(\frac{x}{2}\right)+\left(\frac{x}{2}\right)^2=0$$

i.e., $$y-\frac{x^2}{2}+\frac{x^2}{4}=0$$

or $$4y=x^2.$$

(4) is an equation of the envelope of the family (1).

We check whether (4) is a solution of the given differential equation.

Differentiating (4) $w.r.t.x,$ we get $4\frac{dy}{dx}=2x \, or \, 2p=x.$ (5)

Substituting (4) and (5) into $p^2-px+y,$ we have

$$\frac{x^2}{4}-\frac{x^2}{2}+\frac{x^2}{4}=0$$

Thus (4) is a solution of the given differential equation. This solution does not involve any constant and it cannot be derived from the general solution by giving particular values to c, such a solution is called a singular solution (S.S).

Definition A solution of a differential equation $f(x,y,p)=0$ is called a singular solution (S.S) if

- It is not derived from the general solution by giving any particular value to the arbitrary constants
- At each of its points, it is tangent to some member of the one parameter family of curves represented by the general solution.

We have seen that the envelope, if any, of the one parameter family of curves represented by the general solution of a differential equation is a singular solution. Thus the c- discriminant equation may contain singular solution, if any

The singular solution may also be obtained from the differential equation directly without finding the general solution. Since at ultimate point of intersection of neighboring curves the $p's$ for the intersecting curves become equal, and thus the locus of the points where $p's$ have equal roots will include the envelope. If we eliminate p from

$$f(x,y,p)=0 \, and \, \frac{\partial f}{\partial p}=0$$

The resulting equation is called the p-discriminant. The p-discriminant represents the locus for each point of which $f(x, y, p) = 0$ has equal roots.

If envelope of the general solution of $f(x, y, p) = 0$ exists, it will be contained in the p-discriminant. Thus, the p-discriminant equation may contain (i) singular solution (ii) Solutions that are not singular and (iii) such loci that are not solutions at all.

Example: Solve $xp^2 - 2yp + 4x = 0$ (1)

Solution: $2yp = xp^2 + 4x$

$$2y = xp + \frac{4x}{p}$$

Differentiating $w.r.t.x,$ we get

$$2p = x\frac{dp}{dx} + p + \frac{4}{p} - \frac{4x}{p^2}\frac{dp}{dx}$$

or $$p - \frac{4}{p} = \left(x - \frac{4x}{p^2}\right)\frac{dp}{dx}$$

$$\frac{p^2 - 4}{p} = \frac{x\left(p^2 - 4\right)}{p^2}\frac{dp}{dx}$$

or $$\frac{dp}{p} = \frac{dx}{x}$$

or $p = cx$ (2)

Eliminating p from (1) and (2), we obtain

$$x\left(c^2x^2\right) - 2ycx + 4x = 0$$

or $c^2x^2 - 2yc + 4 = 0$ (3)

From (3), we obtain the c - discriminant as

$4y^2 - 16x^2 = 0$ or $y^2 = 4x^2$

The p-discriminant is from (1)

$4y^2 = 16x^2$ or $y^2 = 4x^2$

Since the c-discriminant and p-discriminant are the same viz., $y^2 = 4x^2$ and it satisfies the given differential equation, it is the singular solution.

Example: Solve $x^2p^2 + yp(2x+y) + y^2 = 0$ (1)

By making substitution $y = u, xy = v$ and find the singular solutions.

Solution: Let $y = u, x = \frac{v}{y} = \frac{v}{u}$

Thus $\frac{dy}{du} = 1,\ \frac{dx}{du} = \frac{u\,.\,\frac{dv}{du} - v}{u^2}$

$$p = \frac{dy}{dx} = \frac{dy}{du}.\frac{du}{dx} = \frac{u^2}{u.\frac{dv}{du} - v}$$

Substituting into (1), we have

$$\frac{v^2}{u^2}\left[\frac{u^2}{u.\frac{dv}{du} - v}\right]^2 + \left(2v + u^2\right).\frac{u^2}{u.\frac{dv}{du} - v}$$

or
$$v^2 + \left(2v + u^2\right)\left(u\frac{dv}{du} - v\right) + \left(u.\frac{dv}{du} - v\right)^2 = 0$$

$$v^2 + 2uv\frac{dv}{du} - 2v^2 + u^3.\frac{dv}{du} - u^2v + u^2\left(\frac{dv}{du}\right)^2 - 2uv\frac{dv}{du} + v^2 = 0$$

$$v = u\frac{dv}{du} + \left(\frac{dv}{du}\right)^2$$

which is a Clairaut's equation. Its solution is

$v = cu + c^2$

i.e., $xy = cy + c^2$ (2)

is the solution of (1).

From (2), the c-discriminant is

$y^2+4xy=0$ i.e., $y(y+4x)=0$

Therefore, $y=0, y+4x=0$ (3)

Differentiating (3), we have $p=-4$. Substituting $p=-4$ into the left-hand member of (1) and using (3), we obtain

$$16x^2-4x(-4)(-2x)+(-4x)^2=32x^2-32x^2=0$$

Thus $y+4x=0$ is also a singular solution.

Example: Solve $(x^2-1)p^2-2xyp-x^2=0$ (1)

Moreover, find the singular solution, if any.

Solution: Solving the equation for y, we have

$$2y=\frac{(x^2-1)p^2-x^2}{xp}=xp-\frac{p}{x}-\frac{x}{p}.$$

Differentiating $w.r.t.x$, we get

$$2p=xp'+p-\frac{xp'-p}{x^2}-\frac{p-xp'}{p^2}, \text{ where } p'=\frac{dp}{dx}$$

or
$$p=p'\left(x-\frac{1}{x}+\frac{x}{p^2}\right)+\frac{p}{x^2}-\frac{1}{p}$$

or
$$p\left(1-\frac{1}{x^2}+\frac{1}{p^2}\right)=xp'\left(1-\frac{1}{x^2}+\frac{1}{p^2}\right)$$

Therefore, $\frac{dp}{p}=\frac{dx}{x}$ or $p=cx.$

Substituting this value of p into (1), we get

$$(x^2-1)c^2x^2-2x^2yc-x=0$$

or $(x^2-1)c^2-2yc-1=0$ (2)

as the general solution of (1).

From (2), the c-discriminant is

$$4y^2+4(x^2-1)=0$$

i.e., $x^2 + y^2 = 1$

which is also envelope of the family (2). We also note that $p - discriminat$ is

$$4x^2y^2 + 4x^2\left(x^2 - 1\right) = 0$$

i.e., $x^2 + y^2 = 1$

Since the envelope and the p-discriminant are same, the singular solution is

$$x^2 + y^2 = 1$$

Example: Find the singular solution of

$$x^3p^2 + x^2yp + a^3 = 0 \tag{1}$$

Solution: The $p - discriminant$ of (1) is

$$x^4y^2 - 4x^3a^3 = 0 \, or \, x^3\left(y^2x - 4a^3\right) = 0 \tag{2}$$

The parts of (2) that satisfy (1) are singular solutions of the given equation. From (2), we have

$$x = 0 \, or \, y^2x - 4a^3 = 0 \tag{3}$$

We rewrite (1) as

$$x^3 + x^2y\frac{dx}{dy} + a^3\left(\frac{dx}{dy}\right)^2 = 0 \tag{4}$$

Clearly, $x = 0$ satisfies (4).

Thus $x = 0$ is a singular solution of (1).

Differentiating $y^2x - 4a^3 = 0$ $w.r.t \, x,$ we have

$$2xyp + y^2 = 0 \, or \, y\left(2xp + y\right) = 0$$

i.e., $p = -\dfrac{y}{2x}.$

Putting this value of p into the left-hand member of (1) and using (3), we obtain

$$x\left(-\frac{y}{2x}\right)^2+x^2y\left(-\frac{y}{2x}\right)+a^3=-\frac{xy^2}{4}+a^3=-\left(xy^2-4a^3\right)=0$$

Thus $xy^2-4a^3=0$ and its derivative satisfy (1) and so it is a solution of (1). Hence $x=0\,and\;xy^2-4a^3=0$ are singular solutions of (1).

1.2.5. The Ricatti Equation

Definition. We have already studied first order linear differential equation

$$y'+P(x)y=R(x) \qquad (1)$$

If we add the term $q(x)y^2$ to the left-hand member of (1), we obtain a nonlinear differential equation

$$y'+P(x)y+Q(x)y^2=R(x) \qquad (2)$$

(2) is called the Ricatti equation.

Note: In many cases, the solution of (2) cannot be expressed in terms of elementary functions. However, the Ricatti equation $y'+Py+Qy^2=R$ (1) can be reduced to a linear equation by the substitution $y=y_1+\frac{1}{u}$, where y_1 is a particular solution of (1) and u is an unknown nonzero function of x.

Proof: Let $y=y_1+\frac{1}{u}$ be as given. Differentiating $w.r.t.x$, we have

$$y'=\frac{dy}{dx}=\frac{dy_1}{dx}-\frac{1}{u^2}\frac{du}{dx}$$

Substituting for $y\,and\,y'$ into (1), we get

$$\frac{dy_1}{dx}-\frac{1}{u^2}\frac{du}{dx}+P\left(y_1+\frac{1}{u}\right)+Q\left(y_1^2+\frac{2y_1}{u}+\frac{1}{u^2}\right)=R$$

or
$$\frac{dy_1}{dx}+Py_1+Qy_1^2-R-\frac{1}{u^2}\left(\frac{du}{dx}-Pu-2Qy_1u-Q\right)=0 \qquad (2)$$

Since y_1 is a solution of (1), we have

$$\frac{dy_1}{dx}+Py_1+Qy_1^2-R=0$$

and so (2) reduces to $\frac{du}{dx}-(P+2Qy_1)u=Q$ which is a linear equation. Thus if a particular solution of (1) is known then its general solution can be found.

Example: Solve $\frac{dy}{dx}-y^2=-1,\ y(0)=3,$ given that $y_1=1$ is a particular solution of the given equation.

Solution: Here $P=0, Q=-1, R=-1.$

Writing $y=1+\frac{1}{u},$ the given equation reduces to

$$\frac{du}{dx}-(0+2(-1)(1))=-1$$

i.e., $$\frac{du}{dx}+2u=-1, \quad (1)$$

which is a linear equation.

I.F. of (1) is $e^{\int 2\,dx}=e^{2x}.$

Multiplying (1) by the I.F., we get

$$\frac{du}{dx}e^{2x}+2u\,e^{2x}=-e^{2x}$$

or $$\frac{d}{dx}\left(ue^{2x}\right)=-e^{2x}$$

Integrating, we obtain

$$u\,e^{2x}=-\int e^{2x}\,dx+c=-\frac{e^{2x}}{2}+c$$

or $$u=-\frac{1}{2}+\frac{c}{e^{2x}} \quad (2)$$

From $y=1+\frac{1}{u},$ we get by the initial condition,

$$y(0)=3=1+\frac{1}{u(0)}\ or\, u(0)=\frac{1}{2}$$

Hence from (2), we have

$$u(0)=\frac{1}{2}=-\frac{1}{2}+c \, or \, c=1$$

Thus $$u=-\frac{1}{2}+\frac{1}{e^{2x}}=\frac{2-e^{2x}}{2e^{2x}}$$

Required solution is $$y=1+\frac{1}{u}=1+\frac{2e^{2x}}{2-e^{2x}}=\frac{2+e^{2x}}{2-e^{2x}}.$$

Example: Solve $$\frac{dy}{dx}-\frac{y}{x}-x^3y^2=-x^5, \qquad (1)$$
by finding a particular solution.

Solution: It is a Ricatti equation with $P=-\frac{1}{x}, Q=-x^3 \text{ and } R=-x^5.$

An obvious solution of (1) is $y_1=x.$ Substituting $y=x+\frac{1}{u}$ into (1), we have

$$\frac{du}{dx}-\left[-\frac{1}{x}+2\left(-x^3\right)x\right]u=-x^3$$

or $$\frac{du}{dx}+\left(\frac{1}{x}+2x^4\right)u=-x^3, \qquad (2)$$
which is a linear equation.

Its $I.F.=e^{\int\left(\frac{1}{x}+2x^4\right)dx}=e^{\ln x}.e^{\frac{2}{5}x^5}=xe^{\frac{2}{5}x^5}$

Multiplying (2) by the I.F. and integrating, we have

$$uxe^{\frac{2}{5}x^5}=-\int x^4.e^{\frac{2}{5}x^5}dx+c'=-\frac{1}{2}e^{\frac{2}{5}x^5}+c'$$

or $$u=\frac{c'-\frac{1}{2}e^{\frac{2}{5}x^5}}{xe^{\frac{2}{5}x^5}}$$

Hence the general solution of (1) is

$$y = x + \frac{x e^{\frac{2}{5}x^5}}{c' - \frac{1}{2}e^{\frac{2}{5}x^5}} = \frac{x\left(c' + \frac{1}{2}e^{\frac{2}{5}x^5}\right)}{c' - \frac{1}{2}e^{\frac{2}{5}x^5}}$$

or

$$\frac{y}{x} = \frac{c' + \frac{1}{2}e^{\frac{2}{5}x^5}}{c' - \frac{1}{2}e^{\frac{2}{5}x^5}}$$

or $\frac{y-x}{y+x} = \frac{e^{\frac{2}{5}x^5}}{2c'} = ce^{\frac{2}{5}x^5}$ is the required solution.

1.3. DIFFERENTIAL EQUATIONS OF HIGHER ORDER

1.3.1. Introduction

In previous sections, we studied methods of solving special types of linear and nonlinear differential equations of the first order. In this chapter, we shall consider systematic methods for the solution of certain classes of differential equations of order more than one.

1.3.2. Linear Differential Equations

Definition. A linear differential equation of order n in the dependent variable y and the independent variable x is of the form

$$a_0(x)\frac{d^n y}{dx^n} + a_1(x)\frac{d^{n-1}y}{dx^{n-1}} + \ldots + a_{n-1}\frac{dy}{dx} + a_n(x)y = F(x), \quad (1)$$

Where $a_0(x), a_1(x), \ldots, a_{n-1}(x), a_n(x) and F(x)$ are functions of the independent variable x only and $a_0(x)$ is not identically zero. Using primes, (1) is also written as

$$a_0 y^n + a_1 y^{n-1} + \ldots + a_{n-1}y' + a_n y = F(x) \quad (2)$$

where $a_0, a_1, \ldots, a_{n-1}, a_n$ are real constants.

Equation (1) is with variable coefficients while (2) is with constant coefficients. In order to solve (2), we shall first consider the equation

$$a_0 y^n + a_1 y^{n-1} + \ldots + a_{n-1} y' + a_n y = 0 \tag{3}$$

The coefficient of y^n may be made 1 by dividing throughout by a_0. The differential equation (3) is called homogeneous linear differential equation of order n. The use of the word homogeneous here is quite different from the one already mentioned in previous chapter. If $F(x)$ is not identically zero then (2) is called non - homogeneous and (3) is called the associated homogeneous equation of (2).

Definition. If $y_1(x), y_2(x), \ldots, y_m(x)$ are m functions of an independent variable x and $c_1, c_2, \ldots, c_m$ are constants, then the expression $c_1 y_1(x) + c_2 y_2(x) + \ldots + c_m y_m(x)$ is called a linear combination of $y_1(x), y_2(x), \ldots, y_m(x)$. We usually write $y_1, y_2, \ldots, y_m$ instead of $y_1(x), y_2(x), \ldots, y_m(x)$ when it is clear from the context that $y_1, y_2, \ldots, y_m$ are functions of x.

As in vector spaces, the m nonzero functions $y_1, y_2, \ldots, y_m$ are called linearly dependent if and only if, there exist constants $c_1, c_2, \ldots, c_m$ at least one of which is nonzero, such that

$c_1 y_1 + c_2 y_2 + \ldots + c_m y_m = 0.$

The functions $y_1, y_2, \ldots, y_m$ are called linear independent if and only if, they are not linearly dependent, i.e., if and only if $c_1 y_1 + c_2 y_2 + \ldots + c_m y_m = 0$

Implies $c_1 = c_2 = \ldots = c_m = 0.$

Now before investigating a solution of (2), we state the following facts:

i. Every homogeneous linear $nth - order$ differential equation $a_0 y^n + a_1 y^{n-1} + \ldots + a_{n-1} y' + a_n y = 0$ has n linearly independent solutions $y_1, y_2, \ldots, y_n$

ii. If $y_1, y_2, \ldots, y\_m$ are n linealy independent solutions of (3), then any linear combination $y_c = c_1y_1 + c_2y_2 + \ldots + c_ny_n$ of $y_1, y_2, \ldots, y_n$ is the general solution of (3), $c_1, c_2, \ldots, c_m$ being arbitrary constants.

iii. Let y_p be any particular solution of (2). i.e., y_p does not contain any constant, then $y_c + y_p$ is the general solution of (2).

(ii) and (iii) can be easily checked by actual substitutions into (3) and (2). Thus, to find the general solution of (2), we have to find a linearly independent set of n solutions of (3). So as to determine y_c and a particular solution y_p of (2) and then obtain $y = y_c + y_p$ (4)

as the general solution of (2). In the general solution (4), y_c is called the Complementary Function (C.F) and y_p is called the Particular Integral (P.I) of (2). These statements are also true for the equation (1) with variable coefficients.

1.3.3. Homogeneous Linear Equation

Consider the equation $a_0\frac{d^ny}{dx^n} + a_1\frac{d^{n-1}y}{dx^{n-1}} + \ldots + a_{n-1}\frac{dy}{dx} + a_n\,y, = 0$ (1)

where $a_0, a_1, \ldots, a_{n-1}$ are real constants. To find a solution of (1) we shall try a judicious guess. The differential equation (1) requires a function y with the property that if y and its successive derivatives are each multiplied by constants $a_j, j = n, n-1, \ldots, 1, 0,$ the resulting products are then added; the sum should equal zero. This can only happen if a function is such that its various derivatives are constant multiples of itself. The exponential function $y = e^{mx}$, m being a constant, has such properties. Here we have

$$\frac{dy}{dx} = me^{mx}, \frac{d^2y}{dx^2} = m^2e^{mx}, \ldots, \frac{d^ny}{dx^n} = m^ne^{mx}.$$

Substituting into (1) we have

$$a_0m^ne^{mx} + a_1m^{m-1}e^{mx} + \ldots + a_{n-1}me^{mx} + a_ne^{mx} = 0$$

or $e^{mx}\,(a_0m^n + a_1m^{n-1}e^{mx} + \ldots + a_{n-1}me^{mx} + a_ne^{mx} = 0$

Since $e^{mx} \neq 0,$ we have

$$a_0 m^n + a_1 m^{m-1} + \ldots + a_{n-1} m + a_n = 0 \qquad (2)$$

Thus, $y = e^{mx}$ is a solution of (1) if and only if m is a solution of (2). Equation (2) is called the characteristics (or auxiliary) equation of the given differential equation (1). Observe that (2) can be obtained from (1) by merely replacing the kth derivative in (1) by $m^k \left(k = n, n-1, \ldots, 3, 2, 1\right)$. Three cases arise according as the roots of (2) are

i. Real and distinct
ii. Real and repeated
iii. Complex

Case I. Distinct Real Roots

Let $m_1, m_2, \ldots, m_n$ be n distinct real roots of (2). Then $e^{m_1 x}, e^{m_2 x}, \ldots, e^{m_n x}$ are n distinct solutions of (1). These n solutions are linearly independent. Hence the general solution of (2) is $y = c_1 e^{m_1 x} + c_2 e^{m_2 x} + \ldots + c_n e^{m_n x}$, where $c_1, c_2, \ldots, c_n$ are arbitrary constants.

Case II. Repeated Real Roots

In equation (1), writing $D \equiv \dfrac{d}{dx}$, we have

$$\left(a_0 D^n + a_1 D^{n-1} + \ldots + a_{n-1} D + a_n\right) y = 0 \quad \text{or} \quad \left[f\left(D\right)\right] y = 0,$$

where $$f\left(D\right) = a_0 D^n + a_1 D^{n-1} + \ldots + a_{n-1} D + a_n$$

Note that if we write D for m in the characterstic equation (2) then it is the same as $f\left(D\right) = 0.$ If $m_1, m_2, \ldots, m_n$ are the roots of $f\left(D\right) = 0,$ then (1) may be written as

$$\left(D - m_1\right)\left(D - m_2\right)\ldots\left(D - m_n\right) y = 0.$$

Let the root be repeated twice say $m_2 = m_1.$

The part of the general solution of (1) corresponding to the twice repeated root m_1 of (2) is solution of $\left(D - m_1\right)^2 y = 0$

i.e., of $(D-m_1)(D-m_1)y=0$ (3)

Let $(D-m_1)y=V$.

Then (3) becomes

$$(D-m_1)V=0$$

or $\frac{dV}{dx}-m_1V=0$ (4)

Separating the variables, we have

$$\frac{dV}{V}=m_1dx.$$

Therefore, in $V=m_1x+k,$ where k is a constant.

$V=c_2e^{m_1x}$, where c_2 is a constant.

Replacing V in (4), we obtain

$$(D-m_1)y=c_2e^{m_1x}$$

$$\frac{dy}{dx}-m_1y=c_2e^{m_1x}$$ (5)

Which is a linear equation of order one. Its I.F. $=\exp\left(\int(-m_1)dx\right)=e^{-m_1x}$

Multiplying (5) by e^{-m_1x}, we get

$$\frac{d}{dx}\left(ye^{-m_1x}\right)=c_2$$

$$ye^{-m_1x}=c_2x+c_1$$

$y=(c_1+c_2x)e^{m_1x}$ is the part of the general solution corresponding to the twice repeated m_1. The general solution of (1) is

$$y=(c_1+c_2x)e^{m_1x}+c_3e^{m_3x}+\ldots+c_ne^{m_nx}$$

In the same manner, if the characteristic equation (2) has the triple repeated root m_1, the corresponding part of the general solution of (1) is the solution of

$$(D-m_1)^3 y = 0$$

Proceeding as before, we can easily find

$$y = \left(c_1 + c_2 x + c_3 x^2\right) e^{m_1 x}$$

As the part of the general solution corresponding to this triple repeated root m_1. If the characteristics equation (2) has the real root m_1 occurring k times then the part of the general solution of (1) corresponding to the $k-fold$ repeated root m_1 is

$$y = \left(c_1 + c_2 x + c_3 x^2 + \ldots + c_k x^{k-1}\right) e^{m_1 x}$$

Case III. Complex Roots

Suppose the characteristic equation has the complex number $a+ib$ as a non- repeated root. Since coefficients of (1) are real, the conjugate complex number $a-ib$ is also a non- repeated root. The corresponding part of the general solution is

$y = k_1 e^{(a+ib)x} + k_2 e^{(a-ib)x}$, where $k_1\ and\ k_2$ are arbitrary constants.

$$y = e^{ax}\left[k_1 e^{ibx} + k_2 e^{-ibx}\right]$$

$$= e^{ax}\left[k_1(\cos bx + i\sin bx) + k_2(\cos bx - i\sin bx)\right]$$

$$= e^{ax}\left[(k_1+k_2)\cos bx + i(k_1-k_2)\sin bx\right]$$

$$= e^{ax}\left[c_1 \sin bx + c_2 \cos bx\right]$$

where $c_1 = i(k_1 - k_2), c_2 = k_1 + k_2$ are two arbitrary constants.

If $a+ib\, and\, a-ib$ are conjugate complex roots, each repeated k times, the the corresponding part of the general solution of (1) may be written as

$$y = \left[\left(c_1 + c_2 x + c_3 x^2 + \ldots + c_k x^{k-1}\right)\sin bx + \left(c_{k+1} + c_{k+2} x + \ldots + c_{2k} x^{k-1}\right)\cos bx\right]$$

In the examples given below we use these concepts to solve homogeneous linear differential equations.

Example: Solve $\left(D^2 + 4D + 3\right)y = 0$

Solution: The characteristic equation is $D^2 + 4D + 3 = 0$ with roots $D = -1, -3$.

Hence the general solution of the given equation is $y = c_1 e^{-x} + c_2\, e^{-3x}$.

Example: Solve $\left(D^3 - 5D^2 + 7D - 3\right)y = 0$.

Solution. The characteristic equation is $D^3 - 5D^2 + 7D - 3 = 0$ (1)

By inspection, $D = 1$ is a solution of this equation. The other two roots can be found from the quadratic factor of $(D-1)\left(D^2 - 4D + 3\right) = 0$

Hence all roots of (1) are $D = 1, 1, 3$.

The general solution is $y = \left(c_1 + c_2 x\right)e^x + c_3 e^{3x}$.

Example. Solve $\left(D^3 - D^2 + D - 1\right)y = 0$.

Solution. The characteristic equation is $D^3 - D^2 + D - 1 = 0$.

By inspection, $D = 1$ is a root of this equation.

Now, $D^3 - D^2 + D - 1 = (D-1)\left(D^2 + 1\right) = 0$.

Hence the other two roots are $D = \pm i = a + ib\, with\, a = 0, b = 1$.

The general solution is

$$y = c_1 e^x + e^{0x}\left(c_2 \sin x + c_3 \cos x\right)$$

$$y = c_1 e^x + c_2 \sin x + c_3 \cos x.$$

Example: Solve $\left(D^2 + D - 12\right)y = 0,\ \ where\, y(2) = 2, y'(2) = 0.$

Solution. The characteristic equation is $D^2+D-12=0$ with roots $D=3,-4$.

Hence the general solution of the given equation is $y=c_1e^{3x}+c_2e^{-4x}$ (1)

We now determine the co-efficient (constants) c_1,c_2 in (1) from the initial conditions as follows. Since the initial conditions are given at $x=2$, we rewrite the general solution in the form

$$y=k_1e^{3(x-2)}+k_2\,e^{-4(x-2)} \qquad (2)$$

where $k_1=c_1e^6,k_2=c_2e^{-8}$.

Applying initial condition, (i.e. replacing c_1,c_2 by the new constants $k_1\,e^{6},k_2\,e^{8}$ respectively in (1), we get

$$y(2)=2=k_1+k_2 \qquad (3)$$

Differentiating (2) $w.r.t.x$, we have

$$y'=\frac{dy}{dx}=3k_1e^{3(x-2)}-4k_2e^{-4(x-2)}$$

Therefore,

$$y'(2)=0=3k_1-4k_2 \qquad (4)$$

Solving (3) and (4), we obtain $k_1=\frac{8}{7},k_2=\frac{6}{7}$.

Hence the solution (2) of the differential equation satisfying the given conditions is

$$y=\frac{8}{7}e^{3(x-2)}+\frac{6}{7}e^{-4(x-2)}.$$

Example. Solve $\left(D^2+4D+5\right)=0,y(0)=1,y'(0)=0.$

Solution. The characteristic equation is $D^2+4D+5=0$ which has roots $D=-2\pm i$.

The general solution is $y=e^{-2x}\left(c_1\sin x+c_2\cos x\right).$ (1)

Applying the given conditions, we have from (1)

$y(0)=1=c_2$

Differentiating (1) $w.r.t.x,$ we get

$y'=-2e^{-2x}\left(c_1 \sin x+c_2 \cos x\right)+e^{-2x}\left(c_1 \cos x-c_2 \sin x\right)$

Therefore,

$y'(0)=0=-2c_2+c_1,$ giving $c_1=2.$

Substituting the values of $c_1\, and\, c_2$ into (1), the required solution is

$y=e^{-2x}\left(2\sin x+\cos x\right).$

Example. Solve $\left(D^3-3D^2+4\right)y=0, y(0)=1, y'(0)=-8, y''(0)=-4.$

Solution. The characteristic equation is $D^3-3D^2+4=0.$

$(D+1)\left(D^2-4D+4\right)=0.$

Therefore, $D=-1,2,2$

The general solution is $y=c_1e^{-x}+e^{2x}\left(c_2+c_3x\right)$ (1)

Now, from (1) $y'=-c_1e^{-x}+2e^{2x}\left(c_2+c_3x\right)+c_3e^{2x}$

$y''=c_1e^{-x}+4e^{2x}\left(c_2+c_3x\right)+4c_3e^{2x}$

Applying initial conditions, we get from the above three equations

$y(0)=1=c_1+c_2$ (2)

$y'(0)=-8=-c_1+2c_2+c_3$ (3)

$y''(0)=-4=c_1+4c_2+4c_3$ (4)

Multiplying (3) by -4 and adding to (4), we have

$28=5c_1-4c_2$ (5)

Multiplying (2) by 4 and adding to (5), we obtain

$9c_1=32\, or\, c_1=\frac{32}{9}.$

Therefore, $c_2 = 1 - \frac{32}{9} = -\frac{23}{9}$ and $c_3 = \frac{6}{9}$.

The required solution is

$$y = \frac{32}{9}e^{-x} + e^{2x}\left(-\frac{23}{9} + \frac{6}{9}x\right) = \frac{1}{9}\left(32e^{-x} - 23e^{2x} + 6xe^{2x}\right).$$

Differential Operators

Let $T: V \to V$, where V is a vector space over a field F, be a linear transformation (or an operator) from V to V. Linear transformation S, T, U defined on V have the following properties.

i. $S + T = T + S$ where $(S+T)v = S(v) + T(v)$, $v \in V$,

$(S+T)+U = S+(T+U)$

$S(T+U) = ST + SU$

$(ST)U = S(TU)$

ii. $ST \neq TS$ in general.

Here $ST: V \to V$ is a linear transformation defined by: $(ST)(v) = S(Tv)$ for all $v \in V$.

Let X be the vector space of all real (or complex) valued real (or complex) functions possessing a kth order derivative for every $k = 1, 2, 3, \ldots$ For each $g \in X$, $y = g(x) \in R(or \in C)$. D^k which associates with each $y = g(x)$, its kth derivative, is a linear operator on X. We write $D^k y$ as $\frac{d^k y}{d^k x}$. Since the sum and product of the linear operator and scalar multiple of a linear operator are also linear operators, the linear combination

$$f(D) = a_0 D^n + a_1 D^{n-1} + \ldots + a_{n-1} D + a_n I \tag{1}$$

Of $D^n, D^{n-1}, \ldots, D, I$ is also a linear operator. Here, I is the identity linear operator defined by $I(y) = y$ for all $y = g(x)$ and $a_0, a_1, \ldots, a_n$ are scalars. The image of a $y = g(x)$ under $f(D)$ is written as

$$f(D)y = a_0 D^n y + a_1 D^{n-1} y + \ldots + a_{n-1} Dy + a_n y$$

$$= a_0 \frac{d^n y}{dx^n} + a_1 \frac{d^{n-1} y}{dx^{n-1}} + \ldots + a_{n-1} \frac{dy}{dx} + a_n y.$$

The equation $f(D)y = 0.$ is then a linear differential equation of order n.

The operator $D, D^2, \ldots, D^n \; and \; f(D),$ given by (1), are called differential operators. We have written $D \; for \frac{d}{dx}, D^2 \; for \frac{d^2}{dx^2}$ and so on $D^n \; for \frac{d^n}{dx^n}.$ Thus $Dy = \frac{dy}{dx}, D^2 y = \frac{d^2 y}{dx^2}$ and $D^n y = \frac{d^n y}{dx^n}.$

The expression $A = f(D) = a_0 D^n + a_1 D^{n-1} + \ldots + a_{n-1} D + a_n I$ is called a differential operator of order n. This may be thought of as an operator which, when applied to any function y, result in

$$Ay = f(D)y = a_0 \frac{d^n y}{dx^n} + a_1 \frac{d^n - 1y}{dx^{n-1}} + \ldots + \frac{a_{n-1} dy}{dx} + a_n Iy.$$

Nonhomogeneous linear Equations

In this section we discuss the solution of nonhomogeneous linear equation $a_0 y^n + a_1 y^{n-1} + \ldots + a_{n-1} y' + a_n y = F(x)$.

Solution of the equation

$$\boldsymbol{f(D)y = F(x)} \qquad (1)$$

where $f(D) = a_0 D^n + a_1 D^{n-1} + \ldots + a_{n-1} D + a_n.$

We have already discussed that the general solution of (1) consist of two parts, namely

i. Complementary Function (C.F)
ii. Particular Solution (P.I)

The C.F is the solution of the homogeneous equation $f(D)y=0$ and we have described different cases of its solution. Now to find the $P.I$ of (1), we can write

$y=\frac{1}{f(D)}F(x)$

And try to evaluate $y=\frac{1}{f(D)}F(x).$

A real number m is said to be a zero of $f(D)$ if $f(m)=0.$

Suppose $f(D)$ has n distinct zeros $m_1, m_2, \ldots, m_n.$

Then $\frac{1}{f(D)}F(x)=\frac{1}{(D-m_1)(D-m_2)\ldots(D-m_n)}F(x)$.

Suppose $\frac{1}{D-m_n}F(x)=y^*.$

Then $(D-m_n)y^*=F(x)$ or $\frac{dy}{dx}-m_n y^*=F(x)$ which is a linear equation of the first order. Its $I.F=\exp\left(\int(-m_n)\right)dx=\exp(-m_n x)$.

Hence

$\frac{d}{dx}\left(ye^{-m_n x}\right)=F(x)e^{-m_n x}$

$y^*=e^{m_n x}\int F(x)e^{-m_n x}dx,$

Omitting the constant of integration, since we are looking for a particular solution, Therefore

$$\frac{1}{f(D)}F(x)=\frac{1}{(D-m_1)\ldots(D-m_{n-1})}e^{m_n x}\int e^{-m_n x}F(x)dx.$$

Next, we evaluate $\frac{1}{D-m_{n-1}}e^{m_n x}\int e^{-m_n x}F(x)dx$

As before and continue the process. This is the required P.I of the equation (1).

Alternatively, we may also write

$$\frac{1}{f(D)}F(x)=\frac{1}{(D-m_1)(D-m_2)\ldots(D-m_n)}F(x)$$

$$=\left[\frac{A_1}{D-m_1}+\frac{A_2}{D-m_2}+\ldots+\frac{A_n}{D-m_n}\right]F(x)$$

after resolving $\frac{1}{f(D)}$ into partial fractions

$$=A_1e^{m_1x}\int e^{-m_1x}F(x)dx+A_2e^{m_2x}\int e^{-m_2x}F(x)dx+\ldots A_ne^{m_nx}\int e^{-m_nx}F(x)dx,$$

which is the required P.I of the equation (1).

Example: Solve $(D^3-D)y=e^x$.

Solution. The characteristic equation is $D^3-D=0$ with roots $0,1,-1$

$C.F=y_c=c_1+c_2e^x+c_3e^{-x}$.

P.I $=$ $y_p=\frac{e^x}{D(D+1)(D-1)}$

Now $\frac{1}{D-1}e^x=e^x\int e^{-x}.e^x dx=xe^x$, by the method already described

Thus

$$\frac{1}{D(D+1)(D-1)}e^x=\frac{1}{D(D+1)}xe^x$$

Again,

$$\frac{1}{D+1}\left(xe^x\right)=e^{-x}\int e^x.xe^x dx=\frac{xe^x}{2}-\frac{e^x}{4}$$

Hence

$$\frac{1}{D(D+1)}\left(xe^x\right)=\frac{1}{D}\left(\frac{xe^x}{2}-\frac{e^x}{4}\right)=\int\frac{xe^x}{2}dx-\int\frac{e^x}{4}dx=\frac{xe^x}{2}-\frac{3}{4}e^x$$

Therefore, the general solution of the equation is

$$y = y_c + y_p = c_1 + c_2 e^x + c_3 e^{-x} + \frac{xe^x}{2} - \frac{3}{4}e^x.$$

$$= c_1 + \left(c_2 - \frac{3}{4} \right) e^x + c_3 e^{-x} + \frac{xe^x}{2}$$

$$= c_1 + c_2' e^x + c_3 e^{-x} + \frac{xe^x}{2}$$

1.3.4. Alternative Method

Consider $f(D)^{-1}$ as the inverse of the operator $f(D)$, so that

$$y_p = \frac{1}{D(D+1)(D-1)} e^x$$

$$y_p = \left(-\frac{1}{D} + \frac{1}{2} \cdot \frac{1}{1+D} + \frac{1}{2} \cdot \frac{1}{D-1} \right) e^x.$$

$$y_p = -\int e^x dx + \frac{1}{2} e^{-x} \int e^x . e^x \, dx + \frac{1}{2} e^x \int e^{-x} . e^x \, dx$$

$$y_p = -e^x + \frac{1}{4} e^x + \frac{1}{2} xe^x = \frac{xe^x}{2},$$

since e^x already occurs in the C.F and so the terms involving e^x are omitted.

1.3.5. Working Rules for finding P.I.

The method of evaluating $\frac{1}{f(D)} F(x)$ involves successive integrations and is quite unwieldy. We list below short methods for finding the P.I. when $F(x)$ is a function of a particular type.

To evaluate $\frac{1}{f(D)} e^{ax}$, where $f(D) = a_0 D^n + a_1 D^{n-1} + \ldots + a_{n-1} D + a_n$.

We know that $D^k e^{ax} = a^k e^{ax}$, where k is a positive integer.

Hence

$$f(D)e^{ax} = \left(a_0D^n + a_1D^{n-1} + \ldots + a_{n-1}D + a_n\right)e^{ax}$$

$$= a_0.D^n e^{ax} + a_1.D^{n-1}e^{ax} + \ldots + a_{n-1}.De^{ax} + a_n.e^{ax}$$

$$= a_0.a^n e^{ax} + a_1.a^{n-1}e^{ax} + \ldots + a_{n-1}.a\,e^{ax} + a_n.e^{ax}.$$

$$= f(a)e^{ax}. \tag{1}$$

Applying $\frac{1}{f(D)}$ to both sides of (1), we obtain

$$\frac{f(D)}{f(D)}e^{ax} = \frac{1}{f(D)}\, f(a)e^{ax}$$

$$e^{ax} = f(a).\frac{1}{f(D)}e^{ax}$$

i.e., $$\frac{1}{f(D)}\,e^{ax} = \frac{e^{ax}}{f(a)},\ if\ f(a) \neq 0$$

if $f(a)=0$, then a is a root of $f(D) = 0$.

Let a be a k-fold root of $f(D) = 0$. so that $(D-a)^k$ is a factor of $f(D)$. Then,

$$f(D) = (D-a)^k \varnothing(D), where \varnothing(a) \neq 0.$$

We first check the effect of $(D-a)^k$ on $e^{ax}p(x)$, for a polynomial $p(x)$ in x.

We have

$$(D-a)\left(e^{ax}p(x)\right) = D\left(e^{ax}p(x)\right) - a\,e^{ax}p(x) = e^{ax}Dp(x)$$

$$(D-a)^2\left(e^{Ax}p(x)\right) = (D-a)\left(e^{ax}Dp(x)\right) = e^{ax}D^2p(x).$$

Continuing in this way, we get

$$(D-a)^k\left(e^{Ax}p(x)\right)=e^{ax}D^k p(x). \tag{3}$$

Setting $p(x)=x^k$ in (3), we are led to

$$(D-a)^k\left(x^k e^{ax}\right)=e^{ax}D^k x^k=k!e^{ax}$$

$$\varnothing(D)(D-a)^k n\left(x^k e^{ax}\right)=\varnothing(D)k!e^{ax}$$

$$=k!\varnothing(D)e^{ax}=k!\varnothing(a)e^{ax},$$ by (1)

Operating on both sides by $\dfrac{1}{\varnothing(D)(D-a)^k}$, *we obtain*

$$x^k e^{ax}=\frac{1}{\varnothing(D)(D-a)^k}k!\varnothing(a)e^{ax}$$

$$\frac{1}{\varnothing(D)(D-a)^k}e^{ax}=\frac{x^k e^{ax}}{k!\,\varnothing(a)} \tag{4}$$

1.3.6. Principle of Superposition

$$f(D)y=\left(a_0D^n+a_1D^{n-1}+\ldots+a_{n-1}D+a_n\right)y=F(x) \tag{1}$$

Be a linear differential equation of order $n.$ if $F(x)=F_1(x)+F_2(x)$, then particular integral of (1) is the sum of particular integrals of

$$F(D)y=F_1(x) \tag{2}$$

and $$F(D)y \quad F(x) \tag{3}$$

Proof: Let y_1 *and* y_2 be particular integrals of (2) and (3) respectively. Then

$$f(D)y_1=F_1(x)$$

and $f(D)y_2=F_2(x)$

Suppose $y = y_1 + y_2$. Setting this value of y into (1), we obtain

$$f(D)y = f(D)(y_1 + y_2)$$

$$= f(D)y_1 + f(D)y_2$$

$$= F_1(x) + F_2(x) = F(x)$$

Showing that y is a solution of (1).

Example: Solve $(D^3+1)y = 1 + e^{-x} + e^{2x}$.

Solution. The characteristic equation is

$D^3 + 1 = 0$ with roots $-1, \dfrac{1 \pm i\sqrt{3}}{2}$

Therefore, $C.F$ is

$$y_c = c_1 e^{-x} + e^{\frac{x}{2}}\left(c_2 \sin\frac{\sqrt{3}}{2}x + c_3 \cos\frac{\sqrt{3}}{2}x\right).$$

The P/I is given by

$$y_p = \frac{1}{D^3+1}\left(1 + e^{-x} + e^{2x}\right)$$

$$= \frac{1}{D^3+1}e^{0x} + \frac{1}{D^3+1}e^{2x} + \frac{1}{(D+1)(D^2-D+1)}e^{-x}$$

$$= 1 + \frac{e^{2x}}{9} + \frac{1}{3(D+1)}e^{-x},$$ using $\dfrac{1}{f(D)}e^{ax} = \dfrac{1}{f(a)}e^{ax}$,

for $\dfrac{1}{D^2-D+1}e^{-x}$, $with\, D = -1\, when\, f(-1) \neq 0$

$$= 1 + \frac{e^{2x}}{9} + \frac{1}{3}xe^{-x},$$

Hence the general solution of the equation is

$y = y_c + y_p$

$$= c_1 e^{-x} + e^{\frac{x}{2}}\left(c_2 \sin\frac{\sqrt{3}}{2}x + c_3 \cos\frac{\sqrt{3}}{2}x \right) + 1 + \frac{e^{2x}}{9} + \frac{1}{3}x.e^{-x}.$$

To Find the P.I when $F(x) = \sin ax \;\; or \cos ax$.

Here we have to evaluate $\frac{1}{f(D)}\sin ax \, and \, \frac{1}{f(D)}\cos ax.$

From Euler's theorem, we have

$e^{iax} = \cos ax + i \sin ax$

Thus $\frac{1}{f(D)}\sin ax \, and \, \frac{1}{f(D)}\cos ax$ are respectively imaginary and real parts of $\frac{1}{f(D)}e^{iax}.$

If $f(D)$ contains only even powers of $D, say\, f(D) = f(D^2),$ it is easy to see that $\frac{1}{f(D^2)}\begin{cases}\sin ax \\ \cos ax\end{cases} = \frac{1}{f(-a^2)}\begin{cases}\sin ax \\ \cos ax\end{cases}, \; provided -a^2 \, is\, not\, a\, zero\, of\, f(D^2).$

Example: Solve $(D^2 - 5D + 6)y = \sin 3x$

Solution: The roots of the characteristic equation

$D^2 - 5D + 6 = 0$ are $2 \, and \, 3.$

$C.F - y_c - c_1 e^{2x} + c_2 e^{3x}$

$$P.I = y_p = \frac{1}{D^2 - 5D + 6}\sin 3x$$

which is imaginary part of $\frac{1}{(D-2)(D-3)}e^{3ix}$

Now

$$\frac{1}{(D-2)(D-3)}e^{3ix}=\frac{e^{3ix}}{(3i-2)(3i-3)},$$

$$=\frac{-1+5i}{78}e^{3ix}$$

$$=\left(-\frac{1}{78}+\frac{5}{78}i\right)(\cos 3x+i\sin 3x)$$

Its imaginary part is $-\frac{1}{78}\sin 3x+\frac{5}{78}\cos 3x=y_p$

Hence the required general solution is

$$y=c_1e^{2x}+c_2e^{3x}-\frac{1}{78}\sin 3x+\frac{5}{78}\cos 3x$$

Theorem. If a is not zero of $f(D)$, then $\frac{1}{f(D-a)}e^{ax}F(x)=e^{ax}\frac{1}{f(D)}F(x)$.

This replacing of $D\,by\,D+a$ is known as exponential shift.

Proof.

We have already shown in previous section that

$$(D-a)^k.\left(e^{ax}p(x)\right)=e^{ax}D^k p(x)$$

Using linearity of differential operators, we conclude that, when $f(D)$ is a polynomial (with constant coefficients), then

$$f(D-a)\left(e^{ax}p(x)\right)=e^{ax}f(D)p(x) \tag{1}$$

Suppose $f(D)p(x)=F(x)$, then

$$p(x)=\frac{1}{f(D)}F(x) \tag{2}$$

From (1) and (2), we have

$$f(D-a)e^{ax}\frac{1}{f(D)}F(x)=e^{ax}F(x)$$

Operating on both sides by $\frac{1}{f(D-a)}$, we get

$$e^{ax}\frac{1}{f(D)}F(x)=\frac{1}{f(D-a)}e^{ax}F(x)$$

or $$\frac{1}{f(D-a)}e^{ax}F(x)=e^{ax}\frac{1}{f(D)}F(x),$$

provided that a is not a zero of $f(D)$.

Example: Solve $\left(D^3+D^2-4D-4\right)y=e^{2x}\cos 3x.$

Solution. The characteristics equation is $D^3+D^2-4D-4=0.$

By inspection $D=2$ is a root of this equation. The other roots are $D=-2,-1$

$$C.F.=y_c=c_1e^{-x}+c_2e^{-2x}+c_3e^{2x}$$

$$P.I=y_p=\frac{1}{(D-2)(D+2)(D+1)}e^{2x}\cos 3x$$

$$=e^{2x}\frac{1}{D(D+4)(D+3)}\cos 3x, \textit{by the exponential shift}$$

$$=e^{2x}\frac{1}{\left(D^3+7D^2+12D\right)}\cos 3x$$

$$=e^{2x}\frac{1}{-9D-63+12D}\cos 3x,\ \textit{putting}\ D^2=-3^2$$

$$=e^{2x}\frac{1}{3(D-21)}\cos 3x$$

$$=e^{2x}\frac{(D+21)}{3\left(D^2-441\right)}\cos 3x$$

$$= e^{2x} \cdot \frac{-1}{3\times 450}\left[\,-3\sin 3x + 21\cos 3x\right]$$

$$= \frac{e^{2x}}{450}\left(\,\sin 3x - 7\cos 3x\right)$$

Alternative Method. Here, for $\frac{1}{D(D+4)(D+3)}\cos 3x,$ we proceed as follow:

$$\frac{\cos 3x}{\left(D^3+7D^2+12D\right)} = Re\,\frac{1}{\left(D^3+7D^2+12D\right)}e^{3ix},$$

$$= Re\frac{1}{\left(-27i-63+36i\right)}$$

$$= Re\frac{e^{3ix}}{9\left(i-7\right)}$$

$$= Re\,\frac{\left(i+7\right)e^{3ix}}{-450}$$

$$= Re\,\frac{\left(i+7\right)(\cos 3x + i\sin 3x)}{-450}$$

$$= -\frac{7}{450}\cos 3x + \frac{1}{450}\sin 3x.$$

Therefore,

$$y_p = \frac{e^{2x}}{450}\left(\,\sin 3x - 7\cos 3x\right)$$ as before.

The general solution is $y = y_c + y_p$

$$= c_1e^{-x} + c_2e^{-2x} + c_3e^{2x} - \frac{7}{450}\cos 3x + \frac{1}{450}\sin 3x.$$

$$= c_1e^{-x} + c_2e^{-2x} + c_3e^{2x} + \frac{e^{2x}}{450}\left(\,\sin 3x - 7\cos 3x\right).$$

1.3.7. Series Expansion of Polynomials or Expressions involving Operators.

This method is useful when $F(x)$ is a polynomial in x.

Let $f(D)y=F(x)$ be such that $F(x)$ is a polynomial in x. To evaluate the particular integral

$$y_p=\frac{1}{f(D)}F(x)$$

It is often useful to expand $\frac{1}{f(D)}$ in a series in D by the Binomial Theorem for negative exponent so that

$$y_p=\left(1+a_1D+a_2D^2+a_3D^3+\ldots\right)F(x)$$

The derivatives on the right will vanish after certain stage since

$D^n x^r=0 \text{ if } n>r.$

The following binomial expansions will be useful in this connection:

$$\frac{1}{1-x}=1+x+x^2+x^3+\ldots.$$

$$\frac{1}{1+x}=1-x+x^2-x^3+\ldots.$$

Example: Solve $\left(D^3-2D+1\right)y=2x^3-3x^2+4x+5$

Solution. The characteristic equation is $D^3-2D+1=0$

$$(D-1)\left(D^2+D-1\right)=0$$

$$D=1,\ \frac{-1+\sqrt{5}}{2},\ \frac{-1-\sqrt{5}}{2}$$

$$y_c=c_1e^x+c_2e^{\frac{-1+\sqrt{5}}{2}x}+c_3e^{\frac{-1-\sqrt{5}}{2}x}$$

Next, $y_p = \frac{1}{1-2D+D^3}\left(2x^3-3x^2+4x+5\right)$

Now $\frac{1}{1-2D+D^3} = \left[1-\left(2D-D^3\right)\right]^{-1}$

$$=1+\left(2D-D^3\right)+\left(2D-D^3\right)^2+\left(2D-D^3\right)^3+\ldots$$

$$=1+2D-D^3+4D^2+8D^3,$$

Neglecting D^4 and higher powers of D in view of the degree 3 of the polynomial $p(x)$ in x therefore

$$y_p=\left(1+2D+4D^2+7D^3\right)\left(2x^3-3x^2+4x+5\right)$$

$$=2x^3-3x^2+4x+5+2\left(6x^2-6x+4\right)+4\left(12x-6\right)+7\left(12\right)$$

$$=2x^3+9x^2+40x+73.$$

The general solution is $y=y_c+y_p$

$$=c_1e^x+c_2e^{\frac{-1+\sqrt{5}}{2}x}+c_3e^{\frac{-1-\sqrt{5}}{2}x}+2x^3+9x^2+40x+73.$$

1.3.8. The Method of Undetermined Coefficients (U. C. Method)

We have studied some special cases in which particular integral can be evaluated by the inverse operator. Now we consider the method of undetermined coefficients which can prove simpler in finding the particular integral of the equation $f(D)y=F(x)$ when $F(x)$ is

a. An exponential function: e^{ax}
b. A polynomial: $\left(b_0x^n+b_1x^{n-1}+\ldots+b_n\right)$
c. Sinusoidal function : $\sin ax$ or $\cos bx$
d. The more general case in which $F(x)$ is sum of a product of terms of the above types, such as $F(x)=e^{ax}\left(b_0x^n+b_1x^{n-1}+\ldots+b_n\right)\begin{Bmatrix}\sin ax\\ \cos bx\end{Bmatrix}$

The P.I, i.e., y_p will be constructed according to the following table:

$F(x)$ *is of the form*	*Take* y_p *as*
$1.a$	$A.x^k$
$2.ax^n$ (n *is a +ve integer*)	$x^k\left(A_0x^n + A_1x^{n-1} + \ldots + A_n\right)$
3. $ax^n e^{rx}$ (n *is a +ve integer*)	$x^k\left(A_0x^n + A_1x^{n-1} + \ldots + A_{n-1}x + A_n\right)e^{rx}$
$4.Cx^n \cos ax$	$x^k\left(A\cos ax + B\sin ax\right)$
$5.Cx^n \sin ax$	
$6.C.x^n.e^{rx} \cos ax$	$x^k\ [\left(A_0x^n + A_1x^{n-1} + \ldots + A_{n-1}x + A_n\right)e^{rx}\cos ax$
$7.Cx^n e^{rx} \sin ax$	$+\left(B_0x^n + B_1x^{n-1} + \ldots + B_{n-1}x + B_n\right)e^{rx}\sin ax$

In x^k, k is the smallest nonnegative integer which will ensure that no terms in y_p is already in the C.F.

If $F(x)$ is sum of several terms, write y_p for each term individually and then add up all of them. The y_p and its derivatives will be substituted into the equation $f(D)y = F(x)$ and coefficients of like terms on the left hand and right-hand sides will be equated to determine the $U.C. A_0, A_1, \ldots, A_n, B_0, B_1, \ldots, B_n$. The method is illustrated by examples as follows.

Example: Solve $y'' - 3y' + 2y = 2x^3 - 9x^2 + 6x$ (1)

Solution: C.F. is easily found as $y_c = c_1e^x + c_2e^{2x}$

For a particular solution, we assume

$y_p = x^k\left(Ax^3 + Bx^2 + Cx + D\right)$.

Since no term of the C.F. is present in y_p, we take $k = 0$ and so

$y_p = Ax^3 + Bx^2 + Cx + D$

$y_p' = 3Ax^2 + 2Bx + C$

$y_p'' = 6Ax + 2B$

Substituting for y_p, y_p' and y_p'' into (1), we have

$(6Ax + 2B) - 3(3Ax^2 + 2Bx + C) + 2(Ax^3 + Bx^2 + Cx + D)$

$= 2x^3 - 9x^2 + 6x$

or $2Ax^3 + x^2(2B - 9A) + x(6A - 6B + 2C) + 2B - 3C + 2D$

$= 2x^3 - 9x^2 + 6x.$

Equating coefficients of like terms, we obtain

Coeff. Of $x^3 : 2A = 2 \quad or\ A = 1.$

Coeff. Of $x^2 : 2B - 9A = -9 \quad or\ 2B = 9A - 9\ giving\ B = 0$

Coeff. Of $x : 6A - 6B + 2c = 6 \quad or\ C = 0.$

Coeff. Of $x^0 : 2B - 3C + 2D = 0 \quad or\ D = 0.$

So, $y_p = x^3$

The required general solution is $y = c_1e^x + c_2e^{2x} + x^3.$

Example: Solve $y'' - 3y' + 2y = x^2e^x.$ (1)

Solution. The C.F. is $y_c = c_1e^x + c_2e^{2x}.$

To construct a particular solution, we use (3) of the table

$y_p = x^k(Ax^2 + Bx + C)e^x.$

Since e^x is already in $C.F.,$ we take $k = 1$ so that no terms of C.F. is in $y_p.$ Hence the modified P.I. is $y_p = (Ax^3 + Bx^2 + Cx)e^x$

$y_p' = (Ax^3 + Bx^2 + Cx)e^x + (3Ax^2 + 2Bx + C)e^x$

$y_p'' = (Ax^3 + Bx^2 + Cx)e^x + 2(3Ax^2 + 2Bx + C)e^x + (6Ax + 2B)e^x.$

Substituting for y_p, y_p' and y_p'' into (1), we have

$$\left(Ax^3+Bx^2+Cx\right)e^x+2\left(3Ax^2+2Bx+C\right)e^x+\left(6Ax+2B\right)e^x+\left(6Ax+2B\right)e^x$$

$$-3\left(Ax^3+Bx^2+Cx\right)e^x-3\left(3Ax^2+2Bx+C\right)e^x+2\left(Ax^3+Bx^2+Cx\right)e^x=x^3e^x$$

$$\left[-3Ax^2+\left(6A-2B\right)x+C\right]e^x=x^3e^x$$

Equating coefficients of like terms, we obtain

$$Coeff.of\ x^3:-3A=1\ \ or\ A=-\frac{1}{3}$$

$$Coeff.of\ x:\ 6A-2B=0\ \ or\ B=-1$$

$$Coeff.of\ x^0:\ C=0$$

So $y_p=-\dfrac{1}{3}x^3e^x-x^2e^x.$

The required solution is

$$y=c_1e^x+c_2e^{2x}-\frac{1}{3}x^3e^x-x^2e^x.$$

Example: Solve $y''+4y=xe^x+x\sin 2x$ (1)

Solution. The C.F., as readily found above, is $y_c=c_1\sin 2x+c_2\cos 2x.$

For the P.I of (1), we find the P.I. of $y''+4y=xe^x$ (2)

And $y''+4y=x\sin 2x$ (3)

Separately. Their sum will be the P.I of (1) (by the Principle of superposition). For a particular solution of (2), we have

$$y_p=x^k\left(Ax+B\right)e^x.$$

Since no term of C.F. is in the y_p, we must take $k=0$. Therefore,

$$y_p=\left(Ax+B\right)e^x$$

$$y_p'=\left(Ax+B\right)e^x+Ae^x$$

$$y_p^{''} = (Ax + B)e^x + 2Ae^x.$$

Substituting for $y_p, y_p^{'}$ *and* $y_p^{''}$ into (2), we have

$$(Ax + B)e^x + 2Ae^x + 4(Ax + B)e^x = xe^x$$

or $5Axe^x + (2A + 5B)e^x = xe^x.$

Equating coefficients of like terms, we obtain

$$Coefficient\ of\ xe^x: \quad 5A = 1\, or\, A = \frac{1}{5}$$

$$Coefficient\ of\ e^x: 2A + 5B = 0\ or\ B = -\frac{2}{25}$$

So, P.I of (2) is $y_p = \frac{1}{5}xe^x - \frac{2}{25}e^x.$

$$y_p = x^k\left[(Cx + D)\sin 2x + (Ex + F)\cos 2x\right].$$

Since $\sin 2x\, and \cos 2x$ are already in the C.F., we must take $k = 1,$ so that the P.I. is

$$y_p = (Cx^2 + Dx)\sin 2x + (Ex^2 + Fx)\cos 2x$$

$$y_p^{'} = (Cx^2 + Dx)(2\cos 2x) + (2Cx + D)\sin 2x$$

$$+(Ex^2 + Fx)(-2\sin 2x) + (2E(x) + F)\cos 2x$$

$$y_p^{''} = (Cx^2 + Dx)(-4\sin 2x) + (2Cx + D)(2\cos 2x) + (2Cx + D)(2\cos 2x)$$

$$+(2Cx + D)(2\cos 2x) + 2C\sin 2x + (Ex^2 + Fx)(-4\cos 2x)$$

$$+(2Ex + F)(-2\sin 2x) + (2Ex + F)(-2\sin 2x) + 2E\cos 2x.$$

Substituting into (3), we have

$$(Cx^2 + Dx)(-4\sin 2x) + 4(2Cx + D)\cos 2x + 2C\sin 2x + (Ex + Fx)(-4\cos 2x)$$

$$-4(2Ex+F)\sin 2x+2E\cos 2x+4\left(Cx^2+Dx\right)\sin 2x+4\left(Ex^2+Fx\right)\cos 2x=x\sin 2x.$$

Equating the coefficients of like terms, we get

Coeff. Of $x\sin 2x$: $-8E=1 \quad or\, E=-\frac{1}{8}$

Coeff. Of $x\cos 2x$: $8C=0\, or\, C=0$

Coeff. Of $\sin 2x: 2C-4F=0\, or\ F=0$

Coeff. Of $\cos 2x: 2E+4D=0\ or\ D=\frac{1}{16}.$

Hence P.I of (3) is

$$y_p=\frac{1}{16}x\sin 2x-\frac{1}{8}x^2\cos 2x. \qquad (5)$$

Sum of (4) and (5) is the required P.I of (I), i.e.,

$y_p=\frac{1}{5}xe^x-\frac{2}{25}e^x+\frac{1}{16}x\sin 2x-\frac{1}{8}x^2\cos 2x$ is the particular solution of (1).

Hence the general solution of (1) is

$$y=c_1\sin 2x+c_2\cos 2x+\frac{1}{5}xe^x-\frac{2}{25}e^x+\frac{1}{16}x\sin 2x-\frac{1}{8}x^2\cos 2x.$$

1.3.9. The Cauchy-Euler Equation

An equation of the form

$$a_0x^n\frac{d^ny}{dx^n}+a_1x^{n-1}\frac{d^{n-1}y}{dx^{n-1}}+\ldots+a_{n-1}x\frac{dy}{dx}+a_ny=F(x)$$

is called Cauchy-Euler (or equidimensional) equation. $a_0,a_1,\ldots,a_{n-1},a_n$ are real constants. The equation can be reduced to a linear differential equation with constant coefficients by the transformation $x=e^t\ or\, t=\ln x$

Then

$$\frac{dy}{dx}=\frac{dy}{dt}.\frac{dt}{dx}=\frac{1}{x}\frac{dy}{dt}$$

$$\frac{d^2y}{dx^2}=\frac{d}{dx}\left(\frac{1}{x}\frac{dy}{dt}\right)=\frac{1}{x}\frac{d}{dx}\left(\frac{dy}{dt}\right)+\frac{dy}{dt}\cdot\frac{d}{dx}\left(\frac{1}{x}\right)$$

$$=\frac{1}{x}\frac{d}{dt}\left(\frac{dy}{dt}\right)\cdot\frac{dt}{dx}+\frac{dy}{dt}\left(-\frac{1}{x^2}\right)=\frac{1}{x^2}\left(\frac{d^2y}{dt^2}-\frac{dy}{dt}\right)$$

$$\frac{d^3y}{dx^3}=\frac{d}{dx}\left[\frac{1}{x^2}\frac{d^2y}{dt^2}-\frac{1}{x^2}\frac{dy}{dt}\right]$$

$$=\frac{1}{x^2}\frac{d}{dx}\left(\frac{d^2y}{dt^2}\right)+\frac{d^2y}{dt^2}\cdot\frac{d}{dx}\left(\frac{1}{x^2}\right)-\frac{1}{x^2}\frac{d}{dx}\left(\frac{dy}{dt}\right)+\frac{dy}{dt}\cdot\frac{d}{dx\left(-\frac{1}{x^2}\right)}$$

$$=\frac{1}{x^2}\frac{d}{dt}\left(\frac{d^2y}{dt^2}\right)\cdot\frac{dt}{dx}-\frac{2}{x^3}\frac{d^2y}{dt^2}-\frac{1}{x^2}\frac{d}{dt}\left(\frac{dy}{dt}\right)\cdot\frac{dt}{dx}+\frac{2}{x^3}\frac{dy}{dt}$$

$$=\frac{1}{x^3}\frac{d^3y}{dt^3}-\frac{3}{x^3}\frac{d^2y}{dt^2}+\frac{2}{x^3}\frac{dy}{dt}$$

$$=\frac{1}{x^3}\left(\frac{d^3y}{dt^3}-3\frac{d^2y}{dt^2}+2\frac{dy}{dt}\right)$$

If we write $D=\frac{d}{dx}\, and\, \Delta=\frac{d}{dt}, then$

$$xD=\Delta$$

$$x^2D^2=\Delta^2-\Delta=\Delta(\Delta-1)$$

$$x^3D^3=\Delta^3-3\Delta^2+2\Delta=\Delta(\Delta-1)(\Delta-2)$$

$$\vdots \qquad\qquad \vdots \qquad\qquad \vdots$$

$$x^nD^n=\Delta(\Delta-1)(\Delta-2)\ldots(\Delta-n+1).$$

Substituting these values of $xD, x^2D^2,\ldots$ into (1), we obtain an equation of nth order with constant coefficients having t as the independendent variable. This equation can be solved by the previous methods.

Example: Solve $x^2\frac{d^2y}{dx^2}-2x\frac{dy}{dx}+2y=x^3.$ (1)

Solution: Let $x=e^t$. Then we have $t=\ln x$

$xD=\Delta.$

$x^2D^2=\Delta(\Delta-1).$

Substituting into (1), we get

$\left[\Delta(\Delta-1)-2\Delta+2\right]y=e^{3t}$

$\left[\Delta^2-3\Delta+2\right]y=e^{3t}.$ (2)

The characteristic equation has roots $2\, and\, 1$. Therefore, C.F. of (2) is

$y_c=c_1e^t+c_2e^{2t}$

P.I. of equation (2) is $y_p=\frac{1}{(\Delta-1)(\Delta-2)}e^{3t}=\frac{1}{2}e^{3t}.$

The general solution of (2) is $y=c_1e^t+c_2e^{2t}+\frac{1}{2}e^{3t}.$

Replacing t by $\ln x\left(or\, e^t by\, x\right)$, we have

$y=c_1x+c_2x^2+\frac{1}{2}x^3$ as the general solution of (1).

Example: Solve $x^4\frac{d^4y}{dx^4}+6x^3\frac{d^3y}{dx^3}+9x^2\frac{d^2y}{dx^2}+3x\frac{dy}{dx}+y-(1+\ln x)^2.$ (1)

Solution: We let $x=e^t$ so that $t=\ln x$

$xD=\Delta$

$x^2D^2=\Delta(\Delta-1)=\Delta^2-\Delta.$

$x^3D^3=\Delta(\Delta-1)(\Delta-2)=\Delta^3-3\Delta^2+2\Delta$

$x^4D^4=\Delta(\Delta-1)(\Delta-2)(\Delta-3)=\Delta^4-6\Delta^3+11\Delta^2-6\Delta.$

Substituting into (1), we get

$$\left[\Delta^4-6\Delta^3+11\Delta^2-6\Delta+6\left(\Delta^3-3\Delta^2+2\Delta\right)+9\left(\Delta^2-\Delta\right)+3\Delta+1\right]y=\left(1+t\right)^2$$

$$\left(\Delta^4+2\Delta^2+1\right)y=1+2t+t^2 \tag{2}$$

The roots of the characteristic equation of (2) are $\pm i, \pm i.$

Therefore, C.F. of (2) is $y_c=\left(c_1+c_2t\right)\sin t+\left(c_3+c_4t\right)\cos t.$

To find a particular integral of (2), we use the method of $U.C.$ Let the particular integral of (2) be

$$y_p=At^2+Bt+C$$

$$y_p'=2At+B$$

$$y_p''=2A.$$

Substituting into (2), we obtain

$$4A+At^2+Bt+C=1+2t+t^2$$

$$At^2+Bt+\left(4A+C\right)=t^2+2t+1.$$

Equating coefficients of like terms, we have

$A=1, B=2, 4A+C=1 \, or \, C=-3.$

Therefore, a particular integral of (2) is

$y_p=t^2+2t-3,$

Hence the general solution of (2) is

$$y=\left(c_1+c_2t\right)\sin t+\left(c^3+c_4t\right)\cos t+t^2+2t-3 \tag{3}$$

Replacing t by $\ln x$ in (3), we get

$$y=\left(c_1+c_2\ln x\right)\sin\left(\ln x\right)+\left(c_3+c_4\ln x\right)\cos(\ln x)+\left(\ln x\right)^2+2\ln x-3$$

as the general solution of (1).

Reduction of Order

If one solution of the second order linear equation $\frac{d^2y}{dx^2}+P\frac{dy}{dx}+Qy=0$ (1)
(where P,Q are not necessarily constants and may be functions of x) is known, then we can use it to find the general solution of

$$\frac{d^2y}{dx^2}+P\frac{dy}{dx}+Qy=F(x) \tag{2}$$

The procedure, due to D' Alembert, is known as the method reduction of order.

Suppose it is known that $y=y_1$ is a solution of (1). We assume that

$$y=vy_1 \tag{3}$$

is a solution of (2), where v is a function of x to be determined. From (3), we get

$$\frac{dy}{dx}=v\frac{dy_1}{dx}+y_1\frac{dv}{dx},$$

$$\frac{d^2y}{dx^2}=v\frac{d^2y_1}{dx^2}+2\frac{dv}{dx}\frac{dy_1}{dx}+y_1\frac{d^2v}{dx^2}$$

Substituting into (2), we have

$$v\frac{d^2y_1}{dx^2}+2\frac{dv}{dx}\frac{dy_1}{dx}+y_1\frac{d^2v}{dx^2}+Pv\frac{dy_1}{dx}+Py_1\frac{dv}{dx}+Qvy_1=F(x)$$

$$y_1\frac{d^2v}{dx^2}+\left(2\frac{dy_1}{dx}+Py_1\right)\frac{dv}{dx}+\left(\frac{d^2y_1}{dx^2}+P\frac{dy_1}{dx}+Qy_1\right)v=F(x) \tag{4}$$

Since $y=y_1$ is a solution of (1)

$$\frac{d^2y_1}{dx^2}+P\frac{dy_1}{dx}+Qy_1=0$$

and, therefore, equation (4) reduces to

$$y_1\frac{d^2v}{dx^2}+\left(2\frac{dy_1}{dx}+Py_1\right)\frac{dv}{dx}=F(x) \tag{5}$$

Setting $\frac{dv}{dx} = u$ in (5), we obtain

$$y_1 \frac{du}{dx} + \left(2\frac{dy_1}{dx} + Py_1 \right) u = F(x)$$

Which is a linear equation of the first order in u and can be solved for From $\frac{dv}{dx} = u,$ we determine v and hence the solution $y = vy_1.$ The method is illustrated by means of example.

Example: Solve $\frac{d^2y}{dx^2} + y = \csc x.$ (1)

Solution. The C.F, of (1) is $y_c = c_1 \sin x + c_2 \cos x.$

We may take any special value of $y_c\ as\ y_1.$ Let us take y_1 to be the value of $y_c\ when\ c_1 = 1\ and\ c_2 = 0$. Then assume that $y = v \sin x$ (2)

is a solution of (1). From (2), we get

$$\frac{dy}{dx} = \sin x \frac{dv}{dx} + v \cos x$$

$$\frac{d^2y}{dx^2} = \sin x \frac{d^2v}{dx^2} + 2\cos x \frac{dv}{dx} - v \sin x.$$

Substituting into (1), we obtain

$$\sin x \frac{d^2v}{dx^2} + 2\cos x \frac{dv}{dx} - v \sin x + v \sin x = \csc x$$

$$\frac{d^2v}{dx^2} + 2\cot x \frac{dv}{dx} = \csc^2 x.$$

Setting $u = \frac{dv}{dx},$ the above equation becomes

$$\frac{du}{dx} + 2\cot x u = \csc^2 x \qquad (3)$$

Which is linear equation of first order. An integrating factor of (3) is

$$exp\left[\int 2\cot x dx\right] = \exp\left[\ln \sin^2 x\right] = \sin^2 x$$

Multiplying (3) by $\sin^2 x$ and integrating, we have

$u \sin^2 x = x,$

Neglecting constant of integration since we seek only a particular solution.

$u = x \csc^2 x$

$$\frac{dv}{dx} = x \csc^2 x$$

$v = \int x \csc^2 x\, dx = -x \cot x + \ln|\sin x|$ (on integrating by parts)

Hence a particular solution of (1) is

$y = v \sin x = -x \cos x + \sin x\ (\ln|\sin x|).$

The general solution of (1) is

$y = c_1 \sin x + c_2 \cos x - x \cos x + \sin x(\ln|\sin x|).$

Example: Solve $(x+2).\frac{d^2y}{dx^2} - (2x+5).\frac{dy}{dx} + 2y = (x+1)e^x$ (1)

Solution: We note that $y = e^{2x}$ makes the left-hand side of (1) zero. Therefore, we put

$y = ve^{2x}$

$$\frac{dy}{dx} = \left(\frac{dv}{dx} + 2v\right)e^{2x}$$

$$\frac{d^2y}{dx^2} = \left(\frac{d^2v}{dx^2} + 4\frac{dv}{dx} + 4v\right)e^{2x}$$

Substituting into (1), we obtain

$$(x+2)\left(\frac{d^2v}{dx^2} + 4\frac{dv}{dx} + 4v\right)e^{2x} - (2x+5)\left(\frac{dv}{dx} + 2v\right)e^{2x} + 2ve^{2x} = (x+1)e^x.$$

$$(x+2)\frac{d^2v}{dx^2}e^{2x}+\left[4(x+2)-(2x+5)\right]\frac{dv}{dx}e^{2x}+\left[4(x+2)-2(2x+5)+2\right]ve^{2x}=(x+1)e^{x}$$

$$(x+2)\frac{d^2v}{dx^2}+(2x+3)\frac{dv}{dx}=(x+1)e^{-x}.$$

Writing $u=\frac{dv}{dx}$, the above equation becomes

$$(x+2)\frac{du}{dx}+(2x+3)u=(x+1)e^{-x}$$

$$\frac{du}{dx}+\frac{3x+3}{x+2}u=\frac{x+1}{x+2}e^{-x} \tag{2}$$

which is linear equation of first order. An integrating factor of (2) is

$$exp\left[\int\frac{2x+3}{x+2}dx\right]=\exp\left[\int\left(2-\frac{1}{x+2}\right)dx\right]$$

$$=\exp\left[2x-\ln(x+2)\right]=\frac{e^{2x}}{x+2}$$

Multiplying (2) by $\frac{e^{2x}}{x+2}$, we have

$$\frac{d}{dx}\left(u\frac{e^{2x}}{x+2}\right)=\frac{x+1}{(x+2)^2}.e^{x}$$

$$u\frac{e^{2x}}{x+2}=\int\frac{x+1}{(x+2)^2}.e^{x}\,dx+c_1$$

$$=\int\frac{e^{x}}{x+2}dx-\int\frac{e^{x}}{(x+2)^2}dx+c_1$$

$$=\frac{e^{x}}{x+2}+c_1$$

$$u=e^{-x}+c_1(x+2)e^{-2x}$$

Therefore, $\dfrac{dv}{dx}=e^{-x}+c_1\left(x+2\right)e^{-2x}.$

Integrating, we get

$$v=-e^{-x}-\frac{1}{4}c_1\left(2x+5\right)e^{-2x}+c_2$$

Hence $y=ve^{2x}=-e^{x}-\dfrac{1}{4}c_1\left(2x+5\right)+c_2e^{2x}.$

is the general solution of (1).

Note. It is easy to see that $y''+Py'+Qy=0$ is satisfied by $y=e^{\pm x}$. If $1\pm P+Q=0$ and by $y=x$ if $P+Qx=0.$ These can be used to find a solution $y''+Py'+Q=0$ by inspection.

1.4. THE WRONSKIAN

Definition. If y_1,y_2 are two differentiable functions of x on $I=[a\ ,b]$ then their Wronskian, denoted by $W=W\left[y_1,y_2\right],$ is obtained by

$$W=W\left[y_1,y_2\right]=y_1\,y_2'-y_1'y_2=\begin{vmatrix} y_1 & y_2 \\ y_1' & y_2' \end{vmatrix}$$

Similarly, the Wronskian of three differentiable functions y_1,y_2,y_3 on $I=[a\ ,b]$ is defined by

$$W-W\left[y_1,y_2,y_3\right]=\begin{vmatrix} y_1 & y_2 & y_3 \\ y_1' & y_2' & y_3' \\ y_1'' & y_2'' & y_3'' \end{vmatrix}$$

The definition can be extended in a similar manner for the Wronskian of n differentiable functions on $I=[a,b].$

Theorem. Let y_1,y_2 be two solutions of the homogeneous linear equation

$$\frac{d^2y}{dx^2}+P\frac{dy}{dx}+Qy=0, \qquad (1)$$

Where P, Q are functions of x and are continuous on $I=[a,b]$. Then their Wronskian $W=W[y_1,y_2]$ is either identically zero or is never zero on I.

Proof. We have $W=y_1 y_2' - y_1' y_2$

Differentiating $w.r.t.x,$ we get

$$W' = y_1 y_2'' + y_1' y_2' - y_1' y_2' - y_1'' y_2$$

$$= y_1 y_2'' - y_1'' y_2.$$

Since y_1, y_2 are solutions of (1), we have

$$y_1'' + Py_1' + Qy_1 = 0 \qquad (2)$$

and $$y_2'' + Py_2' + Qy_2 = 0 \qquad (3)$$

Multiply (2) by y_2 and (3) by y_1 to have an equivalent system

$$y_1'' y_2 + P\, y_1' y_2 + Qy_1 y_2 = 0$$

(4)

$$y_1 y_2'' + P\, y_2' y_1 + Qy_1 y_2 = 0 \qquad (5)$$

Subtracting (4) from (5), we obtain

$$\left(y_1 y_2'' - y_2\, y_1''\right) + P\left(y_1 y_2' - y_1' y_2\right) = 0.$$

$$\frac{dW}{dx} + PW = 0.$$

The general solution of the above equation is readily found as $W = c\exp(-\int P\,dx\,)$.

Since the exponential function is never zero, W is identically zero if $c=0$ or is never zero when $c \neq 0.$

Theorem. If y_1, y_2 are two solutions of $\dfrac{d^2y}{dx^2} + P\dfrac{dy}{dx} + Qy = 0,$

Then their Wronskian $W=W[y_1,y_2]=0$ if and only if y_1, y_2 are linearly dependent.

Proof. If one of the two solutions is identically zero, then the theorem is obviously true. Assume that y_1, y_2 are both nonzero and let y_1, y_2 be linearly dependent. Then $y_1 = cy_2$, where c is a constant.

i.e., $\frac{y_1}{y_2} = c$

or $\frac{d}{dx}\left(\frac{y_1}{y_2}\right) = \frac{y_2 y_1' - y_1 y_2'}{y_2^2} = \frac{dc}{dx} = 0$

i.e., $y_1 y_2' - y_1' y_2 = 0$

$W[y_1, y_2] = 0.$

Conversely, if $W[y_1, y_2] = 0$, then $\frac{d}{dx}\left(\frac{y_1}{y_2}\right) = \frac{W}{y_2^2} = 0.$

$\frac{y_1}{y_2} = c,$ where c is a constant.

Thus $y_1 \; cy_2$ and so y_1, y_2 are linearly dependent.

Corollary. $W[y_1, y_2] \neq 0$ if and only if y_1, y_2 are linearly independent.

Above theorems can only be applied to check the linear dependence or linear independence of differentiable functions which are solutions of a linear homogeneous $O.D.E.$ Arbitrary differentiable functions exist with their Wronskian zero but they are linearly independent.

Consider the functions $y_1 = x^3, y_2 = |x^3|$ on $]-\infty, \infty[$. Here $W(y_1, y_2) = 0$ but y_1 and y_2 are linearly independent.

1.4.1. VARIATION OF PARAMETERS.

We found the solution of the equation $\frac{d^2y}{dx^2} + P\frac{dy}{dx} + Qy = F(x).$ (1)

Where P, Q are functions of x, in the previous section by reduction of order of (1). The solution of (1) can be determined by a procedure known as the method of variation of parameters. This method can be applied even to equations of higher order.

Suppose that linearly independent solutions of $\frac{d^2 y}{dx^2} \text{NP} \frac{dy}{dx} \quad Q$ (2)

Are given by $y = y_1(x)$ and $y = y_2(x)$. Then the complementary function of (1) is

$y_c = c_1 y_1 + c_2 y_2$

Where $c_1 \, and \, c_2$ are arbitrary constants. We replace the arbitrary constants $c_1 \, and \, c_2$ by unknown functions $u_1(x) \, and \, u_2(x)$ and require that

$$y_p = u_1 y_1 + u_2 y_2 \tag{3}$$

be a particular solution of (1). In order to determine the two functions $u_1 \, and \, u_2$ we need two conditions. One condition is that (3) must satisfy (1). A second condition can be imposed arbitrarily.

Differentiating (3) $w.r.t.x,$ we have

$y_p' = (u_1' y_1 + u_2' y_2) + u_1 y_1' + u_2 y_2'.$

If we differentiate the above equation, y_p'' will contain $u_1'' \, and \, u_2''.$ To avoid second derivatives of $u_1 \, and \, u_2.$ we put

$$u_1' y_1 + u_2' y_2 = 0. \tag{4}$$

With this condition, we have $y_p' = u_1 y_1' + u_2 y_2'$

So that $y_p' = u_1 y_1' + u_2 y_2'$

So that $y_p'' = u_1' y_1' + u_2' y_2' + u_1 y_1'' + u_2 y_2''.$

Substituting for y_p, y_p', y_p'' into equation (1), we get

$(u_1' y_1' + u_2' y_2' + u_1 y_1'' + u_2 y_2'') + P(u_1 y_1' + u_2 y_2') + Q(u_1 y_1 + u_2 y_2) = F(x)$

or $u_1(y_1'' + P y_1' + Q y_1) + u_2(y_2'' + P y_2' + Q y_2) + u_1' y_2' = F(x).$

Expressions within the parenthesis are zero since $y_1 \, and \, y_2$ are solutions of (2). Hence

$$u_1' y_1' + u_2' y_2' = F(x) \tag{5}$$

Taking (4) and (5) together, we have two equations in the two unknowns u_1' *and* u_2';

$$u_1' y_1 + u_2' y = 0$$

and $u_1' y_1' + u_2' y_2' = F(x)$,

Solving these, we have

$$\left.\begin{aligned} u_1' &= \frac{-y_2 F(x)}{y_1 y_2' - y_1' y_2} \\ u_2' &= \frac{y_1 F(x)}{y_1 y_2' - y_1' y_2} \end{aligned}\right\} \tag{6}$$

In (6), the Wronskian of y_1, y_2 namely $W(y_1, y_2) = y_1 y_2' - y_1' y_2 \neq 0$, since y_1, y_2 are linearly independent solutions of (2). Integrating (6), we find u_1 *and* u_2 as

$$u_1 = \int \frac{-y_2 F(x)}{y_1 y_2' - y_1' y_2} dx = \int \frac{-y_2 F(x)}{W} dx \tag{7}$$

$$u_2 = \int \frac{y_1 F(x)}{y_1 y_2' - y_1' y_2} dx = \int \frac{y_1 F(x)}{W} dx \tag{8}$$

Where W is the Wronskian of y_1, y_2.

Thus y_p is completely determined. In numerical problems, instead of performing the complete process, formulas (7) will be directly applied to evaluate u_1 *and* u_2.

Example. Find the general solution of $\frac{d^2 y}{dx^2} - 2\frac{dy}{dx} + y = x e^x \ln x$ (1)

Solution. The C.F. of (1) is

$$y_c = c_1 e^x + c_2 x e^x$$

Let

$$y_p = u_1 e^x + c_2 x e^x$$

Here

$$y_1 = e^x, y_2 = x e^x, F(x) = x e^x \ln x$$

$$W = W(y_1, y_2) = y_1 y_2' - y_1' y_2 = e^x\left(e^x + xe^x\right) - xe^x.e^x = e^{2x}.$$

By the formulas of (7) and (8), we have

$$u_1 = \int \frac{-y_2 F(x)}{W} dx = \int \frac{-xe^x.xe^x \ln x}{e^{2x}} dx$$

and $$u_2 = \int \frac{y_1 F(x)}{W} dx = \int \frac{e^x.\ xe^x \ln x}{e^{2x}} dx$$

i.e., $$u_1 = -\int x^2 \ln x dx = -(\ln x).\frac{x^3}{3} + \int \frac{x^3}{3} dx = \frac{x^3}{9} - \frac{x^3}{3} \ln x$$

$$u_2 = \int x \ln x\, dx = \frac{x^2}{2} \ln x - \int \frac{x}{2} dx = -\frac{x^2}{4} + \frac{x^2}{2} \ln x.$$

$$y_p = \left(\frac{x^3}{9} - \frac{x^3}{3} \ln x\right) e^x + \left(-\frac{x^2}{4} + \frac{x^2}{2} \ln x\right) x e^x$$

$$= e^x \left(\frac{x^3}{9} - \frac{x^3}{3} \ln x - \frac{x^3}{4} + \frac{x^3}{2} \ln x\right)$$

$$= e^x \left(-\frac{5x^3}{36} + \frac{x^3}{6} \ln x\right).$$

The general solution of (1) is

$$y = y_c + y_p = c_1 e^x + c_2 x e^x = e^x \left(-\frac{5x^3}{36} + \frac{x^3}{6} \ln x\right).$$

Example. Find the general solution of

$$x^2 \frac{d^2 y}{dx^2} - x(x+2) \frac{dy}{dx} + (x+2) y = x^3 \qquad (1)$$

Given that $y = xe^x$ is a solution of the associated homogeneous equation

$$x^2 \frac{d^2y}{dx^2} - x(x+2)\frac{dy}{dx} + (x+2)y = 0 \qquad (2)$$

Solution. The given equation in the standard form is $\frac{d^2y}{dx^2} - \frac{x+2}{x}\frac{dy}{dx} + \frac{x+2}{x^2}y = x.$

Since $P + Qx = -\frac{x+2}{x} + x\frac{x+2}{x^2} = 0, y = x$ is also a solution of (2). The two solutions, $y = x$ and $y = xe^x$ are linearly independent. The complementary function of (1) is

$y_c = c_1x + c_2xe^x.$

We assume that $y_p = u_1x + u_2xe^x$ (3)

is a particular solution of (1).

Here $y_1 = x, y_2 = xe^x, F(x) = x$

$$W = W(y_1, y_2) = y_1y_2' - y_1'y_2 = x(xe^x + e^x) - xe^x = x^2e^x$$

By the formulas of (7) and (8), we have

$u_1 = \int \frac{-y_2F(x)}{W}dx$ and $u_2 = \int \frac{y_1F(x)}{W}dx$

i.e., $u_1 = -\int \frac{x^2e^x}{x^2e^x}dx = x$ and $u_2 = \int \frac{x^2}{x^2e^x}dx = -e^{-x}$.

Substituting for u_1, u_2 into (3), a particular solution of (1) is

$y - y_c + y_p - c_1x + c_2xe^x - x^2 - x.$

Example. Explain how the method of variation of parameters can be applied to find a particular solution of a non-homogeneous linear third order differential equation whose complementary function is known. Apply the method to find a particular solution of

$$\frac{d^3y}{dx^3}+\frac{dy}{dx}=\cos x.$$

Solution. Suppose the C.F. of a linear third order differential equation

$$\frac{d^3y}{dx^3}+P\frac{d^2y}{dx^2}+Q\frac{dy}{dx}+Ry=F(x) \quad (1)$$

is known to be $y_c=c_1y_1+c_2y_2+c_3y_3; y_1, y_2, y_3$ being linearly independent solutions of the associated homogeneous equation of (1).

We assume that $y_p=u_1y_1+u_2y_2=u_3y_3,$ where u_1,u_2,u_3 are functions of x.

$$y'_p=u_1y'_1+u_3y'_3+u'_1y_1+u'_2y_2+u'_3y_3.$$

We set $u'_1y_1+u'_2y_2+u'_3y_3=0.$

Then

$$y'_p=u_1y'_1+u_2y'_2+u_3y'_3$$

$$y''_p=u_1y''_1+u_2\,y''_2+u_3y''_3$$

$$y'''_p=u_1y'''_1+u_2y'''_2+u_3y'''_3+u'_1y''_1+u'_2y''_2+u'_3y''_3.$$

Substituting for y_p, y'_p, y''_p, y'''_p into equation (1), we have

$$\left(u_1y'''_1+u_2y'''_2+u_3y'''_3+u'_1y''_1+u'_2y''_2+u'_3y''_3\right)+P\left(u_1y''_1+u_2y''_2+u_3y''_3\right)$$

$$+Q\left(u_1y'_1+u_2y'_2+u_3y'_3\right)+R\left(u_1y_1+u_2y_2+u_3y_3\right)=F(x)$$

$$u_1\left(y'''_1+Py''_1+Qy'_1+Ry_1\right)+u_2\left(y'''_2+Py''_2+Qy'_2+Ry_2\right)$$

$$+u_2\left(y'''_2+Py''_2+Qy'_2+Ry_2\right)+u_3\left(y'''_3+Py''_3+Qy'_3+Ry_3\right)+u'_1y''_1+u'_2y''_2+u'_3y''_3=F(x) \quad (4)$$

From the system of equations (2), (3) and (4), we find $u'_1, u'_2\ and\ u'_3$ and on integration we get $u_1, u_2 and\ u_3$.

Now we apply this method to find P.I. of $\frac{d^3y}{dx^3}+\frac{dy}{dx}=\csc x.$

The C.F. of the equation is obtained by solving the equation $D(D+1)=0.$ Thus

$y_c = c_1 + c_2 \cos x + c_3 \sin x$ with $y_1 = 1, y_2 = \cos x, y_3 = \sin x.$

For the particular integral, let $y_p = u_1 y_1 + u_2 y_2 + u_3 y_3,$

$= u_1 + u_2 \cos x + u_3 \sin x.$

Substituting for y_1, y_2, y_3 and their derivatives into equations (2), (3) and (4) above, we have the following system of linear equations in u_1', u_2', u_3'.

$u_1' + u_2' \cos x + u_3' \sin x = 0$

$u_1'.0 - u_2' \sin x + u_3' \cos x = 0$

$u_1'.0 - u_2' \cos x - u_3' \sin x = \csc x.$

We solve this system by the Gaussian Elimination method. Augmented matrix of the system is

$$\left[\begin{array}{ccc|c} 1 & \cos x & \sin x & 0 \\ 0 & -\sin x & \cos x & 0 \\ 0 & -\cos x & -\sin x & \csc x \end{array}\right]$$

$$\left[\begin{array}{ccc|c} 1 & 0 & 0 & \csc x \\ 0 & -\sin x & \cos x & 0 \\ 0 & -\cos x & -\sin x & \csc x \end{array}\right] by\, R_1 + R_3$$

$$\left[\begin{array}{ccc|c} 1 & 0 & 0 & \csc x \\ 0 & -\sin x \cos x & \cos^2 x & 0 \\ 0 & -\cos x \sin x & -\sin^2 x & 1 \end{array}\right] \begin{array}{l} by(\cos x) R_2 \\ and (\sin x) R_3 \end{array}$$

$$\left[\begin{array}{ccc|c} 1 & 0 & 0 & \csc x \\ 0 & -\sin x \cos x & \cos^2 x & 0 \\ 0 & 0 & -1 & 1 \end{array}\right]$$

Therefore,

$-u_3' = 1 \, or \, u_3' = -1,$ from R_3

$(-\sin x \cos x)u_2' + (\cos^2 x)u_3' = 0$ from R_2

or $u_2' = -\cot x.$

$u_1' = \csc x,$ from $R_1.$

Hence

$$u_1 = \int \csc x \, dx = \ln|\csc x - \cot x|$$

$$u_2 = \int \frac{-\cos x}{\sin x} dx = -\ln|\sin x|.$$

and $u_3 = \int -dx = -x.$

$$y_p = u_1 + u_2 \cos x + u_3 \sin x.$$

$$= \ln|\csc x - \cot x| - (\cos x) ln|\sin x| - x \sin x$$

is the required particular solution.

1.4.2. Application of Differential Equations

Differential equations can be used in the solution of many problems in physical, biological and social sciences. In this section, we shall study such problems which can be formulated in terms of first and second order differential equations. The solution of the differential equation will lead to the solution of the given problem.

The following examples illustrate the techniques.

Example. A stone weighting 8 lbs. falls from rest toward the earth from a great height. As it falls it is acted upon by air resistance that is numerically $\frac{1}{2}v(in\ pounds)$. Where v is the velocity (in feet per second). Find the velocity and distance fallen at time t.

Solution. We use Newton's second law $F = ma$ to formulate mathematical model of this problem. The forces acting on the body are

i. F_1, the weight of the stone acting downward and hence is positive.

ii. F_2, the air resistance, numerically equal to $\frac{1}{2}v$, which acts upward and is therefore negative.

Newton's second law becomes $m\frac{dv}{dt} = F_1 + F_3$

i.e., $\frac{1}{4}\frac{dv}{dt} = 9 - \frac{1}{2}v, taking\ g = 32.$

$$\frac{dv}{dt} = 32 - 2v$$

$$\frac{dv}{dt} + 2v = 32$$

which is a linear equation. We now solve this differential equation. Here $I.F = e^{\int 2\,dt} = e^{2t}$.

And so,

$$\frac{d}{dt}\left(ve^{2t}\right) = 32e^{2t}$$

Integrating, we get

$$ve^{2t} = 16e^{2t} + c\ or\ v = 16 + ce^{-2t}$$

Now $v(0) = 0.$

Hence $0 = 16 + c\ or\ c = -16$

Therefore, $v = 16\left\{1 - e^{-2t}\right\}$ (1)

is the velocity after time t. (1) may be written as

$$\frac{dx}{dt} - 16\left(1 - e^{-2t}\right). \quad (2)$$

From (2), we have $dx = 16\left(1 - e^{-2t}\right)dt$

Or $x = 16t + 8e^{-2t} + k.$

Applying the initial condition $x(0)=0,$ we find $k=-8.$

Hence $x=16t+8e^{-2t}-8$ is the distance fallen after time t.

Example. A body of constant mass is projected upward from the earth's surface with an initial velocity $v_0.$ Assuming there is no air resistance, but taking into consideration the variation of the earth's gravitational field with altitude, find the smallest initial velocity for which the body will not return to the earth (This is the so-called escape velocity).

Solution. The general expression for the weight $w(x)$ of a body of mass m is obtained from Newton's inverse- square law of gravitational attraction. If R is the radius of the earth and x is the altitude above sea level, then $w(x)=\frac{k}{(R+x)^2}, k$ being constant. At $x=0, w=mg,$ hence $k=mgR^2$ and $w(x)=\frac{mg\,R^2}{(R+x)^2}.$ The only force acting on the body is its weight which acts downward. Thus, the equation of motion is $m\frac{dv}{dt}=\frac{-mg\,R^2}{(R+x)^2}$ or $v\frac{dv}{dt}=-\frac{mg\,R^2}{(R+x)^2}.$

Separating the variables and integrating, we have

$$\frac{1}{2}v^2=\frac{gR^2}{R+x}+c.$$

But $v=v_0\, at\, t=0$ i.e., at $x=0.$

Therefore, $c=\frac{1}{2}v_0^2-gR$

And so $v^2=v_0^2-2gR+\frac{2gR^2}{R+x}$ (1)

The escape velocity is found by requiring that v given by (1) remains positive for all (positive) value of $x.$ Thus we must have

$v_0^2 \geq n 2g R.$

Hence the escape velocity is

$v_0 = \sqrt{2 gR} = 6.9 miles / sec. taking\ R = 4000\, miles$

1.4.3. Summary and Discussion

In this chapter, we discussed some elementary aspects of differential equation. A very important topic in real life problem.

The study of differential equations is a wide sphere in pure and applied mathematics, physics, and engineering. All of these disciplines are concerned with the estate of differential equations of various types. Pure sum focuses on the life and singularity of solutions, while applied tally emphasizes the rigorous occasion of the methods for approximating solutions. Differential equations pleasure an important role in modeling virtually every physical, technical, or biological process, from celestial motion, to bridge design, to interactions between neurons. Differential equations such as those used to solve real-life problems may not necessarily be directly solvable, i.e. do not have closed pattern solutions. Instead, solutions can be approximated using numerical methods.

Applications. Many fundamental jurisprudences of physics and chemistry can be formulated as differential equations. In biology and economics, differential equations are used to mold the behavior of complex systems. The mathematical opinion of differential equations first developed together with the sciences where the equations had originated and where the results found application. However, diverse problems, sometimes originating in quite distinct scientific fields, may give rise to identical differential equations.

Whenever this happens, mathematical plan seats the equations can be viewed as a unifying principle seat diverse phenomena. As an example, consider dissemination of light and sound in the atmosphere, and of waves on the surface of a pond. All of them may be described by the same second-order partial differential equation, the breaker equation, which allows ourselves to think of brightening and sound as forms of waves, scads like familiar cascade in the water. Conduction of heat, the theory of which was developed by Joseph Fourier, is governed by another second-order partial differential equation, the heat equation. It turns out that many scattering processes, while seemingly different, are described by the same equation;

the Black-Scholes equation in finance is, for instance, related to the heat equation.

Euler-Lagrange equation in classical mechanics, Hamilton's equations in classical mechanics, Radioactive decay in nuclear physics Newton's statute of cooling in thermodynamics, The breaker equation , The heat equation in thermodynamics, Laplace's equation, which defines harmonic functions, Poisson's equation, The geodesic equation, The Navier-Stokes equations in fluid dynamics, The Diffusion equation in stochastic processes, The Convection-diffusion equation in fluid dynamics, The Cauchy-Riemann equations in complex analysis, The Poisson-Boltzmann equation in molecular dynamics, The shallow water equations and Universal differential equation.

Differential Equation. An equation involving one dependent variable and its derivatives with respect to one or more independent variables is called differential equation. For example

$$\frac{dy}{dx} + y\cos x = \sin x$$

Ordinary Differential Equation (O. D. E). A differential equation, in which ordinary derivatives of the dependent variable with respect to a single independent variable occur, is called an ordinary differential equation (O. D. E).

Partial Differential Equation. A differential equation involving partial derivatives of the dependent variable with respect to more than one independent variable is called a partial differential equation.

$$x\frac{\partial z}{\partial x} + y\frac{\partial z}{\partial y} = nx$$

Order of Differential Equation . The order of a differential is the order of the highest derivative that occurs in the equation.

Degree of a Differential Equation. The degree of a differential equation is the greatest exponent of the highest order derivative that appears in the equation. (The dependent variable and its derivatives should be expressed in a form free from radicals and fractions).

- $\frac{dy}{dx} + y\cos x = \sin x$ (order 1, degree 1)

- $\frac{d^2y}{dx^2} + xy\left(\frac{dy}{dx}\right)^2 = 0$ (order 2, degree 1)
- $\left[1+\left(\frac{dy}{dx}\right)^2\right]^{\frac{3}{2}} = \frac{d^2y}{dx^2}$ (order 2, degree 2)

Linear Differential Equation. An ordinary differential equation $F\left(x, y, \frac{dy}{dx}, \frac{d^2y}{dx^2}, \ldots, \frac{d^ny}{dx^n}\right) = 0$ is said to be linear if F is a linear function of the variables $x, \frac{dy}{dx}, \frac{d^2y}{dx^2}, \ldots, \frac{d^ny}{dx^n}$

It should be carefully noted that in a linear ordinary differential equation

- The dependent variable y and its derivative are all of degree one.
- No product of y or any of its derivative appear.
- No transcendental function of y and / or its derivative occur.

Nonlinear Differential Equation. A differential equation that is not linear is called a nonlinear differential equation. Differential equations occur in the mathematical formulation of many problems in science and engineering. Some such problems are

- Determining the motion of projectile, rocket, satellite or planet.
- Finding the charge or current in an electric circuit.
- Study of chemical reactions.
- Determination of curves with given geometrical properties.

Solution of a Differential Equation

A solution (or integral) of a differential equation is a relation between the variables, not containing derivatives, such that this relation and the derivatives obtained from it satisfy the given differential equation identically.

1.4.4. Initial Condition

It is often required to find the solution of a differential equation subject to certain conditions. If the conditions relate to one value of the independent variable such as $y = y_0$ at $x = x_0$ (written as $y(x_0) = y_0$) and $\frac{dy}{dx} = y'(x_0)$

at $x = x_0$, where x_0 belongs to some interval $]\alpha , \beta[$ then they are called initial conditions (or one- point boundary conditions) and x_0 is called the initial point.

1.4.5. Boundary Conditions

The problem of finding the solution of a differential equation such that all the associated constraints relate to two different values of the independent variable is called a two-point boundary value problem (or simply a boundary value problem). The associated supplementary boundary conditions are called two-point boundary conditions.

1.4.6. Separable Equations

Definition. A differential equation of the type $F(x)G(x)dx + f(x)g(y)dy = 0$ is called an equation with separable variables or simply a separable equation. Equation (1) may be written as

$$\frac{F(x)}{f(x)}dx + \frac{g(y)}{G(y)}dy = 0$$

which can be easily integrated.

1.4.7. Homogenous Equations

Definition. A function $f(x, y)$ is called homogeneous of degree n if $f(tx, ty) = t^n f(x, y)$, Where t is a nonzero real number. Thus $\sqrt{xy}$, $\frac{x^{10} + y^{10}}{x^2 + y^2}$ *and* $\sin\left(\frac{x}{y}\right)$ are homogeneous function of degree 1,8 and 0 respectively.

Exact Equation. The expression $M(x, y)dx + N(x, y)dy$ is called an exact differential if there exists a continuously differentiable function $f(x, y)$ of two real variables x and y such that the expression equals the total differential df. We know from calculus that $df = \frac{\partial f}{\partial x}dx + \frac{\partial f}{\partial y}dy.$

Thus, if it is exact then $M(x,y)=\frac{\partial f}{\partial x}=f_x$ and $N(x,y)=\frac{\partial f}{\partial y}=f_y$.

Integrating Factors. If the differential equation $M(x,y)dx+N(x,y)dy=0$ is not exact but when it is multiplied by a function $\mu(x,y)$ and the resulting equation $\mu(x,y)M(x,y)dx+\mu(x,y)N(x,y)dy=0$ is exact, then $\mu(x,y)$ is called an integrating factor (I.F) of the differential equation. The number of integrating factors of an equation may be infinite.

We list below (without proofs) some rules to find the integrating factors of equations of special types.

Rule 1. If $M(x,y)dx+N(x,y)dy=0$ is not exact and $\frac{M_y - N_x}{N}=P$, where P is a function of x only then (1) has an integrating factor $\mu(x)$ which also depends on x. $\mu(x)$ is solution of the differential equation

$$\frac{d\mu}{dx}=P\mu$$

i.e., $\mu(x)=\exp\int P\,dx$

Note that $M_y=\frac{\partial M}{\partial y}$, $N_x=\frac{\partial N}{\partial x}$.

Rule II. If $\frac{N_x-M_y}{M}=Q$, where Q is a function of y only, then the differential equation $Mdx+Ndy=0$ has an integrating factor

$$\mu(y)=\exp\int Q\,dy.$$

Rule III. If $M\,dx+Ndy=0$ (1)

is homogeneous and $xM+yN\neq 0$, then $\frac{1}{xM+yN}$ is an I.F of (1).

Rule IV. If $M\,dx + N dy = 0$ is of the form $y\,f(x,y)\,dx + x\,g(x,y)\,dy = 0$ and $xM - yN \neq 0$, then $\frac{1}{xM - yN}$ is an I.F of (1).

The following differential formulas are useful in the calculation of certain exact equations:

$$d\left(\frac{y}{x}\right) = \frac{x\,dy - y\,dx}{x^2}$$

$$d\left(\frac{x}{y}\right) = \frac{y\,dx - x\,dy}{y^2}$$

$$d(xy) = x\,dy + y\,dx$$

$$d\left(x^2 + y^2\right) = 2\,(x\,dx + y\,dy)$$

$$d\left(\ln\frac{x}{y}\right) = \frac{y\,dx - x\,dy}{xy}$$

$$d\left(arc\tan\frac{x}{y}\right) = \frac{y\,dx - x\,dy}{x^2 + y^2}$$

Linear Equations

A first order ordinary differential equation (ODE) is linear in the dependent variable y and the independent variable x if it is or can be written in the form $\frac{dy}{dx} + P(x)\,y = Q(x)$, where $P\,and\,Q$ are functions of x.

The Bernoulli Equation

Definition. An equation of the form $\frac{dy}{dx} + P(x)\,y = Q(x)\,y^n$ (1)

is called the Bernoulli differential equation. This equation is linear if $n = 0\,or\,1$. If n is not zero or 1, then (1) is reducible to a linear equation.

Clairaut's Equation

Definition. The equation $y = xp + f(p)$, is known as Clairaut's equation.

Envelope

Let $f(x, y, c) = 0$ be a one-parameter family of curves. Suppose all members of the family of curves are drawn for various values of the parameter c arranged in order of magnitude. The two curves that correspond to two consecutive values of c will be designated as neighboring curves. The locus of the ultimate points of intersection of neighboring curves is called the envelope of the family $f(x, y, c) = 0$.

Singular Solution: A solution of a differential equation $f(x, y, p) = 0$ is called a singular solution (S.S) if

i. It is not derived from the general solution by giving any particular value to the arbitrary constants
ii. At each of its points, it is tangent to some member of the one parameter family of curves represented by the general solution.

The Ricatti Equation

Definition. We have already studied first order linear differential equation

$$y' + P(x)y = R(x) \qquad (1)$$

If we add the term $q(x)y^2$ to the left-hand member of (1), we obtain a nonlinear differential equation

$$y' + P(x)y + Q(x)y^2 = R(x) \qquad (2)$$

(2) is called the Ricatti equation.

1.4.8. Linear Differential Equations

Definition. A linear differential equation of order n in the dependent variable y and the independent variable x is of the form

$$a_0(x)\frac{d^n y}{dx^n} + a_1(x)\frac{d^{n-1} y}{dx^{n-1}} + \ldots + a_{n-1}\frac{dy}{dx} + a_n(x)y = F(x), \qquad (1)$$

where $a_0(x), a_1(x), \ldots, a_{n-1}(x), a_n(x) and F(x)$ are functions of the independent variable x only and $a_0(x)$ is not identically zero. Using primes, (1) is also written as

$$a_0y^n + a_1y^{n-1} + \ldots + a_{n-1}y' + a_ny = F(x) \quad (2)$$

where $a_0, a_1, \ldots, a_{n-1}, a_n$ are real constants.

Definition. If $y_1(x), y_2(x), \ldots, y_m(x)$ are m functions of an independent variable x and $c_1, c_2, \ldots, c_m$ are constants, then the expression $c_1y_1(x) + c_2y_2(x) + \ldots + c_my_m(x)$ is called a linear combination of $y_1(x), y_2(x), \ldots, y_m(x)$.

i. Every homogeneous linear $nth-order$ differential equation $a_0y^n + a_1y^{n-1} + \ldots + a_{n-1}y' + a_ny = 0$ has n linearly independent solutions $y_1, y_2, \ldots, y_n$

ii. If $y_1, y_2, \ldots, y\_m$ are n linealy independednt solutions of (3), then any linear combination $y_c = c_1y_1 + c_2y_2 + \ldots + c_ny_n$ of $y_1, y_2, \ldots, y_n$ is the general solution of (3), $c_1, c_2, \ldots, c_m$ being arbitrary constants.

iii. Let y_p be any particular solution of (2). i.e., y_p does not contain any constant, then $y_c + y_p$ is the general solution of (2).

1.4.9. Homogeneous Linear Equation

Consider the equation $$a_0\frac{d^ny}{dx^n} + a_1\frac{d^{n-1}y}{dx^{n-1}} + \ldots + a_{n-1}\frac{dy}{dx} + a_n\, y, = 0 \quad (1)$$

where $a_0, a_1, \ldots, a_{n-1}$ are real constants.

Case I. Distinct Real Roots

Let $m_1, m_2, \ldots, m_n$ be n distinct real roots of (2). Then $e^{m_1x}, e^{m_2x}, \ldots, e^{m_nx}$ are n distinct solutions of (1). These n solutions are linearly independent. Hence the general solution of (2) is $y = c_1e^{m_1x} + c_2e^{m_2x} + \ldots + c_ne^{m_nx}$, Where $c_1, c_2, \ldots, c_n$ are arbitrary constants.

Case II. Repeated Real Roots

$$y=(c_1+c_2x)e^{m_1x}+c_3e^{m_3x}+\ldots+c_ne^{m_nx}$$

In the same manner, if the characteristic equation (2) has the triple repeated root m_1, the corresponding part of the general solution of (1) is the solution of

$$(D-m_1)^3y=0$$

Proceeding as before, we can easily find

$$y=\left(c_1+c_2x+c_3x^2\right)e^{m_1x}$$

Case III. Complex Roots

Suppose the characteristic equation has the complex number $a+ib$ as a non- repeated root. Since coefficients of (1) are real, the conjugate complex number $a-ib$ is also a non- repeated root. The corresponding part of the general solution is

$y=k_1e^{(a+ib)x}+k_2e^{(a-ib)x}$, where $k_1 \; and \; k_2$ are arbitrary constants.

$$y=e^{ax}\left[k_1e^{ibx}+k_2e^{-ibx}\right]$$

$$=e^{ax}\left[k_1(\cos bx+i\sin bx)+k_2(\cos bx-i\sin bx)\right]$$

$$=e^{ax}\left[(k_1+k_2)\cos bx+i(k_1-k_2)\sin bx\right]$$

$$=e^{ax}\left[c_1\sin bx+c_2\cos bx\right]$$

Where $c_1=i(k_1-k_2), c_2=k_1+k_2$ are two arbitrary constants.

If $a+ib \; and \; a-ib$ are conjugate complex roots, each repeated k times, the the corresponding part of the general solution of (1) may be written as

$$y=\left[\left(c_1+c_2x+c_3x^2+\ldots+c_kx^{k-1}\right)\sin bx+\left(c_{k+1}+c_{k+2}x+\ldots+c_{2k}x^{k-1}\right)\cos bx\right]$$

Solution of the equation

- Complementary Function (C.F)
- Particular Solution (P.I)

The C.F is the solution of the homogeneous equation $f(D)y=0$ and we have described different cases of its solution.

Now to find the $P.I$ of (1), we can write

$$y=\frac{1}{f(D)}F(x)$$

and try to evaluate $y=\dfrac{1}{f(D)}F(x).$

To Find the P.I when $\boldsymbol{F(x)}$ $\sin \boldsymbol{ax}$ $\boldsymbol{or} \cos \boldsymbol{ax}$.

Here we have to evaluate $\dfrac{1}{f(D)}\sin ax \, and \, \dfrac{1}{f(D)}\cos ax.$

From Euler's theorem, we have

$$e^{iax}=\cos ax+i\sin ax$$

Thus $\dfrac{1}{f(D)}\sin ax \, and \, \dfrac{1}{f(D)}\cos ax$ are respectively imaginary and real parts of $\dfrac{1}{f(D)}e^{iax}.$

If $f(D)$ contains only even powers of $D, say\; f(D)=f(D^2),$ it is easy to see that

$$\frac{1}{f(D^2)}\begin{cases}\sin ax\\ \cos ax\end{cases}=\frac{1}{f(-a^2)}\begin{cases}\sin ax\\ \cos ax\end{cases}, \quad provided -a^2\, is\, not\, a\, zero\, of\; f(D^2).$$

1.4.10. Series Expansion of Polynomials or Expressions involving Operators.

This method is useful when $F(x)$ is a polynomial in x.

Let $f(D)y=F(x)$ be such that $F(x)$ is a polynomial in x. To evaluate the particular integral

$$y_p = \frac{1}{f(D)} F(x)$$

It is often useful to expand $\frac{1}{f(D)}$ in a series in D by the Binomial Theorem for negative exponent so that

$$y_p = \left(1 + a_1 D + a_2 D^2 + a_3 D^3 + \ldots\right) F(x)$$

The derivatives on the right will vanish after certain stage since

$D^n x^r = 0 \ if \ n > r.$

The following binomial expansions will be useful in this connection:

$$\frac{1}{1-x} = 1 + x + x^2 + x^3 + \ldots.$$

$$\frac{1}{1+x} = 1 - x + x^2 - x^3 + \ldots.$$

1.4.11. The Method of Undetermined Coefficients (U. C. Method)

We have studied some special cases in which particular integral can be evaluated by the inverse operator. Now we consider the method of undetermined coefficients which can prove simpler in finding the particular integral of the equation $f(D)y = F(x)$ when $F(x)$ is

a. An exponential function: e^{ax}

b. A polynomial: $\left(b_0 x^n + b_1 x^{n-1} + \ldots + b_n\right)$

c. Sinusoidal function : $\sin ax \ or \ \cos bx$

The more general case in which $F(x)$ is sum of a product of terms of the above types, such as

$$F(x) = e^{ax}\left(b_0 x^n + b_1 x^{n-1} + \ldots + b_n\right)\begin{Bmatrix} \sin ax \\ \cos bx \end{Bmatrix}$$

The Cauchy-Euler Equation

An equation of the form

$$a_0 x^n \frac{d^n y}{dx^n} + a_1 x^{n-1} \frac{d^{n-1} y}{dx^{n-1}} + \ldots + a_{n-1} x \frac{dy}{dx} + a_n y = F(x)$$

is called Cauchy-Euler(or equidimensional) equation. $a_0, a_1, \ldots, a_{n-1}, a_n$ are real constants. The equation can be reduced to a linear differential equation with constant coefficients by the transformation $x = e^t \; or \; t = \ln x$

Reduction of Order

If one solution of the second order linear equation $\frac{d^2 y}{dx^2} + P\frac{dy}{dx} + Qy = 0$ (1)

(where P, Q are not necessarily constants and may be functions of x) is known, then we can use it to find the general solution of

$$\frac{d^2 y}{dx^2} + P\frac{dy}{dx} + Qy = F(x) \qquad (2)$$

The procedure, due to D' Alembert, is known as the method reduction of order.

The Wronskian

If y_1, y_2 are two differentiable functions of x on $I = [a, b]$ then their Wronskian, denoted by $W = W[y_1, y_2]$, is obtained by

$$W = W[y_1, y_2] = y_1 y_2' - y_1' y_2 = \begin{vmatrix} y_1 & y_2 \\ y_1' & y_2' \end{vmatrix}$$

Similarly, the Wronskian of three differentiable functions y_1, y_2, y_3 on $I = [a, b]$ is defined by

$$W = W[y_1, y_2, y_3] = \begin{vmatrix} y_1 & y_2 & y_3 \\ y_1' & y_2' & y_3' \\ y_1'' & y_2'' & y_3'' \end{vmatrix}$$

The definition can be extended in a similar manner for the Wronskian of n differentiable functions on $I = [a, b]$.

Theorem. If y_1, y_2 are two solutions of $\frac{d^2y}{dx^2}+P\frac{dy}{dx}+Qy=0,$ Then their Wronskian $W=W[y_1,y_2]=0$ if and only if y_1, y_2 are linearly dependent.

Corollary. $W[y_1,y_2]\neq 0$ if and only if y_1, y_2 are linearly independent.

Variation of Parameters.

We found the solution of the equation $\frac{d^2y}{dx^2}+P\frac{dy}{dx}+Qy=F(x).$ (1)

where P,Q are functions of x, in the previous section by reduction of order of (1). The solution of (1) can be determined by a procedure known as the method of variation of parameters. This method can be applied even to equations of higher order.

$$u_1=\int\frac{-y_2F(x)}{y_1y_2'-y_1'y_2}dx=\int\frac{-y_2F(x)}{W}dx$$

$$u_2=\int\frac{y_1F(x)}{y_1y_2'-y_1'y_2}dx=\int\frac{y_1F(x)}{W}dx$$

where W is the Wronskian of y_1, y_2.

CHAPTER 2

FUNDAMENTAL CONCEPTS OF PARTIAL DIFFERENTIAL EQUATIONS

CONTENTS

2.1. INTRODUCTION

Many practical problems in science and engineering , when formulated mathematically, give partial differential equations (often referred to as PDE) , In order to understand the physical behavior of the mathematical model, it is necessary to have some knowledge about the mathematical character properties, and the solution of the governing PDE, An equation which involves several independent variables usually denoted by $(x, y, z, t, \ldots)$, a dependent function u of these variables , and the partial derivatives of the dependent function u with respect to the independent variables such as

$$F\left(x ,y ,z ,t\ldots ,u_x ,u_y ,u_z ,u_t ,\ldots\ldots u_{xx} ,u_{yy} ,\ldots u_{xy} ,\ldots\right)=0 \tag{1.1}$$

is called a partial differential equation. A few well-known examples are:

$(i)\ u_t = k\left(u_{xx} + u_{yy} + u_{zz}\right)$ [linear three-dimensional heat equation]

$(ii) u_{xx} + u_{yy} + u_{zz} = 0$ [Laplace equation in three dimension]

$(iii) u_{xx} + u_{yy} + u_{zz} = 0$ [linear three-dimensional wave equation]

$(iv) u_t = u u_x = \mu u_{xx}$ [one linear one-dimensional Burger equation]

In all these examples, u is the dependent function and the subscripts denote b partial differential with respect to these variable.

Definition 1. The order of the partial differential equation is the order of the highest derivative occurring in the equation. Thus the above examples are partial differential equation of second order, whereas $u_t = uu_{xxx} + \sin x$ is an example for third order partial differential equation.

2.2. CLASSIFICATION OF SECOND ORDER PDE

The most general linear second order PDE, with one dependent function u on a domain Ù of points $X\ \aleph(x_1, x_2, \quad), n \quad 1,$ is

$$\sum_{i,j=1}^{n} a_{ij}\, u_{xi} x_j + \sum_{i=1}^{n} b_i u_{xi} + F(u) = G$$

The classification of PDE depend only on the highest order derivative present.

The classification of PDE is motivated by the classification of the quadratic equation of the form

$$Ax^2+Bxy+Cy^2+Dx+Ey+F=0 \tag{3}$$

which is elliptic parabolic or hyperbolic according as the discriminant B^2-4AC is negative , zero or positive . Thus we have the following second order linear PDE in two variables $x\,and\,y$

$$Au_{xx}+Bu_{xy}+Cu_{yy}+Du_x+Eu_y+Fu=G \tag{4}$$

where the coefficients $A,B,C,\ldots$ may be functions of $x\,and\,y,$ however, for the sake of simplicity we assume then to be constant . Equation is elliptic, parabolic of hyperbolic at a point (x_0,y_0) according as the discriminant. $B^2(x_0,y_0)-4A(x_0,y_0)C(x_0,y_0)$ is negative, zero or positive. If this is true at all points in a domain Ù, then is said to be elliptic, parabolic in that domain. If the number of independent variables is two or three, a transformation can always be found to reduce the given PDE to a canonical form (also called normal form). In general, when the number of independent variables is greater than 3, it is not always possible to find such a transformation excepts in certain special cases. the idea of reducing the given PDE to a canonical form is that the transformed equation assumes a simple form so that the subsequent analysis of solving the equation is made easy.

2.2.1. Canonical Forms

Consider the most general transformation of the independent variables $x\,and\,y$ of to new variables $\xi,\eta,where$

$$\xi=\xi(x,y),=\eta=\eta(x,y) \tag{5}$$

Such that the function $\xi\,and\,\eta\,are$ continuously differentiable and the Jacobian

$$j=\frac{\partial(\xi,\eta)}{\partial(x,y)}=\begin{vmatrix}\xi_x\,\xi_y\\ \eta_x\,\eta_y\end{vmatrix}-(\zeta_x\eta_y-\xi_y\eta_x)\neq 0$$

In the domain Ù Eq. 4 holds, using the chain rule of partial differential . the partial derivative becomes

$$u_x = u_\xi \xi_x + u_\eta \eta_x$$

$$u_y = u_\xi \xi_y + u_\eta \eta_y$$

$$u_{xx} = u_{\xi\xi} \xi_x^2 + 2u_{\xi\eta} \xi_x \eta_x + u_{\eta\eta} \eta_x^2 + u_\xi \xi_{xx} + u_\eta \eta_{xx}$$

$$u_{xy} = u_{\xi\xi} \xi_x \xi_y + u_{\xi\eta} \left(\xi_x \eta_y + \xi_y \eta_x \right) + u_{\eta\eta} \eta_x \eta_y + u_\xi \xi_{xy} + u_\eta \eta_{xy}$$

$$u_{yy} = u_{\xi\xi} \xi_y^2 + 2u_{\xi\eta} \xi_y \eta_y + u_{\eta\eta} \eta_y^2 + u_\xi \xi_{yy} + u_\eta \eta_{yy}$$

Substituting these expression into the original different equation , we get

$$\overline{A} u_{\xi\xi} + \overline{B} u_{\xi\eta} + \overline{C} u_{\eta\eta} + \overline{D} u_\xi + \overline{E} u_\eta + \overline{F} u = \overline{G}$$

where

$$\overline{A} = A\xi_x^2 + B\xi_x \xi_y + C\xi_y^2$$

$$\overline{B} = 2A\xi_x \eta_x + B\left(\xi_x \eta_y + \xi_y \eta_x \right) + 2C\xi_y \eta_y$$

$$\overline{C} = A\eta_x^2 + B\eta_x \eta_y + C\eta_y^2$$

$$\overline{D} = A\xi_{xx} + B\xi_{xy} + C\xi_{yy} + D\xi_x + E\xi_y$$

$$\overline{F} = F, \overline{G} = G \tag{9}$$

It may be noted that the transformed equation has the same form as that of the original equation under the general transformation

Since the classification of Eq. depends on the coefficient $A, B \, and \, C$, we can also rewrite the equation in the form

$$Au_{xx} + Bu_{XY} + Cu_{yy} = H\left(x, y, u, u_x, u_y \right) \tag{10}$$

It can be shown easily that under the transformation (5) Eq. (10) takes one of the following three canonical form :

$$(i) \; u_{\xi\xi} - u_{\eta\eta} = \varnothing\left(\xi, \eta, u, u\xi, u\eta \right) \tag{11a}$$

or $u_{\xi\eta} = \varnothing, \left(\xi, \eta, u, u_\xi, u_\eta \right)$ in the hyperbolic case

$(ii) u_{\xi\xi} + u_{\eta\eta} = \varnothing(\xi, \eta, u, u_\xi, u_\eta)$ in the elliptic case (11b)

$(iii) u_{\xi\xi} = \varnothing(\xi, \eta, u, u\xi, u\eta)$ (11c)

or $u_{\eta\eta} = \varnothing(\xi, \eta, u, u\xi, u\eta)$ in the parabolic case

We shall discuss in detail each of these cases separately .

Using Eq. (9) it can also be verified that

$$\bar{B}^2 - 4\bar{A}\bar{C} = (\xi_x \eta_y - \xi_y \eta_x)^2 (B^2 - 4AC)$$

and therefore we conclude that the transformation of the independent variables does not modify the type of PDE.

2.2.2. Canonical Form for Hyperbolic Equation

Since the discriminant $\bar{B}^2 - 4\bar{A}\bar{C} > 0$ for the hyperbolic case, we set $\bar{A} = 0\, and\, \bar{C} = 0$ in Eq. (9) which will gave us the coordinates $\xi\, and\, \eta\, that$ reduce the given PDE to a canonical form in which the coefficient of $u_{\xi\xi}, u\eta\eta$ are zero. Thus we have

$$\bar{A} = A\xi_x^2 + B\xi_x\xi_y + C\xi_y^2 = 0$$

$$\bar{C} = A\eta_x^2 + B\eta_x\eta_y + C\eta_y^2 = 0$$

which, in rewriting , become

$$A\left(\frac{\xi x}{\xi y}\right)^2 + B\left(\frac{\xi x}{\xi y}\right) + C = 0$$

$$A\left(\frac{\eta_x}{\eta_y}\right)^2 + B\left(\frac{\eta_x}{\eta_y}\right) + C = 0$$

Solving these equation for $(\xi_x / \xi_y)\, and\, (\eta_x / \eta_y)$, we get

$$\frac{\xi_x}{\xi_y} = \frac{B + \sqrt{B^2 - 4AC}}{2A}$$

$$\frac{\eta_x}{\eta_y} = \frac{-B - \sqrt{B^2 - 4AC}}{2A}$$

The condition $B^2 - 4Ac$ implies that the slopes of the curves $\xi(x, y) = C_1, \eta(x, y) = C_2$ are real . Thus, if $B^2 > 4AC$, then at any point (x, y) there exist two real directions given by the two roots (12) along which the PDE (4) reduces of the canonical form. These are called characteristics equation. Though there are two solutions for each quadratic, we have considered only one solution for each. Otherwise we will end up with the same two coordinates. Along the curve $\xi(x, y) = c_1, we\, have$

$$d\xi = \xi_x\, dx\, \xi_y dy = 0$$

Hence,

$$\frac{dy}{dx} = -\left(\frac{\xi_x}{\xi_y}\right)$$

Similarly, along the curve $\eta(x, y) = c_2,$ we have

$$\frac{dy}{dx} = -\left(\frac{\eta_x}{\eta_y}\right)$$

Integrating Eq. (13) and (14), we obtain the equations of family of characteristics $\xi(x, y) = c_1\, and\, \eta(x, y) = c_2,$ which are called the characteristic of the PDE.(4). Now to obtain the canonical form for the given PDE, we substitute the expressions of $\xi\, and\, \eta\, into\, Eq.$ (8) which reduce to Eq.(11a) Though there are two solutions for each quadratic , we have considered only solution for each . Otherwise we will end up with the same two coordinates .

To make the ideas clearer, let us consider the following examples

$$3u_x + 10u_{xy} + 3u_{yy} = 0$$

Comparing with the standard PDE (4), we have $A = 3, B = 10, C = 3, B^2 - 4AC = 64 > 0.$ Hence the given equation is a hyperbolic PDE. The corresponding characteristics are :

$$\frac{dy}{dx} = -\left(\frac{\xi_x}{\xi_y}\right) = \left(\frac{-B + \sqrt{B^2 - 4AC}}{2A}\right) = \frac{1}{3}$$

$$\frac{dy}{dx} = -\left(\frac{\eta_x}{\eta_y}\right) = \left(\frac{-B + \sqrt{B^2 - 4AC}}{2A}\right) = 3$$

To find $\xi\ and\ \eta,$ we first solve for $y\,by$ integrating the above equations. Thus. we get

$$y = 3x + c_1, y = \frac{1}{3}x + c_2$$

Which give the constant as

$$c_1 = y - 3x, c_2 = y - x/3$$

Therefore,

$$\zeta = y - 3x = c_1, \eta = y, -\frac{1}{3}x = c_2$$

These are the characteristics lines for the given hyperbolic equation. In this example, the characteristics are found to be straight lines in the $(x, y) - plane$ along which the initial data, impulse will propagate.

To find the canonical equation, we substitute the expression for $\xi\ and\ \eta\ into\ Eq.$ (9) to get

$$\bar{A} = A\xi_x^2 + B\xi_x\xi_y + C\xi_y^2 = 3(-3)^2 + 10(-3)(1) + 3 = 0$$

$$\bar{B} = 2A\xi_x\eta_x + B(\xi_x\eta_y + \xi_y\eta_x) + 2C\xi_y\eta_y$$

$$= 2(3)(-3)\left(-\frac{1}{3}\right) + 10\left[(-3)(1) + 1\left(-\frac{1}{3}\right)\right] + 2(3)(1)(1)$$

$$= 6 = 10\left(-\frac{10}{3}\right) + 6 = 12 - \frac{100}{3} = -\frac{64}{3}$$

$$C = 0, \bar{D} = 0, \bar{E} = 0\,\bar{F} = 0$$

Hence, the required canonical form is

$$\frac{64}{3}u_{\xi\eta} = 0\,or\,u_{\xi\eta} = 0$$

On integration, we obtain

$$u(\xi,\eta)=f(\xi)+g(\eta)$$

where f *and* g are arbitrary. Going back to the original variables , the general solution is

$$u=(x,y)=f(y-3x)+g\ (y-x/3)$$

2.2.3. Canonical Form for parabolic Equation

For the parabolic equation, the discriminant $\overline{B}^2-4\overline{A}\overline{C}=0$, which can be true if $\overline{B}=0$ and $\overline{A}\ or\ \overline{C}$ is equal to zero n. Suppose we set first $\overline{A}=0\ in\ Eq.(9)$. then we obtain

$$\overline{A}=A\xi_x^2+B\xi_x\xi_y+C\xi_y^2=0$$

$$A\left(\frac{\xi_x}{\xi_y}\right)+B\left(\frac{\xi_x}{\xi_y}\right)+C=0$$

which gives

$$\frac{\xi_x}{\xi_y}=\left(\frac{-B\pm\sqrt{B^2-4AC}}{2A}\right)$$

Using the condition for parabolic case, we get

$$\frac{\xi_x}{\xi_y}=-\frac{B}{2A} \tag{15}$$

Hence, to find the function $\xi=\xi(x,y)$ which satisfies Eq. (15) . we set

$$\frac{dy}{dx}=-\frac{\xi_x}{\xi_y}=\frac{B}{2A}$$

and get the implicit solution

$$\xi(x,y)=C_1$$

In fact, one can verify that $\overline{A}=0$ implies $\overline{B}=0\,as$ follows :

$$B=2A\xi_x\eta_x+B(\xi_x\eta_y+\xi_y\eta_x)+2C\xi_y\eta_x$$

Since $B^2 - 4AC = 0$ the above relation reduces to

$$\bar{B} = 2A\xi_x\eta_x + 2\sqrt{AC}\left(\xi_x\eta_y + \xi_y\eta_x\right) + 2C\xi_y\eta_x$$

$$= 2\left(\sqrt{A}\xi_x + \sqrt{C}\xi_y\right)\left(\sqrt{A}\eta_x + \sqrt{C}\eta_y\right)$$

However,

$$\frac{\xi_x}{\xi_y} = -\frac{B}{2A} = -\frac{2\sqrt{AC}}{2A} = -\sqrt{\frac{C}{A}}$$

We therefore choose ξ in such a way that both $\bar{A}\, and\, \bar{B}$ are zero. Then η can be chosen in any way we like as long as it is not parallel to the ξ coordinates . In other words. We choose $\eta\, such$ that the Jacobian of the transformation is not zero. Thus we can write the canonical equation for parabolic case by simply substituting $\xi\, and\, \eta\, into\, Eq.(8)$ which reduces to either of the forms (11c) .

To illustrate the procedure, we consider the following example:

$$x^2u_{xx} - 2xyu_{xy} + y^2u_{yy} = e^x$$

The discriminant $B^2 - 4AC = 4x^2y^2 - 4x^2y^2 = 0,$ and hence the given PDE is parabolic everywhere. The characteristics equation is

$$\frac{dy}{dx} = -\frac{\xi_x}{\xi_y} = \frac{B}{2A} = -\frac{2xy}{2x^2} = -\frac{y}{x}$$

On integration, we have

And hence $\xi = xy$ will satisfy the characterstics eawuation and we can chosen $\eta = y$. To finds the canonical equation , we substitute the expression for $\xi\, and\, \eta\, into\, Eq.(9)\, to\, get$

$$\bar{A} = Ay^2 + Bxy + cx^2 = x^2y^2 - 2x^2y^2 + y^2x^2 = 0$$

$$\bar{B} - 0.\bar{C} = y^2, \bar{D} - -2xy$$

$$\bar{E} = 0, \bar{F} = 0, \bar{G} = e^x$$

Hence, the transformed equation is

$$y^2 u_{\eta\eta} - 2xyu_\xi = e^x \text{ Or } \eta^2 u_{\eta\eta} = 2\xi u_\xi + e^{\frac{\xi}{\eta}}$$

The canonical form is, therefore

$$u_{\eta\eta} = \frac{2\xi}{\eta^2} u_\xi + \frac{1}{\eta^2} e^{\xi/\eta}$$

2.2.4. Canonical Form for Elliptic Equation

Since the discriminant $B^2 - 4AC < 0, for$ elliptic case, characteristics equation

$$\frac{dy}{dx} - \frac{B - \sqrt{B^2 - 4AC}}{2A}$$

$$\frac{dy}{dx} - \frac{B + \sqrt{B^2 - 4AC}}{2A}$$

Give us complex conjugate coordinates, say $\xi \, and \, \eta.$ Now , we make another transformation from $(\xi, \eta) to (\alpha, \beta)$ so that

$$\alpha = \frac{\xi + \eta}{2}, \beta = \frac{\xi - \eta}{2i}$$

which give us the required canonical equation in the form (11b).

To illustrate the procedure, we consider the following example :

$$u_{xx} + x^2 u_{yy} = 0$$

The discriminant $B^2 - 4AC = -4x^2 < 0.$ Hence the given PDE is elliptic . The characteristics equations are

$$\frac{dy}{dx} = \frac{B - \sqrt{B^2 - 4AC}}{2A} = -\frac{\sqrt{-4x^2}}{2} = -ix$$

Integration of these equation yields

$$iy + \frac{x^2}{2} = c_1, -iy + \frac{x^2}{2} = c_2$$

Hence, we may assume that

$$\xi=\frac{1}{2}\,x^2+iy\,,\eta=\frac{1}{2}x^2-iy$$

Now, introducing the second transformation

$$\alpha=\frac{\xi+\eta}{2}.\beta=\frac{\xi+\eta}{2i}$$

We obtain

$$\alpha=\frac{x^2}{2},\beta=y$$

The canonical form can now be obtained by computing

$$\bar{A}=A\alpha_x^2+\beta\ \alpha_x\,\alpha_y+c\alpha_y^2=x^2$$

$$\bar{B}=2A\alpha_x\,\beta_x+B\left(a_x\,\beta_y+\alpha_y\beta_x\right)+2c\left(\alpha_y\,\beta_y\right)=0$$

$$\bar{C}=A\beta_x^2+B\beta_x\beta_y+c\beta_y^2=x^2$$

$$\bar{D}=Aa_{xx}+Ba_{xy}+ca_{yy}+Da_x+Ea_y=1$$

$$\bar{E}=A\beta_{xx}+B\beta_{xy}+c\beta_{xy}+D\beta_x+E\beta_y=0$$

$$\bar{F}=0,\bar{G}=0$$

Thus required canonical equation is

$$x^2u_{aa}+x^2u_{\beta\beta}+u_\alpha=0$$

or

$$u_{\alpha\alpha}+u_{\beta\beta}=-\frac{u_\alpha}{2\alpha}$$

Example 1. Classify and reduce the relation

$$y^2u_{xx}-2xyu_{xy}+x^2u_{yy}=\frac{y^2}{x}u_x+\frac{x^2}{y}u_y$$

to a canonical form and solve it.

Solution. The discriminant of the given PDE is

$$B^2-4AC=4x^2y^2-4x^2y^2=0$$

Hence the given equation is of a parabolic type, The characteristics equation is

$$\frac{dy}{dx} = -\frac{\xi_x}{\xi_y} = \frac{B}{2A} = \frac{-2xy}{2y^2} = -\frac{x}{y}$$

Integration gives $x^2 + y^2 = c_1. Therefore, \xi = x^2 + y^2$ satisfies the characteristic equation. The $\eta -$ coordinates can be chosen arbitrary so that is not parallel to $\xi,$ i.e. the Jacobian of the transformation is not zero . Thus, we chosen

$$\xi = x^2 + y^2, \eta = y^2$$

To find the canonical equation we compute

$$\bar{A}\, A\xi_x^2 + B\,\xi_x\xi_y + G\xi_y^2 = 4x^2y^2 - 8x^2y^2 + 4x^2y^2 = 0$$

$$\bar{B} = 0\ ,\ \bar{C} = 4x^2y^2,\ \bar{D} = \bar{E} = \bar{F} = \bar{G} = 0$$

Hence, the required canonical equation is

$$4x^2\, y^2 u_{\eta\eta} = 0\, or\, u_{\eta\eta} = 0$$

To solve this equation, we integrate functions of $\xi.$ Now , going back to the original independent variables , the required solution is

$$u = y^2 f\left(x^2 + y^2\right) + g\left(x^2 + y^2\right)$$

Example. 2 Reduce the following equation to a canonical form:

$$\left(1 + x^2\right)u_{xx} + \left(1 + y^2\right)u_{yy} + xu_x + yu_y = 0$$

Solution. The discriminant of the given PDE is

$$B^2 - 4AC = -4\left(1 + x^2\right)\left(1 + y^2\right) < 0$$

Hence the given PDE is an elliptic type. The characteristics equation are

$$\frac{dy}{dx} = \frac{B - \sqrt{B^2 - 4AC}}{2A} = - = \frac{\sqrt{-4\left(1 + x^2\right)\left(1 + y^2\right)}}{2\left(1 + x^2\right)} = -i\sqrt{\frac{1 + y^2}{1 + x^2}}$$

$$\frac{dy}{dx} = \frac{B - \sqrt{B^2 - 4AC}}{2A} = i\sqrt{\frac{1 + y^2}{1 + x^2}}$$

On integrating, we get

$$\xi = In\ (x+\sqrt{x^2+1}) - i\ In\ (y+\sqrt{y^2+1}) = c_1$$

$$\eta = In\ (x+\sqrt{x^2+1}) - i\ In\ (y+\sqrt{y^2+1}) = c_2$$

Introducing the second transformation

$$\alpha = \frac{\xi+\eta}{2}, \beta = \frac{\eta-\xi}{2i}$$

we obtain

$$\alpha = \text{In}\ (\text{x} + \sqrt{x^2+1})$$

$$\beta = \text{In}\ (\text{y} + \sqrt{y^2+1})$$

Then the canonical equation for the given PDE is

$$u_{aa} + u_{\beta\beta} = 0$$

Example. Reduce the following equation top a canonical form and hence solve it.

$$u_{xx} - 2\ \sin x u_{xy} - \cos^2 x u_{yy} - \cos x u_y = 0$$

Solution. Computing with the general second order PDE (4) ,we have

$$A = 1\ ,\ B = -2\sin x,\ C = -\cos^2 x,$$

$$D = 0, E = -\cos x, F = 0, G = 0$$

The discriminant $B^2 - 4AC = 4(\sin^2 x + \cos^2 x) = 4 > 0.$ Hence the given PDE is hyperbolic. The relevant characteristics equations are

$$\frac{dy}{dx} = \frac{B-\sqrt{B^2-4AC}}{2A} = -\sin x - 1$$

$$\frac{dy}{dx} = \frac{B+\sqrt{B^2-4AC}}{2A} = -1 - \sin x$$

On integration, we get

$$y = \cos x - x + c_1,\ y = \cos x + x + c_2$$

Thus, we choose the characteristics line as

$\xi = x + y - \cos x = c_1, \ \eta = -x + y - \cos x = c_2$

In order to find the canonical equation ,we compute

$\bar{A} = A\xi_x^2 + B\xi_x\xi_y + C\xi_y^2 = 0$

$\bar{B} = 2A\xi_x\eta_x + B\left(\xi_x\eta_y + \xi_y\eta_x\right) + 2C\xi_y\eta_y$

$= 2\,(\,\sin x + 1\,)(\,\sin x - 1\,) - 4\sin^2 x - 2C\,\xi_y\eta_y$

$\bar{C} = 0, \ \bar{D} = 0, \bar{E} = 0, \ \bar{F} = 0, \ \bar{G} = 0$

Thus, required canonical equation is

$u_{\xi\eta} = 0$

Integrating with respect to ξ, we obtain

$u_\eta = f(\eta)$

where f is arbitrary. Integrating once again with respect to η, we have

$u = \int f(\eta)\,d\eta + g(\xi)$

or $u = ø\,(\,ç\,) + \ g(\,î\,)$

where $g(\xi)$ is another arbitrary function. Returning to the old variables x, y the solution of the Given PDE is

$u(x, y) = \psi(y - x - cosx)g(y + x - \cos x)$

Example. Find the characteristics of the equation

$u_{xx} + 2u_{xy} + \sin^2(x)u_{yy} + u_y = 0$

when it is of hyperbolic type.

Solution. The discriminant $B^2 - 4AC = 4 - 4\sin^2 x = 4\cos^2 x$ *.Hence* for all $x \neq (2\pi - 1)\pi/2$, the given PDE is of hyperbolic type. The characteristics equation are

$$\frac{dy}{dx} = \frac{B \ \sqrt{B^2 \ \ 4AC}}{2A} = \ \ \cos$$

On integration, we get

$y=x-\sin x+\sin x\ ,\ y=x+\sin x+c_2$

Thus, the characteristics equation are

$\xi=y-x+\sin x,\ \eta=y-x-\sin x$

Example. Reduce the following equation to a canonical form and hence solve it.

$yu_{xx}+(x+y)u_{xy}+xu_{yy}=0$

Solution. The discriminant

$B^2-4AC=(x+y)^2-4xy=(x-y)^2>0$

Hence the given PDE is hyperbolic everywhere except along the line $y=x$; whereas on the line y = x. it is parabolic . when $y\neq x$, the characteristics equations are

$$\frac{dy}{dx}=\frac{B\mp\sqrt{B^2-4AC}}{2A}=\frac{(x+y)\mp(x-y)}{2y}$$

Therefore,

$$\frac{dy}{dx}=1\ ,\frac{dy}{dx}=\frac{x}{y}$$

On integrating, we obtain

$y=x+c_1,y^2=x^2+c_2$

Hence, the characteristics equation are

$\xi=y-x,\eta=y^2-x^2$

These are straight lines and rectangular hyperbolas. The canonical form can be obtained by computing

$\bar{A}=A\xi_x^2+B\xi_x\xi_y+C\xi_y^2=y-x-y+x=0\ ,\bar{B}=-2(x-y)^2$

$\bar{C}-0\ ,\ \bar{D}=0\ ,\ \bar{E}=2(x-y)\ ,\ \bar{F}=\bar{G}=0$

Thus, the canonical equation for the given PDE is

$-2(x-y)^2u_{\xi\eta}+2(x-y)u_{\eta}=0$

or $-2\xi^2 u_{\xi\eta} + 2(-\xi)u_\eta = 0$

or $\xi u_{\xi\eta} + u_\eta = \frac{\partial}{\partial \xi}\left(\xi \frac{\partial u}{\partial \eta}\right) = 0$

Integration yields

$$\xi \frac{\partial u}{\partial \eta} = f(\eta)$$

Again integrating with respect to η, we obtain

$$u = \frac{1}{\xi} \int f(\eta) d\eta + g(\xi)$$

Hence,

$$u = \frac{1}{y-x} \int f(y^2 - x^2) d(y^2 - x^2) + g(y-x)$$

is then general solution.

Example. Classify and transformation the following equation to a canonical form:

$$sin^2(x)u_{xx} + \sin(2x)u_{xy} + \cos^2(x)u_{yy} = x$$

Solution. The discriminant of the given PDE is

$$B^2 - 4AC = \sin^2 2x - 4\sin^2 x \cos^2 x = 0$$

Hence, the given equation is of parabolic type. The characteristics equation

$$\frac{dy}{dx} = \frac{B}{2A} = \cot x$$

Integration gives

$$y = In \sin x + c_1$$

Hence, the characteristic equation are :

$$\xi = y - In \sin x, \eta = y$$

η is chosen in such a way that the Jacobian of the transformation is nonzero. Now the canonical from can be obtained by computing

$\bar{A}=0$, $\bar{B}=0$, $\bar{C}=\cos^2 x$, $\bar{D}=1$, $\bar{E}=0$, $\bar{F}=0$, $\bar{G}=x$

Hence, the canonical equation is

$$\cos^2(x)\, u_{\eta\eta}+u_{\xi}=x$$

or
$$\left[1-e^{2(\eta-\xi)}\right]u_{\eta\eta}=\sin^{-1}\left(e^{\eta-\xi}\right)-u\xi$$

Example. Show that the equation

$$u_{xx}+\frac{2N}{x}u_x=\frac{1}{a^2}u_{tt}$$

where N *and* a are constants, is hyperbolic and obtain its canonical form.

Solution. Comparing with the general PDE (4) and replacing $y\, by\, t$, we have $A-1$, $B=0$, $C=-1/a^2$, $D=2N/x$, and $E=F=G=0$. The discriminant $B^2-4AC=4/a^2>0$. Hence the given PDE is hyperbolic. The characteristics equation are

$$\frac{dt}{dx}=\frac{B\mp\sqrt{B^2-4AC}}{2A}=\mp\frac{\sqrt{4/a^2}}{2}=\mp\frac{1}{a}$$

Therefore,

$$\frac{dt}{dx}=-\frac{1}{a},\frac{dt}{dx}=\frac{1}{a}$$

On integration, we get

$$t=-\frac{x}{a}+c_1\ ,t=\frac{x}{a}+c_2$$

Hence, the characteristic equations arc

$$\xi=x+at\ ,\ \eta=x-at$$

The canonical form can be obtained by computing

$$\bar{A}=A\xi_x^2+B\xi_x\,\xi_t+C\xi_t^2=0\ ,$$

$$\bar{B}=2A\xi_x\eta_x+B\left(\xi_x\eta_t+\xi_t\eta_x\right)+2C\xi_t\eta_t-4\ ,$$

$$\bar{C}=0\ ,\ \bar{D}=D\xi_x+E\xi_t=\frac{2N}{x}\ ,\quad \bar{E}=D\eta_x+E\eta_t=\frac{2N}{x}$$

Thus, the canonical equation for the given PDE is

$$4u_{\xi\eta} + \frac{2N}{x}\left(u_{\xi} + u_{\eta}\right) = 0$$

Example. Transform the following differentiate equation to a canonical form:

$$u_{xx} + 2u_{xy} + 4u_{yy} + 2u_{x} + 3u_{y} = 0$$

Solution. The discriminant $B^2 - 4AC = -12 < 0$. Hence, the given PDE is elliptic. The characteristics equations are

$$\frac{dy}{dx} = \frac{B - \sqrt{B^2 - 4AC}}{2A} = +1 - i\sqrt{3}$$

$$\frac{dy}{dx} = \frac{B + \sqrt{B^2 - 4AC}}{2A} = +1 + i\sqrt{3}$$

Integration of these equations yields

$$y = +\left(1 - i\sqrt{3}\right)x + c_1 \ , y = +\left(1 + i\sqrt{3}\right)x + c_2$$

Hence, we may take the characteristics equations in the form

$$\xi = y - \left(1 - i\sqrt{3}\right)x \ , \eta = y - \left(1 + i\sqrt{3}\right)x$$

In order to avoid calculations with complex variables , we introduced the second transformation

$$a = \frac{\xi + \eta}{2} \ , \beta = \frac{\xi - \eta}{2i}$$

Therefore.

$$a = y - x \ , \beta = \xi + \eta$$

The canonical form can now be obtained by computing

$$\bar{A} = Aa_x^2 + Ba_x a_y + Ca_y^2 = 3$$

$$\bar{B} = 2Aa_x \ \beta_x + B\left(a_x\beta_y + a_y\beta_x\right) + 2Ca_y\beta_y = 0$$

$$\bar{C} = A\beta_x^2 + B\beta_x \ \beta_y + C\beta_y^2 = 3$$

$$\bar{D} = Aa_{xx} + Ba_{xy} + Ca_{yy} + Da_x + Ea_y = 1$$

$$\bar{E} = A\beta_{xx} + B\beta_{xy} + C\beta_{yy} + D\beta_x + E\beta_y = 2\sqrt{3}$$

Thus the required canonical form is

$$3u_{aa} + 3u_{\beta\beta} + u_a + 2\sqrt{3}\,u_\beta = 0$$

or
$$u_{aa} + u_{\beta\beta} = -\frac{1}{3}\,(u_a + 2\sqrt{3u_\beta} = 0$$

2.2.5. Adjoint Operators

Let

$$Lu = \varnothing \tag{16}$$

where L is a differentiate operator given by

$$L = a_0(x)\frac{d^n}{dx^n} + a_1(x)\frac{d^{n-1}}{dx^{n-1}} + \ldots + a_n(x)$$

One way of introducing the adjoint differential operators L^* associated with L is to from the product vLu and integrate it over the interval of interest . Let

$$\int_A^B vLu\,dx = [\ .]_A^B + \int_A^B uL^*\,v\,dx \tag{17}$$

which is obtained after repeated integration by parts. Here L^* is the operator adjoint to L, where the function $u\, and\, v$ are completely arbitrary except that $Lu\, and\, L^*v$ should exist.

Example. Let $Lu = a(x)(d^2u/dx^2) + b(x)(du/dx) + c(x)u$; construct its adjoint L^*.

Solution. Consider the equation

$$\int_A^B vLu\,dx = \int_A^B v\left[a(x)\frac{d^2u}{dx^2} + b(x)\frac{du}{dx} + c(x)u\right]dx$$

$$= \int_A^B (av)\frac{d^2u}{dx^2}dx + \int_A^B (bv)\frac{du}{dx}dx \int_A^B (cv)u\,dx$$

$$\int_A^B (av)\frac{d^2u}{dx^2}du = \int_A^B (av)\frac{d}{dx}(u')dx$$

$$= \left[u'va\right]_A^B - \int_A^B (av)'u'\,dx$$

$$= \left[u'av\right]_A^B - \left[u(av)'\right]_A^B + \int_A^B u(av)''\,dx$$

$$\int_A^B (bv)\frac{du}{dx}dx = \left[u(bv)\right]_A^B - \int_A^B u(bv)'\,dx$$

$$\int_A^B (cv)u\,dx = \int_A^B u(cv)\,dx$$

Therefore,

$$\int_A^B vLu\,dx = \left[u'(av) - u(av)'u(bv)\right]_A^B + \int_A^B u\left[(av)'' - (bv)' + (cv)\right]dx$$

Comparing this equation with Eq. (17), we get

$$L^* v = (av)'' - (bv)' + (cv) = av'' + (2a' - b)v' + (a'' - b' + c)v$$

Therefore,

$$L^* = a\frac{d^2}{dx^2} + (2a' - b)\frac{d}{dx} + (a'' - b' + c)$$

Consider the partial differential equation

$$L(u) = Au_{xx} + 2Bu_{xy} + Cu_{xy} + Du_x + Eu_y + Fu = \phi$$

which is valid in a region S of the $xy-plane$, where $A, B, C, \ldots \phi$ are functions of $x\,and\,y$ .In addition. Linear boundary conditions of the general form

$$au + \beta u_x = f$$

are prescribed over the boundary curve ∂S, the boundary of S, then L^* is called the adjoint operator. In general, a second order linear partial differential operator L is denoted by

$$L(u) = \sum_{i,j=1}^{n} A_{ij} \frac{\partial^2 u}{\partial x_i \partial x_j} + \sum_{i=1}^{n} B_i \frac{\partial u}{\partial x_1} + Cu \tag{19}$$

It's adjoint operator is defined by

$$L^*(v) = \sum_{i,j=1}^{n} \frac{\partial^2}{\partial x_i \partial x_j}\left(A_{ij} v\right) - \sum_{i=1}^{n} \frac{\partial}{\partial x_i}\left(B_i v\right) + Cv \tag{20}$$

Here it is assumed that $A_{ij} \grave{o} C^2$ *and* $B_i \grave{o} C^1$. For any pair of functions $u, v \grave{o} C^2$, it can be shown that

$$vLu(u) - uL^*(v) = \sum_{i=1}^{n} \frac{\partial}{\partial x}\left[\sum_{j=1}^{n} A_{ij}\left(v\frac{\partial u}{\partial x_j} - u\frac{\partial v}{\partial x_j}\right) + uv\left(B_i - \sum_{j=1}^{n} \frac{\partial A_{ij}}{\partial x_j}\right)\right] \tag{21}$$

This is known as Lagrange's identity.

Example. Construct a adjoint to the Laplace operator given by

$$L(u) = u_{xx} + u_{yy} \tag{22}$$

Solution. Comparing Eq. (22) with the general linear PDE (19) , we have $A_{11} = 1, A_{22} = 1$. From $Eq.(20)$ the adjoint of (22) is given by

$$L^*(v) = \frac{\partial^2}{\partial x^2}(v) + \frac{\partial^2}{\partial y^2}(v) = v_{xx} = v_{yy}$$

Therefore,

$$L^*(u) = u_{xx} + u_{yy}$$

Hence, the Laplace operator is a self-adjoint operator

Example. Find the adjoint of the differentiate operator

$$L(u) = u_{xx} + u_i \tag{23}$$

Solution. Comparing Eq. (23) with the general linear PDE (19), we have $A_{11}=1, A_{22}=1$. From Eq (20), the adjoint of (22) is given by

$$L^*(v)\frac{\partial^2}{\partial x^2}(v)+\frac{\partial^2}{\partial y^2}(v)=v_{xx}+v_{yy}$$

Therefore,

$$L^*(u)=u_{xx}+u_t$$

It may be noted that the diffusion operator is not a self-adjoint operator.

2.2.6. Riemann's Method

We have noted with interest that a linear second order PDE

$$L(u)=G(x,y)$$

is classified as hyperbolic if $B^2>4AC,$ and it has two families of real characteristics curves in the $xy-plane$ whose equations are

Here (ξ,η) are the natural coordinates for the hyperbolic system. In the $xy-plane$, the curves $\xi(x,y)=c_1 \, and \, \eta(x,y)=c_2$ are the characteristics of the given PDE as shown in Fig 1(a). while in the $\xi\eta-plane$, the curves $\xi=c_a \, and \, \eta=c_2$ are families of straight lines parallel to the axes as shown in Fig 1(b)

A linear second order partial differential equation in two variables , once classified as a hyperbolic equation, can always be reduce to the canonical form

$$\frac{\partial^2 u}{\partial x \partial y}=G(x,y,u,u_x,u_y)$$

In particular, consider an equation which is already reduced to its canonical form in the variables x, y:

$$L(u)=\frac{\partial^2 u}{\partial x \partial y}=a\frac{\partial u}{\partial x}+b\frac{\partial u}{\partial y}+cu=F(x,y) \tag{24}$$

where L is a linear differential operator and a,bc,F are functions of $x\, and\, y$ only and are differentiable in some domain IR.

Let $v(x,y)$ be an arbitrary function having continuous second order partial derivatives. Let us consider the adjoint operator L^* of L defined by

$$L^*(v)=\frac{\partial^2 v}{\partial x\,\partial y}-\frac{\partial}{\partial x}(av)-\frac{\partial}{\partial y}(bv)+cv \tag{25}$$

Now we introduce

$$M = auv - u\frac{\partial v}{\partial y}, N = buv + v\frac{\partial u}{\partial x} \tag{26}$$

Then

$$M_x + N_y = u_x(av)+u(av)_x - u_x v_y - uv_{xy} + u_y(bv)+u(bv)_y + v_y u_x + vu_{xy}$$

Adding and subtracting $cuv,$ we get

$$M_x + N_y = -u\left[\frac{\partial^2 v}{\partial x\partial y}-\frac{\partial}{\partial x}(av)-\frac{\partial}{\partial y}(bv)+cv\right]+v\left[\frac{\partial^2 u}{\partial x\partial y}+a\frac{\partial u}{\partial x}+b\frac{\partial u}{\partial y}+cu\right]$$

i.e,

$$vLu - ul^*v = M_x + N_y \tag{27}$$

This is knows as Lagrange identity which will be used in the subsequent discussion . The operator L is a self – adjoint if and only if $L = L^*$. Now we shall attempt to solve Cauchy's problem which is described as follows . Let

$$L(u) = F(x,y) \tag{28}$$

with the condition (Cauchy data)

$(i)\ u = f(x) on \Gamma,$ a curve I n the $xy - plane;$

$(ii)\ \frac{\partial u}{\partial n} = g(x) on \Gamma.$

This is $u,$ and its normal derivatives are prescribed on a curve Ã which is not a characteristics line.

Let Ã be a smooth initial curve which is also continuous as shown in Fig 2 since Eq. (24) is in canonical form , x and y are the characteristics coordinates. We also assume that line tangent to $\tilde{A} is\, nowhere$ parallel to the coordinates axes.

Let $P(\xi,\eta)$ be a point at which the solution to the Cauchy problem is sought. Let us draw the characteristics $PQ\, and\, PR$ through P to meet the curve $\tilde{A} at\, Q\, and\, R.$ We assume that $u, u_x\, u_y$ are prescribed along $\tilde{A}$ Let ∂IR be a closed contour $PQRP$ bounding IR. Since Eq. (28) is already in canonical form, the characteristics are lines parallel to $x\, and\, y$ axes . using Green's theorem ,we have

$$\int\int_R \left(M_x + N_y\right) dxdy = \oint_{\partial \text{IR}} \left(M\, dy - N\, dx\right) \tag{29}$$

where ∂IR is the boundary of IR. Applying this theorem to the surface integral of Eq.(27). We obtain

$$\int_{\partial \text{IR}} \left(M\, dy - N\, dx\right) = \int_{\text{IR}} \left[vL(u) - uL^*(v)\right] dx\, dy \tag{30}$$

In other words,

$$\int_{\tilde{A}} \left(Mdy - N\, dx\right) + \int_{RP} \left(M\, dy - N\, dx\right) = \int\int_R \left[vL(v) - uL^*(v)\right] dx\, dy$$

Now using the fact that $dy = 0\, on\, PQ\, and\, dx = 0\, on\, PR,$ we have

$$\int_{\tilde{A}} \left(M\, dy - Ndx\right) + \int_{RP} M\, dy - \int_{PQ} N\, dx = \int\int_{\text{IR}} \left[v\, L(u) - uL^*(v)\right] dx\, dy \tag{31}$$

From Eq. (26), we find that

$$\int_{PQ} N\, dx = \int_P^Q buv\, dx + \int_P^Q vu_x\, dx$$

Integrating by parts the second term on the right –hand side and grouping. the above equation becomes

$$\int_{PQ} N\, dx = \left[nv\right]_P^Q + \int_P^Q u\left(bv - v_x\right) dx$$

Substituting this result into Eq. (31), we obtain

$$[uv]_p = [uv]_Q + \int_{PQ} u\left(bv - v_x\right)dx - \int_{RP} u\left(av - v_y\right)dy$$

$$-\int_{\tilde{A}}\left(M\,dy - N\,dx\right) + \int\int_{IR}\left[vL(u) - uL^*(v)\right]dx\,dy \tag{32}$$

Let us choose $v(x, y;, \xi, \eta)$ to be a solution of the adjoint equation

$$L^*(v) = 0 \tag{33}$$

And at the same time satisfy the following condition :

(i) $v_x = bv$ $when\, u = \eta, i.e\ , on\, PQ$ (34 a)

(ii) $v_y = av$ when $x = \xi, i.e., on\, PR$ (34 b)

$(iii)\, v = 1\, when\, x = \xi, y = \eta$ (34 c)

We call this function $v(x, y; \xi, \eta)$ as the Riemann function or the Riemann- Green function. Since $L(u) = F, Eq.(32)$ reduces to

$$[u]_p = [uv]_Q - \int_{\tilde{A}}\left[u\left(av - v_y\right)dy - v\left(bu + u_x\right)dx\right] + \int\int_{IR}\left(vF\right)dx\,dy \tag{35}$$

This is called the Riemann-Green solution for the Cauchy problem described by Eq.(28) when $u\, and\, u_x$ are prescribed on $\tilde{A}$ Equation (35) can also be written as

$$[u]_p - [uv]_Q - \tilde{A}\int_{\tilde{A}} uv\left(a\,dy - b\,dx\right) + \int_{\tilde{A}}\left(uv_y\,dy - vu_x\,dx\right) + \int_{IR}\int\left(vF\right)dx\,dy \tag{36}$$

This relation gives us the value of at a point p when $u\, and\, u_x$ are prescribed on $\tilde{A}$. But when $u\, and\, u_x\, are$ prescribed on $\tilde{A}$, wc obtain

$$[u]_p = [uv]_R - \int_{\tilde{A}} uv\left(a\,dy - b\,dx\right) - \int_{\tilde{A}}\left(uv_x\,dx + vu_y\,dy\right) + \int\int_{IR}\left(vF\right)dx\,dy \tag{37}$$

By adding Eq. (36) and (37), the value of u at a point P when $u \, and \, u_x$ are prescribed on $\tilde{A}$ But when $u \, and \, u_x$ are prescribed on $\tilde{A}$, we obtain

$$[u]_p = \frac{1}{2}\left\{[uv]_Q + [uv]_R\right\} - \int_{\tilde{A}} uv\left(a\,dy - b\,dx\right) - \frac{1}{2}\int_{\tilde{A}} u\left(v_x\,dx - v_y\,dy\right) + \frac{1}{2}\int_{\tilde{A}} v\left(u_x\,dx - u_y\,dy\right) + \int\int_{IR} \left(vF\right) dx\,dy \tag{38}$$

Thus, we can see that the solution to the Cauchy problem at a point (ξ, η) depends only on the Cauchy data on $\tilde{A}$ the Knowledge of the Riemann-Green function therefore enables us to solve Eq. (28) with the Cauchy data prescribed on a non characteristics curve. Thus, we can see that the solution to the Cauchy problem at a point (ξ, η) depends only on the Cauchy data on $\tilde{A}$ the Knowledge of the Riemann-Green function therefore enables us to solve Eq. (28) with the Cauchy data prescribed on a non characteristics curve.

Example. Let $Lu = a(x)\left(d^2u / dx^2\right) + b(x)\left(du / dx\right) + c(x)u$; construct its adjoint L^*.

Solution. Consider the equation

$$\int_A^B vLu\,dx = \int_A^B v\left[a(x)\frac{d^2u}{dx^2} + b(x)\frac{du}{dx} + c(x)u\right]dx$$

$$= \int_A^B (av)\frac{d^2u}{dx^2}dx + \int_A^B (bv)\frac{du}{dx}dx \int_A^B (cv)u\,dx$$

$$\int_A^B (av)\frac{d^2u}{dx^2}du = \int_A^B (av)\frac{d}{dx}(u')\,dx$$

$$= \left[u'va\right]_A^B - \int_A^B (av)'\,u'\,dx$$

$$= \left[u'av\right]_A^B - \left[u(av)'\right]_A^B + \int_A^B u(av)''dx$$

$$\int_A^B (bv)\frac{du}{dx}dx = \left[u(bv)\right]_A^B - \int_A^B u(bv)' \, dx$$

$$\int_A^B (cv)u \, dx = \int_A^B u(cv) \, dx$$

Therefore,

$$\int_A^B vLu \, dx = \left[u'(av) - u(av)' u(bv)\right]_A^B + \int_A^B u\left[(av)'' - (bv)' + (cv)\right]dx$$

Comparing this equation with we get

$$L^* v = (av)'' - (bv)' + (cv) = av'' + (2a' - b)v' + (a'' - b' + c)v$$

Therefore,

$$L^* = a\frac{d^2}{dx^2} + (2a' - b)\frac{d}{dx} + (a'' - b' + c)$$

Consider the partial differential equation

$$L(u) = Au_{xx} + 2Bu_{xy} + Cu_{xy} + Du_x + Eu_y + Fu = \phi$$

which is valid in a region S of the $xy - plane$, where $A, B, C, \ldots \phi$ are functions of $x and\ y .In$ addition . Linear boundary conditions of the general form

$$au + \beta u_x = f$$

Are prescribed over the boundary curve ∂S, the boundary of S, then L^* is called the adjoint operator. In general , a second order linear partial differential operator L is denoted by

$$L(u) = \sum_{l,j=1}^{n} A_{ij}\frac{\partial^2 u}{\partial x_i \, \partial x_j} + \sum_{i=1}^{n} B_i \frac{\partial u}{\partial x_1} + Cu$$

It's adjoint operator is defined by

$$L^*(v) = \sum_{i,j=1}^{n} \frac{\partial^2}{\partial x_i \partial x_j}(A_{ij}v) - \sum_{i=1}^{n}\frac{\partial}{\partial x_i}(B_i v) + Cv$$

Here it is assumed that $A_{ij} \dot{o} C^2 \; and \; B_i \dot{o} C^1$. For any pair of functions $u, v \dot{o} C^2$, it can be shown that

$$vLu(u) - uL^*(v) = \sum_{i=1}^{n} \frac{\partial}{\partial x}\left[\sum_{j=1}^{n} A_{ij}\left(v\frac{\partial u}{\partial x_j} - u\frac{\partial v}{\partial x_j}\right) + uv\left(B_i - \sum_{j=1}^{n}\frac{\partial A_{ij}}{\partial x_j}\right)\right]$$

Example. Verify that the Green function for the equation

$$\frac{\partial^2 u}{\partial x \partial y} + \frac{2}{x+y}\left(\frac{\partial u}{\partial x} + \frac{\partial u}{\partial y}\right) = 0$$

Subject to $u = 0 \,, \partial u / \partial x \; = 3x^2 \; on \; y = x,$ is given by

$$v(x, y; \xi, \eta) = \frac{(x+y)\{2xy + (\xi - \eta)(x - y) + 2\xi\eta\}}{(\xi + \eta)^3}$$

andobtainthesolutionoftheequationintheform $u = (x - y)(2x^2 - xy + 2y^2)$

.

Solution. In the given problem,

$$L(u) = \frac{\partial^2 u}{\partial x \partial y} + \frac{2}{x+y}\frac{\partial u}{\partial x} + \frac{2}{x+y}\frac{\partial u}{\partial y} = 0 \tag{39}$$

Comparing this equation with the standard canonical hyperbolic equation (24), we have

$$a = b = \frac{2}{x+y} . C = 0 \,, F = 0$$

Its adjoint equation is $L^*(v) = 0$ where

$$L^*(v) = \frac{\partial^2 v}{\partial x \partial y} - \frac{\partial}{\partial x}\left(\frac{2v}{x+y}\right) - \frac{\partial}{\partial y}\left(\frac{2v}{x+y}\right) \tag{40}$$

Such that

$(i) L^* v = 0$ thought the $xy - plane$

$(ii) \frac{\partial v}{\partial x} = \frac{2}{x+y} v \, on \, PQ, i.e, om \; y = \eta$

$$(iii)\ \frac{\partial v}{\partial y} = \frac{2}{x+y} v\, on\, PR, i.e., on\, x = \xi \tag{41}$$

$$(iv) v = 1\ at\ P(\xi, \eta)$$

If v defined by

$$v(x, y; \xi, \eta) = \frac{x+y}{(\xi+\eta)^3}\left[2xy + (\xi - \eta(x-y) + 2\xi\eta\right] \tag{42}$$

Then

$$\frac{\partial v}{\partial x} = \frac{x+y}{(\xi+\eta)^3}\left[2y + (\xi-\eta)\right] + \left[\frac{2xy + (\xi-\eta)(x-y) + 2\xi\eta}{(\xi+\eta)^3}\right]$$

or $$\frac{\partial v}{\partial y} = \frac{1}{(\xi+\eta)^3}\left[4xy = 2x^2 - 2y(\xi-\eta) + 2\xi\eta\right] \tag{43}$$

And

$$\frac{\partial^2 v}{\partial x \partial y} = \frac{4(x+y)}{(\xi+\eta)^3} \tag{44}$$

$$\frac{\partial v}{\partial y} = \frac{1}{(\xi+\eta)^3}\left[4xy = 2x^2 - 2y(\xi-\eta) + 2\xi\eta\right] \tag{45}$$

Using the results described becomes

$$L^*(v)\frac{\partial^2 v}{\partial x \partial y} - \frac{2}{x+y}\left(\frac{\partial v}{\partial x} + \frac{\partial v}{\partial y}\right) + \frac{4v}{(x+y)^2}$$

$$-\frac{4(x+y)}{(\xi+\eta)^3} - \frac{2}{(x+y)(\xi+\eta)^3}\left[4xy + 2\left(x^2 + y^2\right)\right]$$

or $$L^*(v) = \frac{4(x+y)}{(\xi+\eta)^3} - \frac{4(x+y)}{(\xi+\eta)^3} = 0$$

Hence condition (i) is satisfied . Also, on $y = \eta$.

$$\frac{\partial v}{\partial x} = \frac{1}{(\xi+\eta)^3}\left[4x\eta + 2\eta^2 + 2x(\xi-\eta) + 2\xi\eta\right]$$

However,

$$\int_{Q}^{P}\left(\frac{2uv}{x+y}+v\frac{\partial u}{\partial x}\right)dx=\int_{Q}^{P}\frac{2uv}{x+y}dx+\left(uv\right)_{Q}^{P}-\int_{Q}^{P}u\frac{\partial v}{\partial x}dx$$

Now, using the condition $u=0\, on\, y=x$, becomes

$$\int\int_{IR}\left[vL(u)-uL^{*}(v)\right]dx\,dy=\int_{P}^{Q}\left(\frac{2uv}{x+y}-v\frac{\partial u}{\partial x}\right)dx$$

$$-\int_{Q}^{P}\frac{2uv}{x+y}dx-\left(uv\right)_{p}+\left(uv\right)_{Q}+\int_{R}^{Q}\left(\frac{2uv}{x+y}+v\frac{\partial u}{\partial x}\right)dx$$

$$\int_{-Q}^{P}\frac{2uv}{x+y}dy-\int_{P}^{R}\left(u\frac{\partial u}{\partial x}\right)dy$$

Also, using conditions (ii) –(iv) of the above equation simplifies to

$$\left(u\right)_{p}=\left(uv\right)_{Q}-\int_{R}^{Q}v\frac{\partial u}{\partial x}dx$$

Now using the given condition, viz.

$$\frac{\partial u}{\partial x}=3x^{2}\, on\, RQ$$

We obtain

$$\left(u\right)_{p}=\left(uv\right)_{Q}-3\int_{R}^{Q}x^{2}\left[\frac{2x\left[2x^{2}+2\xi\eta\right]}{\left(\xi+\eta\right)^{3}}\right]dx$$

$$=-\frac{12}{\left(\xi+\eta\right)^{3}}\int_{\xi}^{\eta}\left(x^{5}+x^{3}\,\xi\eta\right)dx$$

$$=-\frac{12}{\left(\xi+\eta\right)^{3}}\left[\frac{1}{6}\left(\eta^{6}-\xi^{6}\right)+\frac{1}{4}\xi\eta\left(\eta^{4}-\xi^{4}\right)\right]$$

$$=\frac{\xi^{2}-\eta^{2}}{\left(\xi+\eta\right)^{3}}\left[2\left(\xi^{4}+\xi^{2}\eta^{2}+\eta^{4}\right)+3\xi\eta(\xi^{2}+\eta^{4}\right]$$

$$=(\xi-\eta)\left(2\xi^2-\xi\eta+2\eta^2\right)$$

Therefore,

$$u(x,y)=(x-y)\left(2x^2-xy+2y^2\right)$$

hence the result.

2.3. SUMMARY AND DISCUSSION

Many practical problems in science and engineering, when formulated mathematically, give partial differential equations (often referred to as PDE). An equation, which involves several independent variables usually, denoted by $(x,y,z,t,\ldots)$, a dependent function u. of these variables, and the partial derivatives of the dependent function u with respect to the independent variables such as

$$F\left(x\ ,y\ ,z\ ,t\ldots\ ,u_x\ ,u_y,u_z\ ,u_t\ ,\ldots\ldots u_{xx},u_{yy}\ ,\ldots.u_{xy}\ ,\ldots\right)=0$$

is called a partial differential equation. A few well-known examples are:

$(i)\ u_t=k\left(u_{xx}+u_{yy}+u_{zz}\right)$ [linear three-dimensional heat equation]

$(ii)u_{xx}+u_{yy}+u_{zz}=0$ [Laplace equation in three dimension]

Order of PDE. The order of the partial differential equation is the order of the highest derivative occurring in the equation. Thus the above examples are partial differential equation of second order, whereas $u_t=uu_{xxx}+\sin x$ is an example for third order partial differential equation.

2.4. CLASSIFICATION OF SECOND ORDER PDE

The classification of PDE depend only on the highest order derivative present. The classification of PDE is motivated by the classification of the quadratic equation of the form

$$Ax^2+Bxy+Cy^2+Dx+Ey+F=0$$

which is elliptic parabolic or hyperbolic according as the discriminant B^2-4AC is negative , zero or positive . Thus we have the following second order linear PDE in two variables $x\ and\ y$

$$Au_{xx} + Bu_{xy} + Cu_{yy} + Du_x + Eu_y + Fu = G$$

where the coefficients $A, B, C, \ldots$ may be functions of $x \, and \, y,$ however, for the sake of simplicity we assume then to be constant . Equation is elliptic , parabolic of hyperbolic at a point (x_0, y_0) according as the discriminant. $B^2(x_0, y_0) - 4A(x_0, y_0)C(x_0, y_0)$ is negative , zero or positive. If this is true at all points in a domain $\grave{\text{U}},$ then is said to be elliptic, parabolic in that domain . if the number of independent variables is two or three , a transformation can always be found to reduce the given PDE to a canonical form (also called normal form).

Adjoint Operators

Let

$$Lu = \varnothing$$

where L is a differentiate operator given by

$$L = a_0(x)\frac{d^n}{dx^n} + a_1(x)\frac{d^{n-1}}{dx^{n-1}} + \ldots + a_n(x)$$

One way of introducing the adjoint differential operators L^* associated with L is to from the product vLu and integrate it over the interval of interest . Let

$$\int_A^B vLu \, dx = [\ .]_A^B + \int_A^B uL^* \, v \, dx$$

which is obtained after repeated integration by parts. Here L^* is the operator adjoint to $L,$ where the function $u \, and \, v$ are completely arbitrary except that $Lu \, and \, L^* v$ should exist.

$$vLu(u) - uL^*(v) = \sum_{i=1}^{n} \frac{\partial}{\partial x}\left[\sum_{j=1}^{n} A_{ij}\left(v\frac{\partial u}{\partial x_j} - u\frac{\partial v}{\partial x_j}\right) + uv\left(B_i - \sum_{j=1}^{n}\frac{\partial A_{ij}}{\partial x_j}\right)\right]$$

This is known as Lagrange's identity

CHAPTER 3

APPLICATION OF DIFFERENTIAL EQUATIONS IN MECHANICS

CONTENTS

3.1. INTRODUCTION

The motion of a particle along a straight line is called *rectilinear motion.* Let the particle start from point O along a line. We take the line as $x-axis$. Let after time t particle be at point P at a distance x from O. Let $\bar{r}$ be the position vector of point $p\,w.r.t.$ origin O. then

$$\bar{r} = \overline{OP} = x\hat{i}$$

$$\bar{v} = \frac{d\bar{r}}{dt} = \frac{dx}{dt}\hat{i}$$

Let $v = |\bar{v}|$ and $a = |\bar{a}|, then$

$$v = \frac{dx}{dt}$$

$$a = \frac{d^2x}{dt^2}$$

Also,

$$a = \frac{dv}{dt} = \frac{dv}{dx}\frac{dx}{dt} = \frac{dv}{dx}v = v\frac{dv}{dx}$$

3.1.1. Motion with Constant Acceleration

Let the particle start form O with velocity u at time t = 0 with constant acceleration a. Let the particle be at distance x from O after time t. Then

$a = \dfrac{dv}{dt}$ where a is constant.

Integration will yield

$$v = at + A$$

Since velocity $v=u$ at $t=0,$ therefore, $A=u$, thus

$$v = at + u = u + at$$

Now, since $v = \frac{dx}{dt}$, therefore eq. (1) gives

$$\frac{dx}{dt} = u + at$$

By integrating

$$x = ut + a\left(\frac{t^2}{2}\right) + B$$

Using $x = 0, at\, t = 0,$ we get, $B = 0,$ hence

$$x = ut + \frac{at^2}{2}$$

Also, $a = v\frac{dv}{dx}$, where a is constant, then

$$ax = v\,dv$$

Integrating

$$\frac{v^2}{2} = ac + C$$

Since v = u, at x = 0, therefore, C = $\frac{u^2}{2}$, and hence

$$v^2 - u^2 = 2ax$$

The equation and are called equation of motion. If the particle moves with constant retardation the acceleration is taken as negative.

3.1.2. Vertical Motion under Gravity

For a failing body, the acceleration is constant. It is called acceleration due to gravity and is denoted by 'g'. In FPS system g = $32\, ft / sec^2$.In CGS system g =981 cm / sec^2 and in MKS system $g = 9.81 m / sec^2$. If the body is projected vertically upward then acceleration g is negative. For falling bodies equation of motion is

$$v = u + gt$$

$$x = ut + \frac{1}{2}gt^2$$

$$v^2 - u^2 = 2gx$$

Example

A particle is projected vertically upward with velocity $\sqrt{2gh}$ and another is let fall from a height h at the same time. Find the height of the point where they meet each other.

Solution.

Let both particles meet at point P at height x. Then for the first particle

$$x = ut - \frac{1}{2}gt^2$$

Put $u = \sqrt{2gh}$, then

$$x = \sqrt{2gh}t - \frac{1}{2}gt^2 \quad (1)$$

For the second particle

$$x = ut + -gt$$

$$x = ut + \frac{1}{2}gt^2$$

Put $u = 0, x = h - x$, then

$$h - x = 0 + \frac{1}{2}gt^2 \quad (2)$$

Using $eqs.$ (1) and (2), we get

$$u = \sqrt{2gh}t$$

$$or \qquad t = \sqrt{2gh}t$$

Thus equation (1) gives

$$x=\sqrt{2gh}t\left(\frac{h}{\sqrt{2gh}}\right)-\frac{1}{2}g\left(\frac{h^2}{2gh}\right)=h-\frac{h}{4}=\frac{3h}{4}$$

Example

A particle is projected vertically upward. After time t another particle is sent up from the same point with the same velocity and meets the first at height h during the downward flight of the first. Find the velocity of the projection.

Solution

Let u be the velocity of the projection and v *be* the velocity at height h, then

$$v^2-u^2=-2gh$$

or $v=\sqrt{u^2-2gh}$

Now, the time taken by the first particle from height h to the highest point and back to the height h is t, therefore, the time taken from the highest point is t / 2 .

Velocity at highest point $=o$

Velocity at height $h=v$

Here we use the formula

$$v=u-gt$$

Put $v=0, u=v, t=$ t/2, then

$$0=v-\frac{gt}{2}$$

From (1) and (2)

$$\sqrt{u^2\quad 2gh}=\frac{gt}{2}$$

$$u^2-2gh=\frac{g^2t^2}{4}$$

$$u^2=2gh+\frac{g^2t^2}{4}$$

$$u = \sqrt{\frac{8gh + g^2t^2}{4}}$$

3.1.3. Distance Traveled in nth Second

Let $x_1\ and\ x_2\ b$ e the distance travelled in the first n and n – 1 seconds respectively, then from the equation of motion

$$x = ut + \frac{1}{2}at^2$$

Then $x_1 = un + \frac{1}{2}an^2$

And $x_2 = u(n-1) + \frac{1}{2}a(n-a)^2$

Distance travelled in the nth second = $x_1 - x_2$

$$x_1 - x_2 = un + \frac{a}{2}n^2 - \left\{u(n-a) + \frac{a}{2}(n-1)^2\right\}$$

$$= un + \frac{a}{2}n^2 - \left\{un - u + \frac{a}{2}\left(n^2 - 2n + 1\right)\right\}$$

$$= u + \frac{1}{2}a(2n-1)$$

3.1.4. Motion with variable Acceleration

(i) **Time Dependent Acceleration**

As the acceleration is time dependent, then it is function of time Therefore we can write:

$$a = f(t)$$

or $\frac{dv}{dt} = f(t)$

or $dv = f(t)dt$

Integrating

$\int dv = \int f(t)\,dt$

or $v = g(t) + A$

where g(t) = $\int f(t)\,dt,$ and A is constant of integration, which gives the velocity. Also since

$v = \frac{dx}{dt},$

Therefore

$\frac{dx}{dt} = g(t) + A$

or $dx = g(t)\,dt + A\,dt$

By integrating, we get

$x = \int g(t)\,dt + At + B$

where B is another constant which gives the distance travelled by the particle· By using the initial conditions we can find the constants $A\, and\, B$.

Example 3

Discuss the motion of the particle moving in a straight line if it stars from rest t = 0 and its acceleration is equal to $accost + bsint.$

Solution

Give acceleration is

$a - accost + b\,sint$

By integrating, we have

$v = asint - b\,cost + A$

Since $v = 0\, at\, t = 0$, therefore $A = b,$ thus

$C = asint - bcost + b$

Also, v = $\frac{dx}{dt} = asint - bcost + b$

Again integrating

$x = -acost - bsint + bt + B$

where B is constant of integration.

Since x = 0 , at t = 0, therefore B = a. Thus

$x = -acost - bsint + bt + a$

$= a(1 - cost) + b(t - sint)$

(ii) Velocity Dependent Acceleration

Let the acceleration a is function of velocity v, then

$a = f(v$

Since $a = v\frac{dv}{dx}, therefore$

$v\frac{dv}{f(v)} = dx$

Integrating

$\int v\frac{dv}{f(v)} = \int dx$

or $x = g(v) + A$

where $g(v) = \int v\frac{dv}{f(v)} and$ A is constant of integration which gives the distance travelled by the particles.

Again consider a = f(v). since

$a = \frac{dv}{dt},$

Therefore

$\frac{dv}{f(v)} = dt \quad or$

or $\int\frac{dv}{f(v)} = \int dt$

or $t = h(v) + B$

where $h(v) = \int \frac{dv}{f(v)}$ and B is constant of integration and gives the time .

Example: A particle moves in a straight line with an acceleration kv'. If its initial velocity is u, find the velocity and the time spend when the particle has travelled a distance of x.

Solution.

Given that

$a = kv^3$

Since $a = v\frac{dv}{dx}$, therefore

$$v\frac{dv}{dx} = kv^3$$

$$or \quad v^{-2}\, dv = k\, dx$$

$$or \quad \int v^{-2}\, dv = kdx$$

$$or\ v^{-1} = kx + A$$

Given $v = u, x = 0\, at\, t = 0, therefore\ A = -\frac{1}{u}$

Thus $\frac{1}{v} = \frac{1}{u} - kx \quad or\ c = \frac{u}{1 - kux}$

Again since $v = \frac{dx}{dt}$, then

$$\frac{dx}{dt} = \frac{u}{1 - kux}$$

$$or \quad (1 - kux)\, dx = u\, dt, thus$$

$$\int (1 \quad kux)\, dx - u \int dt$$

$$x - ku\left(\text{—} \right) = ut + B$$

Using $x = 0, at\, t = 0, we\, get\, B = 0.$ Hence

$$x - ku\left(\frac{x^2}{2}\right) = ut\, or \quad t = \frac{x}{2u}(2 - kux)$$

Example. A Particle starts with a velocity u and move in a straight line. If the suffers a retardation equal to the square of velocity, find the distance travelled by the particle in time t.

Solution. Given retardation $= v^2, therefore$

$$a = -vt^2$$

We know that

$$v\frac{dv}{dx} = a \text{ ,there fore}$$

$$\frac{dv}{dx} = -v$$

$$or \qquad \frac{dv}{v} = -dx$$

$$or \qquad \int\frac{dv}{v} = -\int dx$$

$$or \qquad In\, v = -x + A$$

where a is constant of integration .

Using $v = u, x = 0\, at\, t = 0, we\, have$

$$A = In\, u, thus$$

$$\ln\left(\frac{u}{v}\right) = x \quad or \quad \frac{u}{v} = e^x \quad or \quad v = \frac{u}{e^x}$$

Since $v = \frac{dx}{dt}$, therefore

$$\frac{dx}{dt} = \frac{u}{e^x}$$

$$or \quad e^x = u\, dt, then$$

$e^x = ut + B$

where B is constant of integration .

Using $x=0\,at\,t=0,$ we get $B=1$. Hence

$e^x = ut + 1 \;\; or \;\; x = in(ut+1)$

(iii) Distance Dependent Acceleration

Let the acceleration be a function of distance x . Then a = f(x).

Since $v\frac{dv}{dx} = f(x)dx$

$or \;\; \int v dv = \int f(x)dx$

$$\frac{v^2}{2} = g(x) + A$$

where g(x) $\int f(x)dx\, and\, A$ is the constant of integration

$or \;\; v = \pm\sqrt{2g(x)+2A}$

We know that

$$\frac{dx}{dt} = v, therefore$$

$$\frac{dx}{dt} = \pm\sqrt{2g(x)+2A}$$

$$dt = \pm\frac{dx}{\sqrt{2g(x)+2A}}$$

Integrating, we get

$$t = \pm\int\frac{dx}{\sqrt{2g(x)+2A}} + B$$

where B is constant of integration.

Example

Discuss the motion of the particle moving in a straight line if it starts at a distance 'a' from a point O and move with an acceleration equal to times its distance from the O.

Solution

Let x be the distance of particle from O, then

Acceleration $= a = \mu x$

Since $v\frac{dv}{dx} = a \;\; or \;\; v\frac{dv}{dx} = \mu x \;\; or \;\; vdv = \mu xdx$

$$or \quad \int vdv = \mu \int xdx$$

$$\frac{v^2}{2} = \mu\frac{x^2}{2} + A$$

Given $v = 0, x = a \, at \, t = 0,$ so that

$$A = -\mu\frac{a^2}{2}, thus$$

$$\frac{v^2}{2} = \mu\left(\frac{x^2}{2} - \frac{a^2}{2}\right)$$

$$or \quad v = \pm\sqrt{\mu}\sqrt{x^2 - a^2}$$

$$Take \, v = \sqrt{\mu}\sqrt{x^2 - a^2} \;, since \, v = \frac{dx}{dt}, then$$

$$\int\frac{dx}{\sqrt{x^2 - a^2}} = \sqrt{\mu}\int dt$$

$$or \qquad \cosh^{-1}\left(\frac{x}{a}\right) = \sqrt{\mu}\int dt$$

Using $x = a \, at \, t = 0, B = 0, thus$

$$\cosh^{-1}\left(\frac{x}{a}\right) = \sqrt{\mu}\, t$$

or $x = a\cosh(\sqrt{\mu}t)$

If we take $v = -\sqrt{\mu}\sqrt{x^2 - a^2}, since\, v = \frac{dx}{dt}, then$

$$\int \frac{dx}{\sqrt{x^2 - a^2}} = -\sqrt{\mu}\int dt + C$$

$$\cosh^{-1}\left(\frac{x}{a}\right) = -\sqrt{\mu}\,t + C$$

Using $x = a\, t = 0, C = 0,$ thus

$$\cosh^{-1}\left(\frac{x}{a}\right) = -\sqrt{\mu}\,t$$

or $= a\cosh\left(-\sqrt{\mu}\,t\right) = a\cosh\left(\sqrt{\mu}\,t\right)$

3.1.5. Graphical Solution of Rectilinear Problem

Consider the graph of the curve $y = f(x)$ in the $xy-$ plane, then

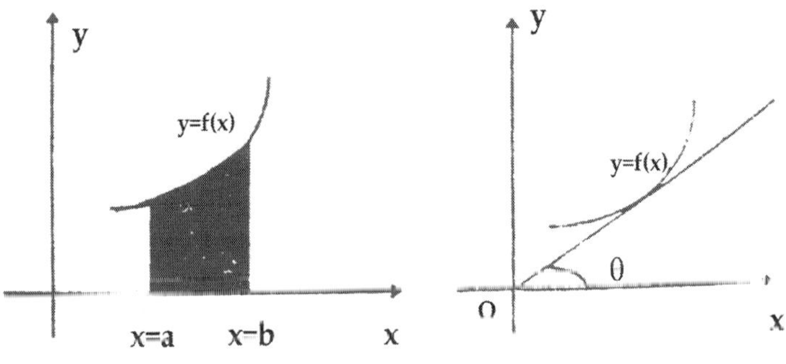

$(a)\ \frac{dy}{dx} =$ slope of the tangent at any point

(b) Area bounded by the curve $y = f(x), from\, x = a\, to\, x = b\, is = \int_a^b y\,dx$

(c) **(i) Velocity- Time Curve**

Consider the graph of velocity time graph v = v(t), then acceleration = a =

$$\frac{dv}{dt}$$

or $a =$ slope of velocity – time curve $= \tan\theta$

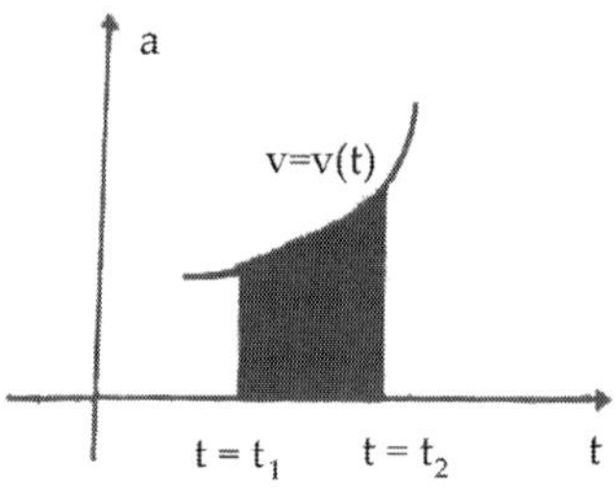

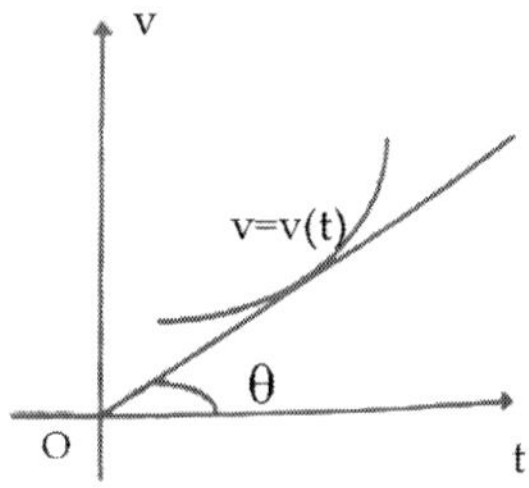

We know that $v = \frac{dx}{dt}$, thus

$dx = vdt$

or $\int_{t=t_1}^{t=t_2} dx = \int_{t=t_1}^{t=t_2} vdt$

or $\left|x\right|_{t=t_1}^{t=t_2} = \int_{t=t_1}^{t=t_2} vdt$

or $x(t_1) - x(t_2) = \int_{t=t_1}^{t=t_2} vdt$

Change in distance $= area$ bounded by velocity – time curve, t-axis $t = t_1 \; and \; t = t_2$

(ii) Space – Time curve

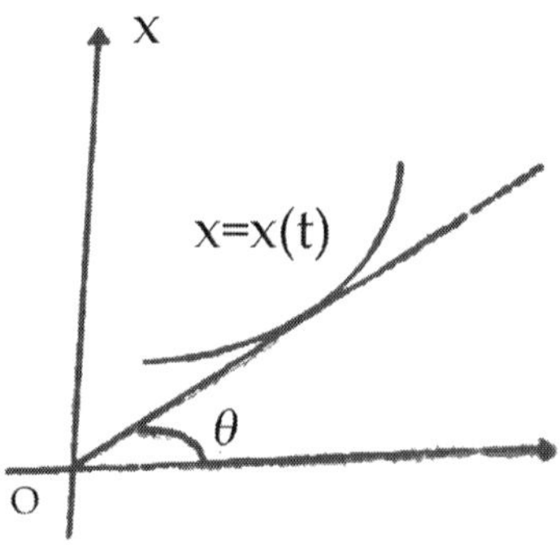

Consider the space- time curve $x = x(t)$, as

$$velocity = v = \frac{dx}{dt}$$

$velocity =$ slope of the space-time curve

$= tan\theta$

(ii) Acceleration-time graph

Consider the acceleration – time graph $a = a(t)$, then, since

$$a = \frac{dv}{dt} \quad or \quad dv = adt$$

or $$\int_{t=t_1}^{t=t_2} dv = \int_{t=t_1}^{t=t_2} adt$$

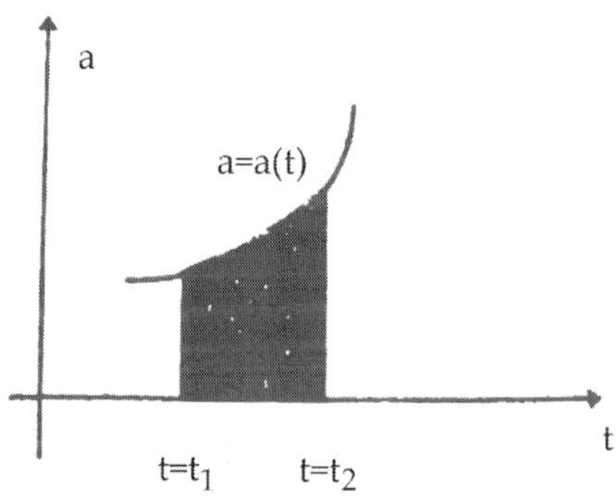

or $$|v|_{t=t_1}^{t=t_2} = \int_{t=t_1}^{t=t_2} adt$$

or $$v(t_1) - v(t_2) = \int_{t=t_1}^{t=t_2} adt$$

Change in velocity = area bounded be the acceleration – time curve, $t - axis\, t = t_1\, and\, t = t_2$

(iv) Velocity-space curve

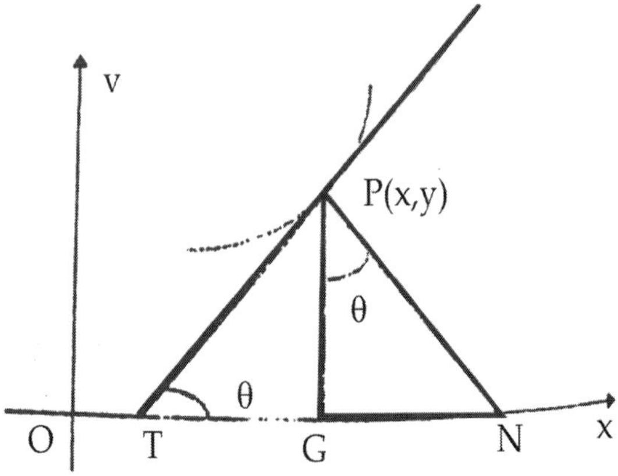

Consider the velocity- space curve $v = v(x)$. Now since

$$a = v\frac{dv}{dx}$$

where

Slope of tangent = $\frac{dv}{dx}$

Therefore

$$tan\theta = \frac{dv}{dx}$$

In triangle PNG

$$\tan\theta = \frac{GN}{GP} = \frac{GN}{v}$$

From (1) and (2), we have

$$\frac{dv}{dx} = \frac{GN}{v} \quad \text{or} \quad v\frac{dv}{dx} = GN$$

Therefore

Acceleration $= a = Lenght\, of\, subnormal\, at\, P$

Example. A particle starts from rest with a constant acceleration a. when its velocity acquires a certain value v, it moves uniformly and then its velocity start decreasing with constants retardation 2a till it comes to rest .

Find the distance traveled by the particle. If the time taken from rest to rest is t.

Solution

Let t_1 ,$and\, t_2\, and\, t_3$ be the times for accelerated, uniform motion and retorted motion respectively , then

$t = t_1 + t_2 + t_3$

Acceleration = a = slope of OA $= \frac{v}{t_1}$

or $t_1 = \frac{v}{a}$

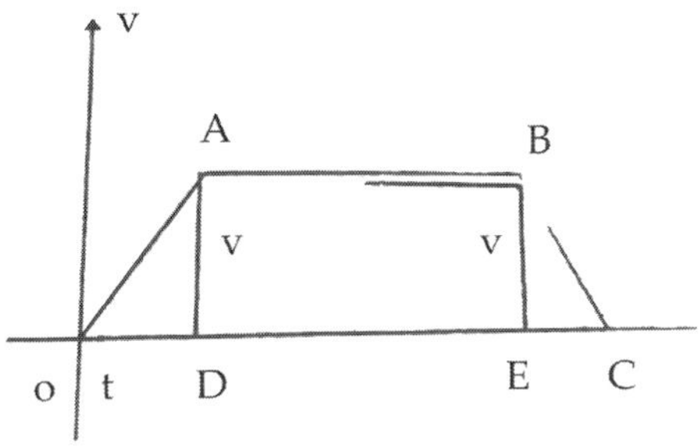

Similarly retardation = $2a = \frac{v}{t^3}$

or $t_3 = \frac{v}{2a}$

Thus $t = t_1 + t_2 + t_3$

$= \frac{v}{a} + t_2 + \frac{v}{2a}$

This implies that

$t_2 = t \quad \frac{3v}{2a}$

Distance = area under the velocity- time curve

= area of trapezium OABC

$$=\frac{1}{2}(OC+AB)(AD)$$

$$=\frac{1}{2}(t_1+t_2+t_3+t_2)v$$

$$=\frac{1}{2}(t+t_2)v$$

$$=\frac{1}{2}\left(t+t-\frac{3v}{2a}\right)v=vt-\frac{3v^2}{4a}$$

Example

A particle moving along a straight-line stars from rest and is accelerated uniformly till it attains the velocity v. The motion is the retorted and the particle come to rest after traversing a total distance x. if the acceleration is f, find the retardation and the total time take y the particle from rest to rest.

Solution

Let t_1 and t_2 be the time for acceleration and retorted motion respectively, then

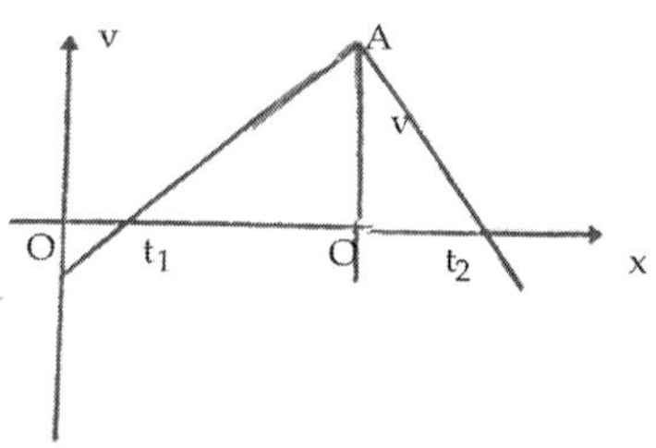

Acceleration = f

$=slope\, of\, OA$

Or $f=\frac{v}{t_1}\quad or\, t_1=\frac{v}{f}$

Let g be the retardation, then

g = slope of AB

$$=\frac{v}{t^2}$$

or $t^2 = \frac{v}{g}$

Distance = x = area of triangle OAB $= \frac{1}{2}(t_1 + t_2)v$

or $t_1 + t_2 = \frac{2x}{v}$

Thus total time = $\frac{2x}{v}$

Put $t_1 = \frac{v}{f} \, and \, t_2 = \frac{v}{g} in(1),$ then

$$x = \frac{1}{2}\left(\frac{v}{f} + \frac{v}{g}\right)v = \frac{1}{2}\left(\frac{1}{f} + \frac{1}{g}\right)v^2$$

or $\frac{2x}{v^2} = \frac{1}{f} + \frac{1}{g}$

or $\frac{1}{g} = \frac{2x}{v^2} - \frac{1}{f} = \frac{2xf - V^2}{fv^2}$

or $g = \frac{fv^2}{2xf - v^2}$

3.2. PROJECTILE MOTION

An object thrown into the space with certain velocity from a gun or dropped from a moving plane under the action of gravity is called a projectile. Thus, a projectile move with a constant horizontal velocity and at the same falls freely under the action of gravity. The path of projectile is called the *trajectory*

3.2.1. Equation of the Trajectory of a projectile

Let the particle m be projected from point O with an initial velocity v_o making an angle α with the horizontal. Take O as the origin and the horizontal and vertical lines through O as $x-axis$ and $y-axis$ respectively.

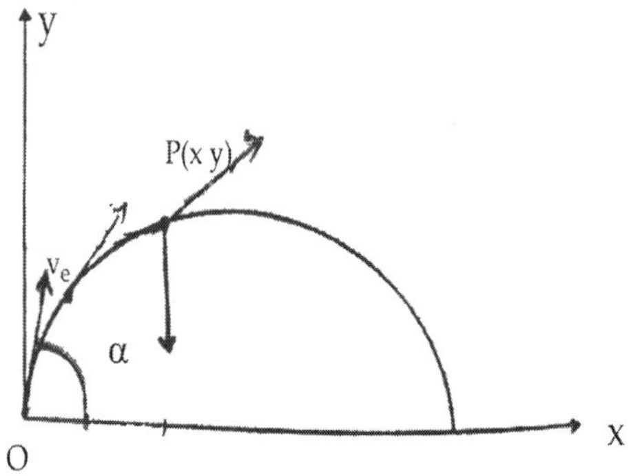

Suppose that after time t, the particle is at point $p(x, y)$ whose position vector is $\bar{r}$. *i.e*

$$\bar{r} = x\hat{i} + y\hat{j}$$

$$\dot{\bar{r}} = \dot{x}\hat{i} + \dot{y}\hat{j}$$

$$\ddot{\bar{r}} = \ddot{x}\hat{i} + \ddot{y}\hat{j}$$

The gravitational force $\bar{F}$ acting on the particle at P is

$$\bar{F} = mg\hat{j}$$

$$m\ddot{\bar{r}} = -mg\hat{j}$$

or $\ddot{\bar{r}} = g\hat{j}$

$$\ddot{x}\hat{i} + \ddot{y}\hat{j} = -g\hat{j},$$

Then

$$\ddot{x} = 0 \quad and \quad \ddot{y} = -g$$

Integrating, we get

$$\dot{x} = A \quad and \quad \dot{y} = -gt + B$$

at $t = 0,$ the particle is at O, therefore $at\ t = 0,\ \dot{x} = v_0 \cos\alpha, \dot{y} = v_o \sin\alpha$

Therefore

$$A = v_o\, cos\alpha, B = v_o\, sin\alpha$$

Hence

$$\dot{x} = v_0 \cos\alpha,\ \dot{y} = -gt + v_o\, sin\alpha$$

which gives the horizontal and vertical component of velocity at any time t. again integrate

$$x = (v_0 \cos\alpha)t,\ and\ \ y = -g\frac{t^2}{2} + t + D$$

$$at\ t = 0, x = 0, y - 0, therfore\, C = 0, D = 0,$$

Hence,

$$x = (v_0 \cos\alpha)t$$

$$y = v_0\, sin\alpha - \frac{1}{2}gt^2$$

The eqs (1) and (2) are parametric equation of trajectory. To find the Cartesian equation, eliminate t from the $eqs.$ (1) and (2), we get

which is the Cartesian equation of trajectory $\cdot$ from $(3),$ we have

$$\frac{gx^2}{2v_o^2 \cos^2\alpha} - x\frac{sin\alpha}{cos\alpha} = -y$$

or

$$x^2 - x\frac{sin\alpha}{cos\alpha}\left(\frac{2v_o^2 \cos^2\alpha}{g}\right) = -y\frac{2v_o^2 \cos^2\alpha}{g}$$

or

$$x^2 - \frac{2xv_0^2\, sin\alpha \cos\alpha}{g} = -2y\frac{v_o^2 \cos^2\alpha}{g}$$

or

$$x^2 - 2x\frac{v_0^2 sin\alpha \cos\alpha}{g} + \left(\frac{v_0^2 sin\alpha \cos\alpha}{g}\right)^2$$

$$= -2y\frac{v_o^2 \cos^2\alpha}{g} + \left(\frac{v_0^2 sin\alpha \cos\alpha}{g}\right)^2$$

or $\left(x-\frac{v_0^2 \sin\alpha\cos\alpha}{g}\right)^2=\frac{2v_o^2\cos^2\alpha}{g}\left(y-\frac{v_o^2\sin^2\alpha}{2g}\right)$

Now

(i) **Vertex**

$$\left(\frac{v_0^2 \sin\alpha\cos\alpha}{g},\frac{v_o^2\sin^2\alpha}{2g}\right)$$

is the verte3x of the

(ii) Latus Rectum

Latus Rectum = $|4a|=\frac{2v_o^2\cos^2\alpha}{g}$

(iii) Maximum Height

Maximum height attained above x- axis

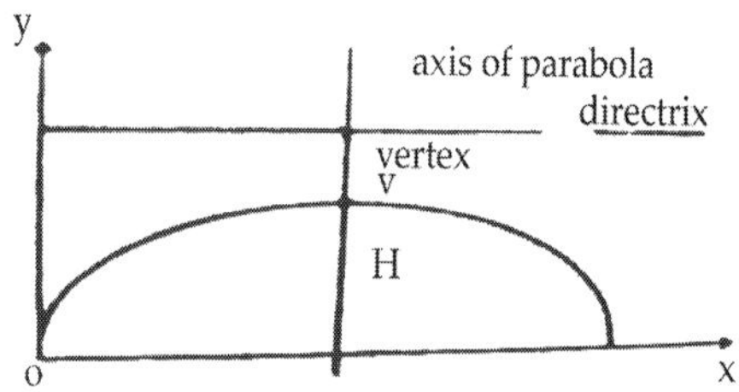

H = Maximum height attained above x – axis

$= y-$ Coordinate of vertex

$$=\frac{v_o^2\sin^2\alpha}{2g}$$

This equation can also be obtained by putting $\frac{dy}{dx}=0$ in the equation of trajectory.

(iv) Equation of Directrix

Height of Directrix above $x-axis$

$$=H+\frac{1}{4}(latus\,rectum)$$

$$= \frac{v_o^2 \sin^2 \alpha}{2g} + \frac{v_o^2 \cos^2 \alpha}{2g}$$

$$= \frac{v_o^2}{2g}\left(\sin^2 \alpha + \cos^2 \alpha\right) = \frac{v_o^2}{2g}$$

Equation of Directrix is $y = \frac{v_o^2}{2g}$

(v) Focus

x – Coordinate of focus = x – coordinate of vertex

$$= \frac{v_0^2 sin\alpha \cos\alpha}{g}$$

y – Coordinate of focus = H - $\frac{1}{4}(latus\, rectum)$

$$= \frac{v_o^2 \sin^2 \alpha}{2g} + \frac{v_o^2 \cos^2 \alpha}{2g}$$

$$= \frac{v_o^2}{2g}\left(\sin^2 \alpha + \cos^2 \alpha\right)$$

$$= -\frac{v_o^2}{2g} cos2\alpha$$

Thus, the focus

$$= \left(\frac{v_0^2 sin\alpha \cos\alpha}{g}, -\frac{v_o^2}{2g} cos2\alpha\right) = \left(\frac{v_o^2}{2g} sin2\alpha - \frac{v_o^2}{2g} cos2\alpha\right)$$

(vi) Time of Flight

The time taken by the particle to reach the horizontal plane through the point of projection is ca;;ed its time of flight. In this case put y = 0, in the equation

$$y = g\frac{t^2}{2} + (v_o \sin\alpha)t$$

Then $0=-g\frac{t^2}{2}+(v_o\sin\alpha)t$

$since\,t\neq 0,$therefore

$$v_o\sin\alpha-\frac{1}{2}gt=0$$

or $t=\dfrac{2v_o\,sin\alpha}{g}$ is the time of flight.

(vii) Range

(a) Horizontal Range

The horizontal range of the projectile is the horizontal distance described by the particle during the time of flight. Therefore

Range R = (horizontal velocity) (time of flight)

$$v_o\cos\alpha\left(\frac{2v_o\,sin\alpha}{g}\right)$$

$$=\frac{v_o^2\,2(\sin\alpha\,cos\alpha)}{g}=\frac{2v_o^2sin2\alpha}{g}$$

This range can also be obtained by putting y = 0 in the Cartesian equation of trajectory $i.e$ putting y = 0 in the equation

$$y=xtan\alpha-\frac{gx^2}{2v_o^2}\sec^2\alpha$$

Also, it is notable that this range will remain same if we replace $\alpha by\,\alpha=\frac{\pi}{2}-\alpha$

(b) Maximum Range

Range R = $\dfrac{v_o^2sin2\alpha}{g}$

For maximum range $\dfrac{dr}{d\alpha}=0$

$$\Rightarrow \quad \frac{2v_o^2}{g}\cos 2\alpha = 0$$

$$\Rightarrow \quad cos 2\alpha = 0$$

$$\Rightarrow \quad 2\alpha = \frac{\pi}{2}$$

$$\Rightarrow \quad \alpha = \frac{\pi}{4}$$

Therefore

$$R_{max} = \frac{v_o^2}{g}$$

3.2.2. Range in inclined plane

Let β the inclination of the inclined plane. Let a particle is projected from point O with velocity v_o by making an angle α to the horizontal where $\beta < \alpha$. Let the projectile meet the inclined plane at $P(x, y)$ then OP = r is called the range of the inclined plane.

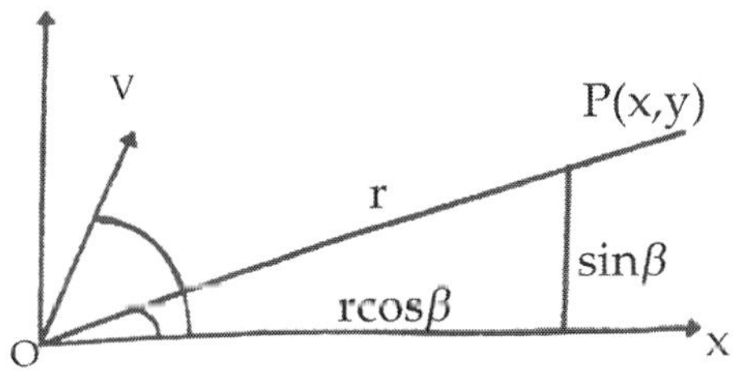

$$x = rcos\,\beta \quad , \; y = rsin\beta$$

$$P(x, y) = (rcos\beta, rsin\beta)$$

Equation of trajectory is

$$y = \tan\alpha - \frac{1}{2} g \frac{x^2}{v_o^2} \sec^2 \alpha$$

The point $p(rcos\beta, rsin\beta)$ lies on it, then

$$r\,sin\beta = r\beta tan\alpha - \frac{1}{2}g\frac{r\cos^2\beta}{v_o^2\cos^2\alpha}$$

or $$\frac{grcos^2\beta}{2v_o^2\cos^2\alpha} = \cos\beta\frac{sin\alpha}{cos\alpha} - sin\beta$$

$$= \frac{sin\alpha\,cos\beta - cos\alpha\,sin\beta}{cos\alpha}$$

$$= \frac{\sin(\alpha-\beta)}{cos\alpha}$$

Thus

$$r = \frac{2v_o^2\cos^2\alpha}{g\cos^2\beta} \times \frac{\sin(\alpha-\beta)}{cos\alpha}$$

$$= \frac{2v_o^2\cos^2\alpha\sin(\alpha-\beta)}{g\cos^2\beta}$$

For range down the inclined plane, change $\beta\,to - \beta$.

Maximum Range

Equation (1) can be written as, by using

$$\sin(\alpha+\beta) - \sin(\alpha-\beta) = 2\cos\alpha sin\beta$$

$$r = \frac{v_o^2}{gcos^2\beta}(\sin(2\alpha-\beta) = 1\ ,so\,that$$

$$2\alpha - \beta = \frac{\pi}{2}\ \ or\ \ \alpha = \frac{\beta}{2} + \frac{\pi}{4}$$

Thus

$$r_{max} = \frac{v_o^2}{gcos^2\beta}(1 - \sin\beta)$$

$$\frac{v_o^2(1-\sin\beta)}{g(1-\sin^2\beta)} = \frac{v_o^2}{g(1-\sin\beta)}$$

Or for maximum range put $\frac{dr}{dx}$ 0

or $\cos(2\alpha-\beta) = 0$

or $2\alpha-\beta = \frac{\pi}{2}$

Time of Flight

Time of flight = $\frac{Horizontol\ distance}{Horizontol\ velocity}$

$$= \frac{x}{v_o\cos\alpha} = \frac{r\cos\beta}{v_o\cos\alpha}$$

Using (1), we get

Time of flight = $\frac{2v_o^2\cos^2\alpha\sin(\alpha-\beta)}{g\cos^2\beta}\frac{\cos\beta}{v_o\cos\alpha}$

$$= \frac{2v_o^2\sin(\alpha-\beta)}{g\cos\beta}$$

For time of flight down the plane, change β *to* $-\beta$.

Example: Show that the speed at a point P of the projectile is the same as attained by a particle of same mass at P falling freely from the point vertically above P on the Directrix of the trajectory.

Let v_o be The Velocity of projection and v be the velocity at point $p(x, y)$ of trajectory, then

$$\bar{v} = \dot{x}\hat{i} + \dot{y}\hat{j}$$

$$\bar{v} = (v_0\cos\alpha)\hat{i} + (v_o\sin\alpha - gt)\hat{j}$$

$$v = |\bar{v}| = \sqrt{(v_0\cos^2\alpha) + (v_o\sin\alpha - gt)^2}$$

$$v = \sqrt{v_o^2 \cos^2 \alpha + v_o^2 \sin^2 \alpha - 2gtv_o\, sin\alpha + g^2 t^2}$$

$$= \sqrt{v^2 - 2g\left(v_o t\, sin\alpha - \frac{1}{2}gt^2\right)}$$

Since

$$y = v_o t \sin\alpha - \frac{1}{2}gt^2$$

Therefore

$$v = \sqrt{v_0^2 - 2gy}$$

Let N be the point on the Directrix vertically above $P(x, y)$. From fig.

$$MN = MP + PN$$

$$\frac{V_0^2}{2g} = y + PN$$

Let v' be the velocity of the particle in failing freely a distance NP. Using the equation of motion

$$v^2 - U^2 = 2gx$$

where u is initial velocity and v is final velocity

Here $u = 0, x = PN, v = v'$, then

$$v'^2 - 0 = 2g(PN)$$

$$v'^2 = 2g\left(\frac{v_o^2}{2g} - y\right)$$

$$v'^2 = v_o^2 - 2gy$$

$$v' = \sqrt{v_o^2 - 2gy}\ .$$

From $(1)\, and\, (2)$, we have

$$v' = v$$

3.2.3. Parabola if Safety

As the equation of the projectile is

$$y = x\,tan\alpha - \frac{1}{2}g\frac{x^2}{v^2}\sec^2\alpha$$

which shows that for different values of α we have different trajectories. So, we define that the locus of all the points on different trajectories which lie at maximum distance from the point of projection O is called the parabola of safety. All the trajectories from O must touch the parabola of safety at some point. Thus, it is envelope of the family of all the trajectories from with same velocity of protestation v_o, α being regarded as parameter. Equation of trajectory is

$$y = x\tan\alpha - \frac{1}{2}g\frac{x^2}{v_o^2}\sec^2\alpha$$

$$y = x\tan\alpha - \frac{1}{2}g\frac{x^2}{v^2}$$

$$\frac{gx^2}{2v_o^2}\tan^2\alpha - x\tan\alpha + \left(\alpha + \frac{gx^2}{2v_o^2}\right) = 0$$

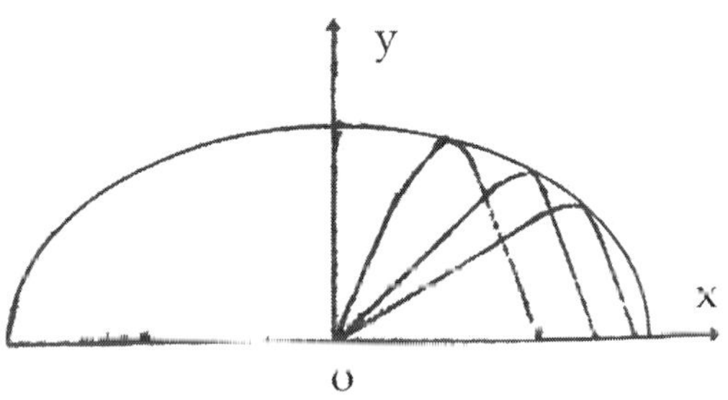

It is quadratic equation in $tan\alpha$. For the equation of envelope , put discriminate = 0, $i.e$

$$x^2 - 4\left(\frac{gx^2}{v_o^2}\right)\left(y + \frac{gx^2}{v_o^2}\right) = 0$$

Since $x \neq 0$, dividing it by x^2 , we get

$$-2\left(\frac{g}{v_o^2}\right)\left(y+\frac{gx^2}{v_o^2}\right)=0$$

$$-\frac{2gy}{v_o^2}-\frac{g^2x^2}{v_o^2}=0$$

$$\frac{g^2x^2}{v_o^2}=\frac{2gy}{v_o^2}+1$$

$$x^2=-2\frac{v_o^2}{g}\left(y-\frac{v_o^2}{2g}\right)$$

which is the equation of the parabola of safety having vertex

$$\left(0,\frac{v_o^2}{2g}\right)$$

Latus Rectum = $\frac{2v_o^2}{g}$

Equation of Directrix is

$$y=\frac{v_o^2}{2g}+\frac{1}{4}(latus\,rectum)$$

$$y=\frac{v_o^2}{2g}+\frac{1}{4}\left(\frac{2v_o^2}{g}\right)$$

$$y=\frac{v_o^2}{2g}+\frac{v_o^2}{2g}=\frac{v_o^2}{g}$$

$$x-coordinate\,of\;focus=x-coordinate\,of\;vertex=0$$

$$y-coordinate\;of\;focus\;=y-\;coordinate\;of\;vertex\;=0$$

$$-\frac{1}{4}(latus\,rectum)$$

$$=\frac{v_o^2}{2g}-\frac{1}{4}\frac{2v_o^2}{g}=0$$

Therefore, focus

3.2.4. Projection with Air Resistance

Let a particle of mass m be projected with an initial velocity v_o making an angle α with the horizontal as shown in the figure.

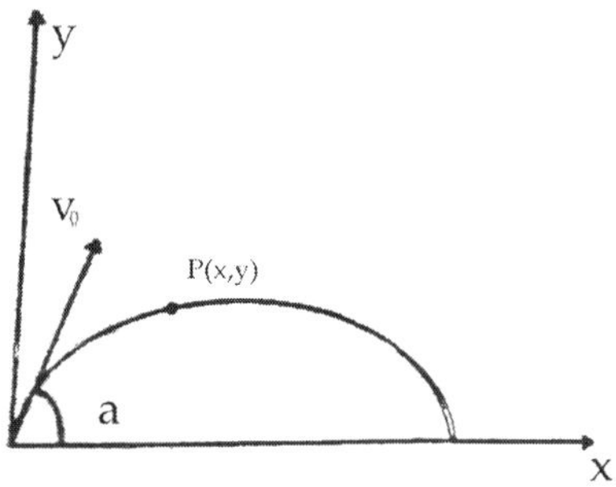

Here we restrict ourselves to the simple case of air resistance proportional to the first power of velocity. We suppose that after some time t the particle is at P(x, Y). The equations of motion are, with mass canceled,

$$\ddot{x}=-k\dot{x}$$

$$\ddot{y}=-g-k\dot{y}$$

From (1)

$$\frac{dx}{dt}=-kx$$

$$\frac{dx}{x}=-kdt$$

Integrate

$$In(x)=-kt+c_1$$

At $t=0\, x=v_o\cos\alpha\ ,therefore\, c_1=in(v_o cos\alpha)$

In(x) $=-kt+in\left(v_o cos\alpha\right)$

$$x=v_o\cos\alpha\, e^{-kt} \tag{3}$$

$$x=-\frac{v_o}{k}cos\alpha\, e^{-kt}+c_2$$

At $t=0$,$x=0$, there fore

$$c_2=\frac{v_o}{k}cos\alpha, thus$$

$$x=\frac{v_o}{k}\cos\alpha\, e^{-kt}+\frac{v_o}{k}cos\alpha$$

or $$x=-\frac{v_o}{k}\left(1-e^{-kt}\right)cos\alpha \tag{4}$$

Now, consider equation (2)

$$\frac{dy}{dt}=-g-ky$$

or $$\frac{dy}{g+ky}=-dt$$

Integrate

$$in\left(g+ky\right)=-kt+c_3$$

At $t=0, y=v_o sina,$ that implies that

$$c_3=in\ (g+kv_o\sin\alpha)$$

In $\left(g+ky\right)=-kt+In\left(g+kv_o sin\alpha\right)$

In $$In\frac{g+ky}{g+kv_o\sin\alpha}=-kt$$

$$g+ky=\left(g+kv_o\sin\alpha\right)e^{-kt}$$

$$y=-\frac{g}{k}+\left(\frac{g}{k}+kv_o\, sin\alpha\right)e^{-kt} \tag{5}$$

Again integrating

$$y = -\frac{g}{k} - \frac{1}{k}\left(\frac{g}{k} + v_o sin\alpha\right)e^{-ky} + c_4$$

$at\ t = 0, y = 0,$ this implies that

$$c_4 = \frac{1}{k}\left(\frac{g}{k} + v_o\ sin\alpha\right)$$

Thus

$$y = -\frac{g}{k} - \frac{1}{k}\left(\frac{g}{k} + v_o\ sin\alpha\right)e^{-kt} + \frac{1}{k}\left(\frac{g}{k} + v_o\ sin\alpha\right)$$

$$y = -\frac{g}{k}t - \frac{1}{k}\left(\frac{g}{k} + v_o\ sin\alpha\right)\left(1 - e^{-kt}\right) \qquad (6)$$

The equation of the path is obtained by eliminating t from equation $(4) and (6)$ as

$$y = -\frac{xcos\alpha}{v_o}\left(\frac{g}{k} + v_o\ sin\alpha\right) + \frac{g}{K^2} In\left(1 - \frac{xcos\alpha}{v_o}\right)$$

Example

A shell burst on contact with the ground and pieces from it fly in all directions with all speed up to 80 ft. / sec. Prove that a man 100 ft. is in danger for $\frac{5}{\sqrt{2}}$ seconds.

Solution

We know that the range of the particle is

$$R = \frac{v_o^2 sin2\alpha}{g}$$

Here $R - 100\ ft., v_o - 80\ ft\ /\ sec., g = 32\ ft\ /\ sec^2.,$ therefore

$$100 = \frac{6400 sin2\alpha}{32}$$

or $\sin 2\alpha = \frac{1}{2}$

or $\alpha = 15^o, 75^o$

For the range $100\,ft.$ there are two directions of projection. Let t_1 and t_2 be the times of flight respectively, then

$$t_1 = \frac{2v_o sin15^o}{g}$$

and $$t_2 = \frac{2v_o sin75^o}{g}$$

The man is in danger for time = $t_2 - t_1$

$$= \frac{2v_o sin15^o}{g} - \frac{2v_o sin75^o}{g}$$

$$= \frac{2v_o}{g}\left(sin15^o - sin75^o\right)$$

$$= \frac{160}{32}\left(2\cos\left(\frac{75+15}{2}\right)\sin\left(\frac{75-15}{2}\right)\right)$$

$$= 5\left(2cos45^o sin30^o\right) = \frac{5}{\sqrt{2}}\,sec.$$

Example Find the range of a rifle bullet when α is the elvation of projection and v_o is speed. show that, if the rifle is field with the same elevation and speed from a car travelling with v towards the target, the range will be increase by $\frac{2vv_o sin\alpha}{g}$

Solution

We know that the range is given by

$$R = (horizontal\,velocity)(time\,of\;flight)$$

$$R = v_o cos\alpha\left(\frac{2v_o sin\alpha}{g}\right) = \frac{2v_0^2 \sin\alpha cos\alpha}{g}$$

When the shell is fired from a car moving with speed v towards the target the horizontal velocity is increased by v.

Horizontal velocity = $v_o\, cos\alpha + v$

New $range + R'$

$= (new\, horizontal\, time\, of\ flight)$

$$= (v_o cos\alpha + v)\left(\frac{2v_o sin\alpha}{g}\right)$$

$$R' = \frac{2v_0^2 \sin\alpha cos\alpha}{g} + \left(\frac{2v_o v\, sin\alpha}{g}\right)$$

Increase in range = $r' - R = \left(\frac{2v_o v\, sin\alpha}{g}\right)$

Example

A battleship is steaming ahead with speed v and a gun is mounted on a battleship so as to point straight backward and is set at an angle of elevation α if v_o is the speed of projection (relative to gun), show that the range is $\left(\frac{2v_o sin\alpha}{g}\right)(v_o\, cos\alpha - v)$

Prove also that the angle of elevation for maximum range is $\cos^{-1}\left(\frac{v + \sqrt{v^2 + 8v_o^2}}{4v_o}\right)$

Solution

The ship is moving with speed v away from the target, so the horizontal velocity of the projectile will be decreased by v.

Actual horizontal velocity = $v_o\, cos\,\alpha - v$

Actual range = R = (horizontal velocity) (time of flight)

$$(v_o cos\alpha - v)\left(\frac{2v_o sin\alpha}{g}\right)$$

$$\left(\frac{2v_o sin\alpha}{g}\right)\left(v_o\, cos\alpha - v\right)$$

Since

$$R = \left(\frac{2v_o sin\alpha}{g}\right)\left(v_o cos\alpha - v\right)$$

$$\frac{v_o}{g}\,(v_o\,\sin 2\alpha - 2v\,sin\alpha)$$

Therefore, for maximum range

$$\frac{dR}{d\alpha} = \frac{2v_o}{g}\left(v_o\,cos2\alpha - v\,cos\alpha\right) = 0$$

$$v_o\,\cos 2\alpha - vcos\alpha = 0$$

$$v_o\,\,(2\cos^2\alpha - 1) - vcos\alpha = 0$$

$$v_o(2\cos^2\alpha - 1) - v\,cos\alpha = 0$$

$$2v_0\,\cos^2\alpha - vcos\alpha - v_0 = 0$$

$$cos\alpha = \frac{v \pm \sqrt{v^2 + 8v_o^2}}{4V_o}$$

$$\frac{d^2R}{d\alpha^2} = \frac{2v_o}{g}\left(-2v_o\,sin2\alpha + vsin\alpha\right)$$

$$= -\left(\frac{2v_o sin\alpha}{g}\right)4v_o\,cos\alpha -$$

$$\cos\alpha = \frac{v + \sqrt{v^2 + 8v_o^2}}{4V_o}$$

when

$$\frac{dR}{d\alpha} = \frac{2v_o sin\alpha}{g}\,(\sqrt{v^2 + 8v_0^2}) = -ve$$

For 0 < $\alpha < 90, sin\alpha is\ positive$.

Thus, R is maximum when

$$cos\alpha = \frac{v+\sqrt{v^2+8v_o^2}}{4V_o}$$

$$\alpha = \cos^{-1}\left(\frac{v+\sqrt{v^2+8v_o^2}}{4V_o}\right)$$

or

$$\alpha = \cos^{-1}\left(\frac{v+\sqrt{v^2+8v_o^2}}{4V_o}\right)$$

or

Example

Prove that the speed required to project a particle from a height h to fall a horizontal distance a from the point of projection is at least

$\sqrt{g(\sqrt{a^2+h^2-h)}}$

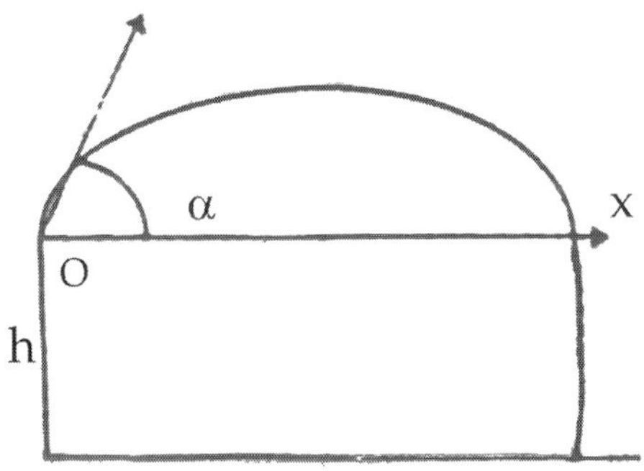

Solution

Equation of trajectory is

$$y = xtan\alpha - \frac{gx^2}{2v_o^2}\sec^2\alpha$$

Point $(a,-h)$ lies on it, then

$$-h = a\tan\alpha - \frac{ga^2}{2v_0^2}\sec^2\alpha$$

$$-h = a\tan\alpha - \frac{ga^2}{2v_0^2}(1+\tan^2\alpha)$$

$$ga^2\tan^2\alpha - 2av_o \, tan\alpha + ga^2 - 2hv_o^2) \geq 0$$

Since $tan\alpha$ is real therefore, discriminant $\geq 0, i.e.$

$$4a^2v_o^4 - 4ga^2\left(ga^2 - 2hv_o^2\right) \geq 0$$

$$\left(v_0^2 + gh\right)^2 \geq g^2\left(a^2 + h^2\right)$$

$$v_0^2 + gh \geq g\sqrt{a^2 + h^2}$$

$$v_0^2 \geq g\left(\sqrt{a^2 + h^2} - h\right)$$

$$v_o \geq \sqrt{g\left(\sqrt{a^2 + h^2} - h\right)}$$

Therefore, least value of v_o is $\sqrt{g\left(\sqrt{a^2 + h^2} - h\right)}$

Example

A particle is launched at an angle a from a cliff of height H above sea level. If it falls into the sea at a distance D from the base of the *cliff*, prove that the maximum height above the sea level is

$$H + \frac{D^2\tan^2\alpha}{4(H + D\tan\alpha)}$$

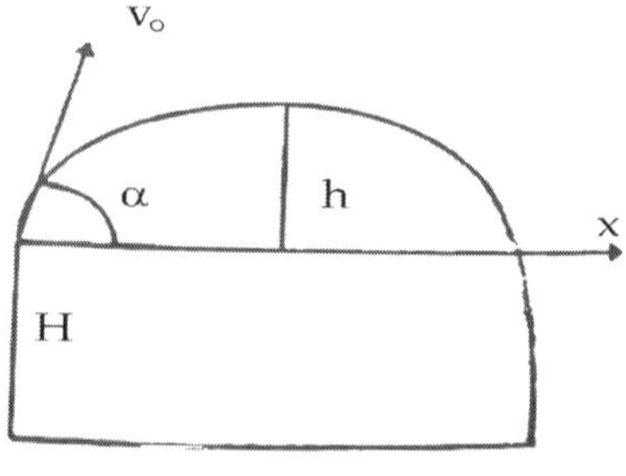

Solution

Equation of the trajectory is

$$y = xtan - \frac{gx^2}{2v_0^2}\sec^2\alpha$$

Point (D, -H) lies on it, therefore

$$-H = D\tan\alpha - \frac{gD^2}{2v_o^2}\sec^2\alpha$$

$$H + D\tan\alpha - \frac{gD^2}{2v_o^2\cos^2\alpha}$$

$$v_o^2 = \frac{gD^2}{2\cos^2\alpha(H + D\tan\alpha)}$$

Now, $h = height\, above\, x - axis,$ so that

$$h = \frac{v_o^2\sin^2\alpha}{2g}$$

Putting the value of v_o^2 ,then

$$h = \frac{gD^2}{2\cos^2\alpha\left(H + Dtan\alpha\right)}\frac{\sin^2\alpha}{2g}$$

$$= \frac{D^2\tan^2\alpha}{4\left(H + Dtan\alpha\right)}$$

Height above sea level $= H + h$

$$= H + \frac{D^2\tan^2\alpha}{4\left(H + Dtan\alpha\right)}$$

Example

A front and a slip are both armed with guns which give their projectile a muzzle velocity $\sqrt{2gk}\, and$ guns in the fort are at height h above the guns in the ship. If d_1 and d_2 are the greatest horizontal ranges at which the fort and ship respectively can engage, prove that

$$\frac{d_1}{d_2} = \sqrt{\frac{k+h}{k-h}}$$

Solution

Let S be the ship and F be the fort.

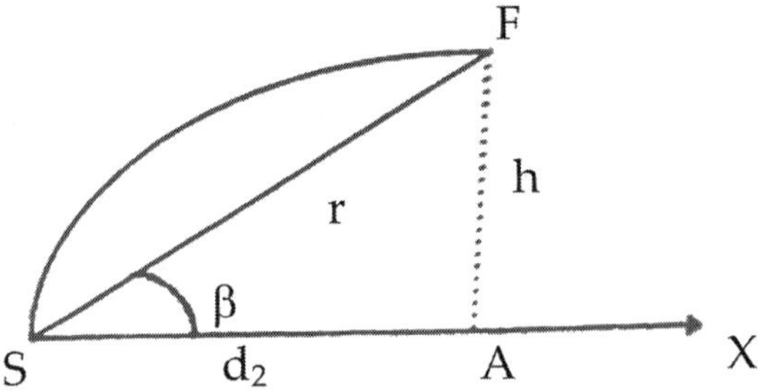

Let $\angle XSF = \beta$ and $SF = r$

d_2 is greatest horizontal range for gun in ship, so r is maximum range on inclined plane with inclination β. Now,

$$r = \frac{v_o^2}{g(1+\sin\beta)}$$

Put $v_o = \sqrt{2gk}$ in the above equation, then

$$r = \frac{2gk}{g(1+\sin\beta)} = \frac{2k}{1+\sin\beta}$$

$$r + r\sin\beta = 2k$$

Since form the figure $h = r \sin$, therefore

$$r + h = 2k$$

Also from right angle triangle SAF, we have

$$r^2 = h^2 + d_2^2$$

So that

$$h^2 + d_2^2 = (2k - h)^2 = 4k^2 - 4kh + h^2$$

$$d_2^2 = 4k^2 - 4kh = 4k(k - h) \qquad (1)$$

For gun in fort, change h to – h, then

$$d_1^2 = 4k(K + h) \qquad (2)$$

Dividing (2) by (1), then

$$\frac{d_1}{d_2} = \sqrt{\frac{k+h}{k-h}}$$

Example

The horizontal rang of a ball for two different projection is same and equal to R. if $h_1 \, and \, h_2$ are the greatest height in the two parts for which this is possible prove that

$$R^2 = 16 h_1 h_2$$

Solution

Let R be the range, when α is angle of the projection and v_o is velocity of projection , then

$$R = \frac{2v_o^2 \, sin\alpha cos\alpha}{g}$$

Let R' be the range when angle of projectile is $\frac{\pi}{2} - \alpha,$ then

$$R' \frac{2v_o^2 \sin\left(\frac{\pi}{2} - \alpha\right) cos\left(\frac{\pi}{2} - \alpha\right)}{g}$$

or

$$R' = \frac{2v_o^2 \, sin\alpha cos\alpha}{g}$$

which show that R is the same when angle of projection are α and $\frac{\pi}{2} - \alpha$.

Let $h_1 and \, h_2$ be the maximum height in these cases respectively, therefore

$$h_1 = \frac{v_o^2 \sin^2 \alpha}{2g}$$

$$h_2 = \frac{v_o^2 \sin^2\left(\frac{\pi}{2} - \alpha\right)}{2g} = \frac{v_o^2 \cos^2 \alpha}{2g}$$

Hence

$$16h_1h_2 = 16\frac{v_o^2 \sin^2 \alpha}{2g}\left(\frac{v_o^2 \cos^2 \alpha}{2g}\right)$$

$$= 4\frac{v_o^4 \sin^2 \alpha \cos^2 \alpha}{g} = R^2$$

$$i.e \;\; R^2 = 16h_1h_2$$

Example

Prove that if time of flight of a bullet over a horizontal range R is T seconds then the inclination of direction to the horizontal is

$$tan^{-1}\left(\frac{gt^2}{2r}\right)$$

Solution

We know that

$$T = \frac{2_o \sin \alpha}{g}$$

$$R = \frac{2v_o^2\, sin\alpha cos\alpha}{g}$$

$$\frac{gT^2}{2R} = \frac{4v_o^2 \sin^2 \alpha}{g} \frac{g}{4v_o^2\, sin\alpha cos\alpha} = tan\alpha$$

Hence

$$\alpha = \tan^{-1}\left(\frac{gT^2}{2R}\right)$$

Example

A cricket ball thrown from height of 6 feet at an angle of 30° with the horizontal speed of 60 $ft./sec.$ is caught by another player at height 2 feet from the ground. How far apart were the two men?

Solution

Let two men are x feet apart. As the ball is caught at a height of 2 feet from the ground, so the distance it falls vertically is 6-2 = 4ft. The point $(x,-4)$ lies on the trajectory. Now

$$y = (v_o \, sin\alpha)t - \frac{1}{2}gt^2$$

Here $y = -4$, $v_0 = 6 - ft \, / sec.,\ \alpha = 3 =^0, t = ?$

$$-4 == (60sin30)t - \frac{1}{2}(32)t^2$$

$$8t^2 - 15t - 2 = 0$$

or (t-2) (8t+1) = 0

$$t = 2 \ or \quad t = -\frac{1}{8}$$

Since $t \neq -\frac{1}{8}$, therefore $t = 2$.

$$x = Horzontal\ distance$$

$$= (Horizontal\ velocity)(time)$$

$$= (v_o cos\alpha)t = (60 \, cos30^0)(2)$$

$$x = 60\sqrt{3}\ ft.$$

3.3. SUMMARY AND DISCUSSION

Rectilinear Motion. The motion of a particle along a straight line is called *rectilinear motion.*

Vertical Motion under Gravity

For a failing body, the acceleration is constant. It is called acceleration due to gravity and is denoted by 'g'. In FPS system g = $32\,ft/sec^2$.*In* CGS system g =981 cm/sec^2 and in MKS system $g = 9.81m/sec^2$. If the body is projected vertically upward then acceleration g is negative. For falling bodies equation of motion is

$$v = u + gt$$

$$x = ut + \frac{1}{2}gt^2$$

$$v^2 - u^2 = 2gx$$

Distance traveled in nth Second

Let $x_1\ and\ x_2\ b$ e the distance travelled in the first n and n – 1 seconds respectively, then from the equation of motion

$$x = ut + \frac{1}{2}at^2$$

then $$x_1 = un + \frac{1}{2}an^2$$

and $$x_2 = u(n-1) + \frac{1}{2}a(n-a)^2$$

Distance travelled in the nth second = $x_1 - x_2$

$$x_1 - x_2 = un + \frac{a}{2}n^2 - \left\{u(n-a) + \frac{a}{2}(n-1)^2\right\}$$

$$= un + \frac{a}{2}n^2 - \left\{un - u + \frac{a}{2}(n^2 - 2n + 1)\right\}$$

$= u+\frac{1}{2}a(2n-1)$

Motion with variable Acceleration

(i) Time Dependent Acceleration

(ii) Velocity Dependent Acceleration

(iii) Distance Dependent Acceleration

Projectile

An object thrown into the space with certain velocity from a gun or dropped from a moving plane under the action of gravity is called a projectile. Thus, a projectile move with a constant horizontal velocity and at the same falls freely under the action of gravity. The path of projectile is called the *trajectory.*

Horizontal and vertical component of velocity at any time t are

,

are the parametric equation of trajectory.

(i) Vertex

$$\left(\frac{v_0^2 sin\alpha \cos\alpha}{g}, \frac{v_o^2 \sin^2\alpha}{2g}\right)$$

(ii) Latus Rectum

Latus Rectum = $|4a| = \frac{2v_o^2 \cos^2\alpha}{g}$

(iii) Maximum Height

$$= \frac{v_o^2 \sin^2\alpha}{2g}$$

(iv) Equation of Directrix

$$y = \frac{v_o^2}{2g}$$

(v) Focus

$$= \left(\frac{v_0^2 sin\alpha \cos\alpha}{g}, -\frac{v_o^2}{2g} cos2\alpha \right) = \left(\frac{v_o^2}{2g} sin2\alpha - \frac{v_o^2}{2g} cos2\alpha \right)$$

The time taken by the particle to reach the horizontal plane through the point of projection is called its time of flight.

$t = \frac{2v_o\, sin\alpha}{g}$ is the time of flight.

(vii) Range

(a) Horizontal Range

The horizontal range of the projectile is the horizontal distance described by the particle during the time of flight. Therefore

Range R = (horizontal velocity) (time of flight)

$$v_o \cos\alpha \left(\frac{2v_o\, sin\alpha}{g} \right)$$

$$= \frac{v_o^2\, 2(\sin\alpha\, cos\alpha)}{g} = \frac{2v_o^2 sin2\alpha}{g}$$

This range can also be obtained by putting y = 0 in the Cartesian equation of trajectory $i.e$ putting y = 0 in the equation

$$y = xtan\alpha - \frac{gx^2}{2v_o^2} \sec^2\alpha$$

Also, it is notable that this range will remain same if we replace $\alpha by \alpha = \frac{\pi}{2} - \alpha$

(b) Maximum Range

Range R = $\frac{v_o^2 sin2\alpha}{g}$

For maximum range $\frac{dr}{d\alpha} = 0$

$$\Rightarrow \quad \frac{2v_o^2}{g}\cos 2\alpha = 0$$

$$\Rightarrow \quad cos2\alpha - 0$$

$$\Rightarrow \quad 2\alpha = \frac{\pi}{2}$$

$$\Rightarrow \quad \alpha = \frac{\pi}{4}$$

Therefore $R_{max} = \dfrac{v_o^2}{g}$

Time of Flight

Time of flight = $\dfrac{Horizontol\ distance}{Horizontol\ velocity}$

$$= \frac{x}{v_o \cos\alpha} = \frac{rcos\beta}{v_o \cos\alpha}$$

Using (1), we get

Time of flight = $\dfrac{2v_o^2 \cos^2\alpha \sin(\alpha - \beta)}{g \cos^2\beta} \dfrac{cos\beta}{v_o cos\alpha}$

$$= \frac{2v_o^2 \sin(\alpha - \beta)}{g \cos\beta}$$

For time of flight down the plane, change β to $-\beta$.

Parabola of Safety

As the equation of the projectile is

$$y - x\,tan\alpha - \frac{1}{2} g \frac{x^2}{v^2} \sec^2 \alpha$$

which shows that for different values of α we have different trajectories. So we define that the locus of all the points on different trajectories which lie at maximum distance from the point of projection O is called the parabola of safety.

CHAPTER 4

ELLIPTIC DIFFERENTIAL EQUATION

CONTENTS

4.1. INTRODUCTION

In previous topics, we have seen the classification of second order partial differential equation into elliptic, parabolic and hyperbolic types. In this chapter we shall consider various properties and technique for solving Laplace and Poisson equations which are elliptic in nature.

Various physical phenomena are governed by the well-known Laplace and poison equations. A few of them, frequently encountered in applications are: study heat condition, magnetic potential, torsion of prismatic shaft, bending of prismatic beams, distribution of gravitational potential, etc. In the following two sub-section, we shall give the derivation of Laplace and Poisson equations in relation to the most frequently occurring physical situation , namely the gravitational potential.

4.1.1. Derivation of Laplace Equation

Consider two particles of masses $m\, and\, m_1$ situated at Q and P separated by a distance r as shown in . According to Newton's universal law of gravitational, the magnitude of the force, proportional to the product of their masses and inversely proportional to the square b of the distance. Between is given

$$F = G\frac{mm_1}{r^2} \tag{1}$$

where $G is$ the gravitational constant. It $\boldsymbol{r}$ represent the vector $\overrightarrow{PQ},$ assuming unit mass at Q and $G = 1,$ the force at Q due to the mass at P is given by

$$F = -\frac{m_1 r}{r^3} = \nabla\left(\frac{m_1}{r}\right) \tag{2}$$

which is called the intensity of the gravitational force. Suppose a particle of unit mass moves under the attraction of a particle of mass m_1 at P infinity up to $Q;$ then the work done by the force $\boldsymbol{F}$ **is**

$$\int_{\infty}^{r} \boldsymbol{F}\,.d\boldsymbol{r} = \int_{\infty}^{r} \nabla\left(\frac{m_1}{r}\right).d\boldsymbol{r} = \frac{m_1}{r} \tag{3}$$

This is defined as the potential V at Q due to a particle at P and is denoted by

$$V = \frac{m_1}{r} \tag{4}$$

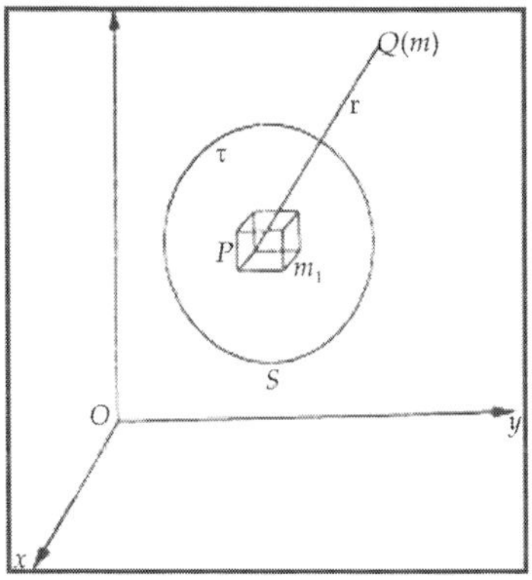

Figure 1

From Eq. 2, the intensity of the force at P is

$$\boldsymbol{F} = -\nabla V \tag{5}$$

Now, if we consider a system of particles of masses $m_1, m_2, \ldots, m_n$ which are at distances $r_1, r_2, \ldots, r_n$ respectively, then the force of attraction per unit mass at Q due to the system is

$$= \sum_{i-1}^{n} \nabla \frac{m_i}{r_i} = \nabla \sum_{i-1}^{n} \frac{m_i}{r_i} \tag{6}$$

The work done by the force acting on the particle is

$$\int_{\infty}^{r} \boldsymbol{F}.dr = \sum_{i=1}^{n} \frac{m_i}{r_i} = -V \tag{7}$$

Therefore,

$$\nabla^2 \mathrm{V} = -\nabla^2 \sum_{\mathrm{i=1}}^{\mathrm{n}} \frac{\mathrm{m_i}}{\mathrm{r_i}} = \sum_{\mathrm{i=1}}^{\mathrm{n}} \nabla^2 \frac{m_i}{r_i} = 0,\ \mathrm{r_i} \neq 0 \tag{8}$$

where

$\nabla^2 = div\nabla = \dfrac{\partial^2}{\partial x^2}+\dfrac{\partial^2}{\partial y^2}+\dfrac{\partial^2}{\partial z^2}$ is called the Laplace operator.

In the case of continuous distribution of matter of density p in a volume r, we have

$$V(x,y,z)=\iiint_r \frac{p(\xi,\eta,\zeta)}{r}dr \tag{9}$$

where $r=\left\{(x-\xi)^2+(y-\eta)^2+(z-\zeta)^2\right\}^{1/2}$ and Q is outside the body. It can be verified that

$$\nabla^2 V = 0 \tag{10}$$

which is called the Laplace equation□A2I.

4.1.2. Derivation of Poisson Equation

Consider a closed surface S consisiting of particle of masses $m_1, m_2, \ldots m_n$. Let Q be any point on S. Let $\sum_{i=1}^{n} m_i = M$ be the total mass inside S, and let $g_1, g_2, \ldots, g_n$ be the gravity field at Q due to the presence of $m_1, m_2 \ldots, m_n$ respectively within S. Also, let $\sum_{i=1}^{n} g_t = g,$ the entire gravity field at Q. Then, according to Gauss law, we have

$$\iint_S \mathbf{g}.d\mathbf{S} = -4\pi GM \tag{11}$$

where $M = \iiint_r p\,d\tau, p$ is the mass density function and τ is the volume in which the masses are distributed through. Since the gravity field is conservative. We have

$$g = \nabla V \tag{12}$$

where V is a scalar potential. But the Gauss divergence theorem states that

$$\iint_S g.dS = \iiint_\tau \nabla.g\, d\tau \tag{13}$$

Also. Eq. (11) gives

$$\iint_S g\hat{\alpha}dS = -4\pi G \iiint_\tau \text{pd} \tag{14}$$

Combining Eq. (13) and (14), we have

$$\iiint_\tau (\nabla.g + 4\pi Gp)d\tau = 0$$

Implying

$$\nabla.g = -4\pi Gp = \nabla.\nabla\ v$$

Therefore,

$$\nabla^2 V = -4\pi Gp \tag{15}$$

This equation is known as *poisson's equation*.

4.2. BOUNDARY VALUE PROBLEM (BVPS)

The function V, whose analytical form we seek for the problems stated in section in addition to satisfying the Laplace and Poisson equation in a bounded region IR in R^3, should also satisfy certain boiundary conditions on the boundary ∂IR of this region Such problems are referred to as boundary value problems (BVPs) for Laplace and Poisson equations. We shall denote the set of all boundary points of IR by ∂IR. By the closure of IR, we mean the set of all interior points of IR together with its boundary points and is denoted by IR, we mean the set of all interior point together Symbolically, $\overline{IR} = IR \cup \partial IR$

If a function $f \in c^n \left(f \text{ belongs to } c^n \right)$ then all its derivative of order n are continuous.

If it belong to c^0, then we mean f is continuous.

There are mainly three types of boundary value problems for Laplace equation. If $f \in c^0$ and is specified on the boundary ∂IR of some finite region IR, the problem of determining a function $\psi(x, y, z)\, such\, that\, \nabla^2\psi = 0$ within IR and satisfying $\psi = f\, on\, \partial IR$ is called the boundary value problem of first kind, or the Dirichlet problem. For example, finding the steady state temperature within the region IR when no heat sources or sinks are present and when the temperature is prescribed on the boundary ∂IR, is Dirichlet problem. Another example would be to find the potential inside the region IR when the potential is specified on the boundary $\partial IR. These$ two examples correspond to the interior Dirichlet problem.

Similarly, if $f \in c^0$ and is prescribed on the boundary ∂IR of a finite simply connected region IR, determinanting function $\psi(x, y, z)$ which satisfies $\nabla^2\psi = 0$ outside IR and is such that $\psi = f$ on ∂IR, is called an exterior Dirichlet problem. For example, determination of the distribution of the potential outside a body whose surface potential is prescribed , is an exterior Dirichlet problem. The second type of BVP is associated with Von Neumann. The problem is to determine the function $\psi(x, y, z)$ so that $\nabla^2\psi = 0\, within\, IR\, while\, \partial\psi\partial n.$ is specified at every point of $\partial IR, where\, \partial\psi / \partial n$ denotes the normal derivative of the field variable ψ. This problem is called the Neumann problem. If ψ is the temperature, $\partial\psi / \partial n$ is the heat flux representing the amount of heat crossing per unit volume per unit time along the normal direction, which is zero when insulated . The third type of BVP is concerned with the determination of the function $\psi(x, y, z)$ such that $\nabla^2\psi = 0$ within IR, while a boundary condition of the form $\partial\psi / \partial n$ + $h\psi = f$, where $h \geq 0$ is specified at every point of ∂IR. This is called a mixed BVP or Churchill's problem. If we assume Newton's law of cooling, the heat lost is $h\psi$, where ψ is the temperature difference from the surrounding medium and $h > 0$ is a constant depending on the medium. The heat f supplied at a point of the boundary is partly conducted into the medium and partly lost by radiation to the surrounding. Equating these amounts, we get the third boundary condition. The third type of BVP is concerned with the determination of the function $\psi(x, y, z)$ such that $\nabla^2\psi = 0$ within $IR,$ while a boundary condition of the form $\partial\psi / \partial n + h\psi = f,$ where $h \geq 0$ is specified at every point of $\partial IR.$ This is called a mixed BVP or Churchill's problem. If we assume Newton's law of cooling, the heat lost is $h\psi,$ where ψ is the temperature difference from the surrounding medium and $h > 0$ is a constant depending on the medium. The heat f supplied at a point of the boundary is partly conducted into the medium and partly lost by radiation to the surrounding. Equating these amounts, we get the third boundary condition

4.3. SOME IMPORTANT MATHEMATICAL TOOLS

Among the mathematical tools, we employ in deriving many important results. The Gauss divergence theorem plays a vital role, which can be stated as follows : Let ∂IR be a closed surface in the $xyz-space$ and IR denote the bounded region enclosed by ∂IR in which **F** is a vector belonging to $c^1\, in\, IR$ and continuous on $IR.$ Then

$$\iint_{\partial IR} F.\hat{n}\, dS = \iiint_{IR} \nabla.F\, dV$$

where dV is an element of volume, dS is an element of surface area . and $\hat{n}$ the outward drawn normal.

Green's identities which follow form divergence theorem are also useful and they can be derived as follows : Let $F = f\phi,$ where f is a vector function of position and ϕ is a scalar function of position. Then

$$\iiint_{IR} \nabla.(f\phi)\, dV = \iint_{\partial IR} \hat{n}.\, f\, \phi\, dS$$

Using the vector identity

$$\nabla.(f\phi) = f.\nabla\phi + \phi\nabla.\boldsymbol{f}$$

we have

$$\iiint_{IR} f.\nabla.\phi dV = \iint_{\partial IR} \hat{n}.f\phi dS \iiint_{IR} \phi \nabla.\text{f dV}$$

If we choose $f = \nabla\psi$, the above equation yields

$$\iiint_{IR} \nabla\phi.\nabla\psi\, dV = \iint_{\partial IR} \phi\hat{n}.\nabla\psi\, dS - \iiint_{IR} \phi\nabla^2\psi dV \qquad (17)$$

Noting that $\hat{n}.\nabla\psi$ is trhe derivative of ψ in the direction of $\hat{n}$, we introduced the notation

$\hat{n}.\nabla\psi = \partial\psi / \partial n$

Into Eq. (17) to get

$$\iiint_{IR} \nabla\phi.\nabla\psi dV = \int_{IR} \phi\frac{\partial\psi}{\partial n} dS - \iiint_{IR} \psi\nabla^2\phi \text{ dV} \qquad (18a)$$

This equation is known as Green's first identity. Of course, it is assumed that both ϕ and ψ possess continuous second derivatives.

Interchanging the role of ϕ and ψ, we obtain form relation (18 a) the equation

$$\iiint_{IR} \nabla\psi.\nabla\phi dV = \int_{IR} \psi\frac{\partial\phi}{\partial n} dS - \iiint_{IR} \psi\nabla^2\phi \text{ dV} \qquad (18b)$$

Now, subtracting Eq. (18b) from Eq. (18) we get

$$\iiint_{IR} \left(\phi\nabla^2\psi - \psi\nabla^2\phi\right) dV = \iint_{\partial IR} \left(\phi\frac{\partial\psi}{\partial n} - \psi\frac{\partial\phi}{\partial n}\right) dS \qquad (19)$$

This is known as Green' second identity. If we set $\phi = \psi$ in Eq. (18a) we get

$$\iiint_{IR} (\nabla\phi)^2 dV = \iint_{\partial IR} \phi\frac{\partial\phi}{\partial n} dS - \iiint_{IR} \phi\nabla^2\phi dV \qquad (20)$$

which is a special case of Green's first identity.

4.4. PROPERTIES OF HARMONIC FUNCTIONS

Solutions of Laplace equation are called harmonic functions which possess a number of interesting properties, and they are presented in the following theorems.

Theorems. If a harmonic function vanishes everywhere on the boundary , then it is identically zero everywhere.

Proof. If ϕ is a harmonic function, then $\nabla^2 \phi = 0$ in IR . Also, if $\phi = 0$ on $\partial IR,$ we shall show that $\phi = 0 \, in \, \overline{IR} = IR \, U \, \partial IR.$ Recalling Green's first identity, i.e., Eq. (20)

$$\int\int\int_{IR} (\nabla \phi)^2 \, dV = \int\int_{\partial IR} \phi \frac{\partial \phi}{\partial n} dS - \int\int\int_{IR} \phi \, \nabla^2 \phi dV$$

And using the above facts we have, at once, the relation

$$\int\int\int_{IR} (\nabla \phi)^2 \, dV = 0$$

Since $(\nabla \phi)^2$ is positive, it follows that the integral will be satisfied only if $\nabla \phi = 0.$ This implies that ϕ is a constant in IR . Since ϕ is continuous in $\overline{IR} \, and \, \phi$ is zero on $\partial IR,$ it follows that $\phi = 0 \, in \, IR.$

Theorem. If ϕ is a harmonic function in IR and $\partial \phi / \partial n = 0$ on $\partial IR,$ then ϕ is a constant in $\overline{IR}$.

Proof. Using Green's first identity and the data of theorem, we arrive at

$$\int\int\int_{IR} (\nabla \phi)^2 \, dV = 0$$

Implying $\nabla \phi = 0, i.e., \phi$ is a constant in $IR.$ Since the value of ϕ is not known on the boundary $\partial IR \, while \, \partial \phi / \partial n = 0,$ it is implied that ϕ is a constant on ∂IR and hence on $\overline{IR}$.

Theorem. If the Dirichlet problem for a bounded region has a solution, then it is unique .

Proof. If $\phi_1 \, and \, \phi_2$ are two solutions of the interior Dirichlet problem, then

$$\nabla^2\phi_1 = 0 \, in \, IR; \ \phi_1 = f \, on \, \partial IR$$

$$\nabla^2\phi_2 = 0 \, in \, IR; \ \phi_2 = f \, on \, \partial IR$$

Let $\psi = \phi_1 - \phi_2$. Then

$$\nabla^2\phi = \nabla^2\phi_1 - \nabla^2\phi_2 = 0 \ in \, IR;$$

$$\psi = \phi_1 - \phi_2 = f - f = 0 \ on \, \partial IR;$$

Therefore,

$$\nabla^2\psi = 0 \, in \, IR, \ \psi = 0 \, on \, \partial IR$$

Now using, we obtain $\phi = 0 \, on \, \overline{IR},$ which implies that $\phi_1 = \phi_2$. Hence the solution of the Dirichlet problem is unique.

Theorem. If the Neumann problem for a bounded region has a solution, then it is either unique or it differ from one another by a constant only.

Proof. Let $\phi_1 \, and \, \phi_2$ be two distinct solution of the Neumann problem. Then we have

$$\nabla^2\phi_1 = 0 \, in \, IR; \frac{\partial\phi_1}{\partial n} = f \, on \, \partial IR,$$

$$\nabla^2\phi_2 = 0 \, in \, IR; \frac{\partial\phi_2}{\partial n} = f \, on \, \partial IR,$$

Let $\psi = \phi_1 - \phi_2$. Then

$$\nabla^2\psi = \nabla^2\phi_1 - \nabla^2\phi_2 = 0 \, in \, IR$$

$$\frac{\partial\phi}{\partial n} = \frac{\partial\phi_1}{\partial n} - \frac{\partial\phi_2}{\partial n} = 0 \, on \, \partial IR$$

Hence from theorem ψ is a constant on $\overline{IR}, i.e., \phi_1 - \phi_2 =$ constant. Therefore, the solution of the Neumann problem is not unique. Thus, the solutions of a certain Neumann problem can differ from one another by a constant only.

4.4.1. The Spherical Mean

Let IR be a region bounded by ∂IR and let $p(x, y, z)$ be any point in IR. Also, let $S(p,r)$ represent a sphere with centre at P and radius r such that it lies entirely within the domain IR as depicted in. Let u be a continuous function in IR. Then the spherical mean of u denote byu $\overline{u}$ is defined as

$$u(r)=\frac{1}{4\pi r^2}\int\int_{S(p,r)} u(Q)\,dS \qquad (21)$$

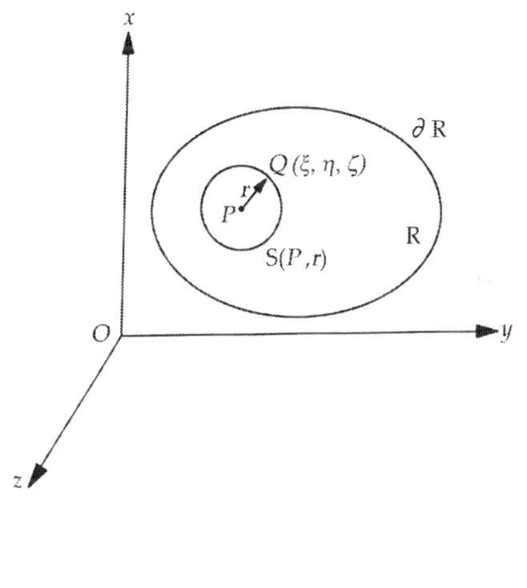

Figure 2

where $Q(\xi,\eta,\zeta)$ is any variable point on the surface of the sphere $S(p,r)$ and dS is the surface element of integration. For a fixed radius r, the value $\overline{u}(r)$ is the average of the value of u taken over the spehere $S9p,r)$, and hence it is called the spherical mean. Taking the origin at P, in terms of spherical polar coordinates, we have

$\xi = x + r\sin\theta\,\cos\phi$

$\eta = y + r\sin\theta\, \sin\phi$

$\zeta = z + r\cos\theta$

Then the spherical mean can be written as

$$\bar{u}(r) = \frac{1}{4\pi r^2} \int_{\theta=0}^{2\pi} \int_{\theta=0}^{\pi} u(x + r\sin\theta\, \cos\phi, y + r\, sin\theta\, sin\phi, z + r\, cos\theta) r^2 \sin\theta\, d\theta\, d\phi$$

Also since u is a continuous on $S(p,r), \bar{u}\ too$ is a continuous function of r on some interval $0 < r \le R,$ which can be verified as follows.

$$\bar{u}(r) = \frac{1}{4\pi r^2} \int_{\phi=0}^{2\pi} \int_{\theta=0}^{\pi} u(Q) sin\theta\, d\theta\, d\phi = \frac{u(Q)}{4\pi} \int_0^{2\pi} \int_0^{\pi} \sin\theta\, d\,\theta\, d\phi = u(Q)$$

Now, taking the limit as $r \to 0, Q \to p,$ we have

$$Lt_{r\to 0}\, \overline{u}(r) = u(p) \qquad (22)$$

Hence, u is continuous in $0 \le r \le R.$

4.4.2. Mean Value Theorem for harmonic Functions

Theorem. Let u be harmonic in a region $IR.$ Also, let $p(x,y,z)$ be a given point in IR and $S(p,r)$ be a sphere with centre at p such that $S(P,R)$ is completely contained in the domain of harmonicas of $u.$ Then

$$u(p) = \bar{u}(r) = \frac{1}{4\pi r^2} \iint_{s(p,r)} u(Q)\, dS$$

Proof. Since u is harmonic in IR, it spherical mean $\bar{u}(r)$ is continuous in IR and is given by

$$\bar{u}(r) = \frac{1}{4\pi r^2} \iint_{s(p,r)} u(Q)\, dS = \frac{1}{4\pi r^2} \int_0^{2\pi} \int_0^{\pi} u(\xi, \eta, \zeta) r^2\, \sin\theta d\theta \backslash d\phi$$

Therefore,

$$\frac{d\overline{u}(r)}{dr} = \text{—} \int_0^{\pi}\int_0^{\pi}\left(u_{\xi}\xi_r + u_{\eta}\eta_r + u_{\zeta}\zeta_r\right) sin\theta \, d\theta \, d\phi$$

$$= \frac{1}{4\pi r^2}\int_0^{2\pi}\int_0^{\pi}\left(u_{\xi}\ \sin\theta\, cos\phi + u_{\tau}\ sin\theta\ \sin\phi + u_{\zeta}\ cos\theta\right)\sin\theta\, d\,\theta\, d\phi \qquad (23)$$

Nothing that $\sin\theta\, cos\phi, \sin\theta \sin\phi\, and \cos\theta$ are the direction cosines of the normal $\hat{n}\, on\, S(p, r)$,

$\nabla u = iu_{\xi} + ju_{\eta} + ku_{\xi}$, $\hat{n} = (in_1, jn_2, kn_3)$,

The expression within the parenthesis of the integrand of eq. (23) can be written as $\nabla.\hat{n}$. Thus

$$\frac{d\overline{u}(r)}{dr} = \frac{1}{4\pi r^2}\iint_{s(p,r)} \nabla u.\ \hat{n}r^2 \sin\,\theta\, d\theta\, d\phi$$

$$= \frac{1}{4\pi r^2}\iint_{s(p,r)} \nabla u.\hat{n}\, dS$$

$$= \frac{1}{4\pi r^2}\iint_{V(p,r)} \nabla.\nabla u\, dV\,(by\, divergence\, theorem)$$

$$= \frac{1}{4\pi r^2}\iint_{V(p,r)} \nabla^2 u\, dV = 0\,(since\, u\, is\, harmonic)$$

Therefore, $\frac{d\overline{u}}{dr} = 0,$ implying $\overline{u}$ is constant.

Now the continuity of $\overline{u}\ at\ r = 0$ gives, from Eq. (22), the relation

$$\overline{u}(r) - u(p) = \frac{1}{4\pi r^2}\iint_{V(p,r)} \mathrm{u(Q)d\,S} \qquad (24)$$

4.4.3. Maximum-Minimum Principal and Consequences

Theorem. Let IR be a region bounded ∂IR. Also. Let u be a function which is continuous in a closed region IR and satisfies the Laplace equation

$\nabla^2 u = 0$ in the interior of IR. Further, if u is not constant everywhere on $\overline{IR}$, then the maximum and minimum values of u must occur only on the boundary ∂IR.

Proof. Suppose u is a harmonic function but not constant everywhere on $\overline{IR}$. if possible , let u attain its maximum value M at some interior point $p\, in\, IR. Since\, M$ is the maximum of u which is not a constant , there should exist a spehere $S(p,r)$ about p such that some of the values of u on $S(p,r)$, must be less than M. But by the mean value property, the value of u at P is the average of the values of $u\, on\, S(p,r)$, and hence it is less than M. This contradicts the assumption that $u = M\, at\, p$. Thus x u must be constant over the entire sphere $S(p,r)$.

Let Q be any other point inside IR which can be connected to P by an arc lying entirely within the domain IR. By convergence this arc with spheres and suing the Henie-Borel Theorem to choose a finite number of covering spheres and repeating the argument given above , we can arrive at the conclusion that u will have the same constant value at Q as at P. Thus u cannot attain a maximum value at any point inside the region IR. Therefore, u can attain its maximum value only on the boundary ∂IR. By convergence this arc with spheres and suing the Henie-Borel Theorem to choose a finite number of covering spheres and repeating the argument given above , we can arrive at the conclusion that u will have the same constant value at Q as at P. Thus u cannot attain a maximum value at any point inside the region IR. Therefore, u can attain its maximum value only on the boundary ∂IR.

Some important consequence of the maximum-minimum principle are given in the following theorems.

Theorem.

Proof. Let $u_1\, and\, u_2$ be two solutions of the Dirichlet problem and let $f_1\, and\, f_2$ be a given continuous functions on the boundary ∂IR such that

$\nabla^2 u_1 = 0\, in\, IR\, ; u_1 = f_1\, on\, \partial IR$,

$\nabla^2 u_2 = 0 \, in \, IR \; ; \; u_2 = f_2 \, on \, \partial IR$,

Let $u = u_1 - u_2$. Then,

$\nabla^2 u = \nabla^2 u_1 - \nabla^2 u_2 = 0 \;\; in \;\; IR \; ; \quad u = f_1 - f_2 \;\; on \, \partial IR$

Hence, u is a solution of the Dirichlet problem with boundary condition $u = f_1 - f_2 \, on \, \partial IR.$ By the maximum-minimum principle , u attains the maximum and minimum values on $\partial IR.$ Thus at any interior point in $IR,$ we shall have , for a given $\varepsilon > 0,$

$-\varepsilon < u_{min} \leq u \leq u_{max} < \varepsilon$

Therefore,

$|u| < \varepsilon . in \, IR, implying \, |u_1 - u_2| < \varepsilon \;\; on \, IR$

Thus, small changes in the initial data bring about an arbitrary small change in the solution. This completes the proof of the theorem.

Theorem. Let $\{f_n\}$ be a sequence of the functions, each of which is continuous on $\overline{IR}$ and harmonic on $IR.$ if the sequence $\{f_n\}$ convergence uniformly on $\partial IR,$ then it converges uniformly on $\overline{IR}$.

Proof. Since the sequence $\{f_n\}$ converges uniformly on $\partial IR,$ for a given $\varepsilon > 0$, we find an integer N such that

$|f_n - f_m| \langle \varepsilon \;\; for \, n, m \rangle N$

Hence, form stability theorem, for all $n, m > N$ it follows immediately that

$|f_n - f_m| < \varepsilon \, In \, IR$

Therefore, $\{f_n\}$ convergence uniformly on $\overline{IR}.$

Example. Show that if the two-dimensional Laplace equationc §X $\nabla^2 u = 0$ is transformed by introducing plane polar coordinates r, θ defined by the relations $x = r\cos\theta, y = r\sin\theta,$ it takes the form

$$\frac{\partial^2 u}{\partial r^2}+\frac{1}{r}\frac{\partial u}{\partial r}+\frac{1}{r^2}\frac{\partial^2 u}{\partial \theta^2}=0$$

Solution. In many practical problems, it is necessary to write the Laplace equation-« X in the various coordinates systems. For instance, if the boundary of the region ∂IR is circle, then it is natural to use polar coordinates defined by $x = r\,cos\theta, y = r\sin\theta$. Therefore,

$$r^2 = x^2 + y^2,\ \theta = \tan^{-1}\left(\frac{y}{x}\right).$$

$$r_x = cos\theta,\quad r_y = sin\theta,\quad \theta_x = -\frac{\sin\theta}{r},\quad \theta_y = \frac{cos\theta}{r}$$

Since

$$u=u(r,\theta)u_x = u_r r_x + u_\theta \theta_x = \left(u_r \cos\theta - u_\theta \frac{sin\theta}{r}\right)$$

Similarly,

$$u_y = u_r r_y + u_\theta \theta_y = \left(u_r\, sin\theta + u_\theta \frac{cos\theta}{r}\right)$$

Now for the second order derivatives,

$$u_{xx} = (u_x)_x = (u_x)_r r_x + (u_x)_\theta \theta_x = \left(u_r cos\theta - u_\theta \frac{sin\theta}{r}\right)_r \cos\theta + \left(u_r \cos\theta - u_\theta \frac{sin\theta}{r}\right)_\theta \left(-\frac{sin\theta}{r}\right)$$

Therefore,

$$u_{xx} = \left(u_{rr} \cos\theta - u_{\theta r} \frac{sin\theta}{r} + u_\theta \frac{sin\theta}{r^2}\right) cos\theta$$

$$+\left(u_{r\theta}\, cos\theta - u_r\, sin\theta - u_{\theta\theta} \frac{sin\theta}{r} - u_\theta \frac{cos\theta}{r}\right)\left(-\frac{sin\theta}{r}\right) \qquad (25)$$

Similarly, we can show that

$$u_{yy} = \left(u_{rr} \sin\theta + u_{r\theta} \frac{cos\theta}{r} - u_\theta \frac{cos\theta}{r^2}\right) sin\theta$$

$$+\left(u_{r\theta}\, sin\theta + u_r\, cos\theta + u_{\theta\theta} \frac{cos\theta}{r} - u_\theta \frac{sin\theta}{r}\right)\left(\frac{cos\theta}{r}\right) \qquad (26)$$

By adding Eq. (25) and (26) and equating to zero, we get

$$u_{xx}+u_{yy}=u_{rr}+\frac{1}{r}u_r+\frac{1}{r_2}u_{\theta\theta}=0 \qquad (27)$$

which is the Laplace equation in polar coordinates. One can observe that the Laplace equation in the Cartesian coordinates has constant coefficients only, whereas in polar coordinates , it has variable coefficients.

Example. Show that in cylindrical coordinates r,θ,z defined by the relations $x=r\,cos\theta, y=r\sin\theta, z=z,$ the Laplace equation $\nabla^2 u=0$ takes the form

$$\frac{\partial^2 u}{\partial r^2}+\frac{1}{r}\frac{\partial u}{\partial r}+\frac{1}{r^2}\frac{\partial^2 u}{\partial \theta}+\frac{\partial^2 u}{\partial z^2}=0$$

Solution. The Laplace equation in Cartesian coordinates is

$$\nabla^2 u=\frac{\partial^2 u}{\partial x^2}+\frac{\partial^2 u}{\partial y^2}+\frac{\partial^2 u}{\partial z^2}=0$$

The relations between Cartesian and cylindrical coordinates gives

$$r^2=x^2+y^2, \theta=\tan^{-1}(y/x),\ z=z$$

Since

$$u=u(r,\theta,z)$$

$$u_x=u_x r_x+u_\theta\theta_x+u_z z_x=u_r\cos\theta-u_\theta\left(\frac{\sin\theta}{r}\right)$$

$$u_y=u_r r_y+u_\theta\theta_y+u_z z_y=u_r\sin\theta-u_\theta\left(\frac{\cos\theta}{r}\right)$$

$$u_z=u_r r_z+u_\theta\theta_z+u_z=u_z$$

For the second order derivatives, we find

$$u_{xx}=(u_x)_x=(u_x)_r r_x+(u_x)_\theta\theta_x+(u_x)_z z_x$$

$$=\left[u_r\,cos\theta\quad u_\theta\left(\frac{sin\theta}{r}\right)\right]_r\cos\theta\ \text{I}\left[u_r\,cos\theta-u_\theta\left(\frac{sin\theta}{r}\right)\right]_\theta\left(-\frac{sin\theta}{r}\right)$$

$$=\left(u_{rr}\,cos\theta-u_{r\theta}\frac{sin\theta}{r}+u_\theta\frac{sin\theta}{r^2}\right)cos\theta$$

$$+\left(u_{r\theta}cos\theta-u_r\ \sin\theta-u_{\theta\theta}\frac{\sin\theta}{r}-u_\theta\frac{cos\theta}{r}\right)\left(-\frac{sin\theta}{r}\right) \tag{28}$$

Similarly

$$u_{yy}=\left(u_y\right)y=\left(u_y\right)_r r_y+\left(u_y\right)_\theta\theta_y+\left(u_y\right)_z z_y$$

$$=\left(u_r\ sin\theta+u_\theta\frac{cos\theta}{r}-u_\theta\frac{cos\theta}{r^2}\right)\sin\theta$$

$$+\left(u_{r\theta}\ sin\theta++u_r\ cos\theta+u_{\theta\theta}\frac{cos\theta}{r}-u_\theta\frac{sin\theta}{r}\right)\left(\frac{\cos\theta}{r}\right) \tag{29}$$

$$u_{zz}=u_{zz} \tag{30}$$

Adding Eq. (28)-(30), we obtain

$$\nabla^2u=u_{rr}+\frac{1}{r}u_r+\frac{1}{r^2}u_{\theta\theta}+u_{zz} \tag{31}$$

Example. Show that in spherical polar coordinates r,θ,ϕ defines by the relations $x=r\ \sin\theta\cos\phi, y=r\,sin\theta\sin\phi, z=r\ \cos\theta,$, the Laplace equations $\nabla^2u=0$ takes the form

$$\frac{\partial}{\partial r}\left(r^2\frac{\partial u}{\partial r}\right)+\frac{1}{sin\theta}\frac{\partial}{\partial\theta}\left(\sin\ \theta\frac{\partial u}{\partial\theta}\right)+\frac{1}{\sin^2\theta}\frac{\partial^2u}{\partial\phi^2}=0$$

Solution. In Cartesian coordinates , the Laplace equation is

$$\nabla^2 u=u_{xx}+u_{yy}+u_{xx}+u_{zz}=0$$

In spherical coordinates,

$$u=u\left(r,\theta,\phi\right), r^2=x^2+y^2+z^2, \cos\theta=\frac{z}{r}, \tan\phi=y/x$$

.

It can be easily verified that

$$\theta_x=\frac{cos\theta\,cos\phi}{r},\ \theta_y=\frac{cos\theta\,sin\phi}{r},\theta_z=-\frac{\sin\theta}{r}$$

$$\phi_x=\frac{sin\phi}{r\,sin\phi},\phi_y=\frac{cos\phi}{r\,sin\phi}\phi_z=0$$

Now,

$$u_x = u_x r_x + u_\theta \theta_x + u_\phi \phi_x = u_r \sin\theta \cos\phi + u_\theta \frac{\cos\theta \cos\phi}{r} - u_\phi \frac{sin\phi}{rsin\phi}$$

$$u_y = u_y r_y + u_\theta \theta_y + u_\phi \phi_y = u_r \sin\theta \sin\phi + u_\theta \frac{\cos\theta \sin\phi}{r} + \frac{u_\phi \, sin\phi}{r\, sin\phi}$$

$$u_z = u_r r_z + u_\theta \theta_y + u_\phi \phi_y = u_r \cos\theta + u_\theta \left(-\frac{sin\theta}{r} \right)$$

For the second order derivatives,

$$u_{xx} = (u_x)_r r_x + (u_x)_\theta \theta_x + (u_x)_\phi \phi_x$$

$$= \left(u_r \sin\theta \cos\phi + u_\theta \frac{cos\theta\, cos\phi}{r} - u_\phi \frac{sin\phi}{r\sin\theta} \right)_r (\sin\theta \cos\phi)$$

$$= \left(u_r \sin\theta \cos\phi + u_\theta \frac{\cos\theta \cos\phi}{r} - u_\phi \frac{\sin\phi}{r\sin\theta} \right)_\theta \left(\frac{\sin\theta \cos\phi}{r} \right)$$

$$= \left(u_r \sin\theta \cos\phi + u_\theta \frac{\cos\theta \cos\phi}{r} - u_\phi \frac{\sin\phi}{r\sin\theta} \right)_\phi \left(-\frac{\sin\theta}{r\sin\theta} \right)$$

$$= \left(\sin^2\theta \cos^2\phi\right)_{urr} + \frac{\cos^2\theta \cos^2\phi}{r^2} u_{\theta\theta} + \frac{\sin^2\phi}{r^2\, sin^2\theta} u_{\phi\phi}$$

$$+ u_{r\theta} \left(\frac{2\sin\theta\, cos\,\theta \cos^2\phi}{r} \right) + u_{r\phi} \left(-\frac{2\sin\phi \cos\phi}{r} \right)$$

$$+ u_{\theta\phi} \left(-\frac{2\cos\theta \cos\phi \sin\phi}{r^2 \sin\theta} \right) + u_r \left(\frac{\cos^2\theta \cos^2\phi}{r} + \frac{\sin^2\phi}{r} \right)$$

$$+ u_\phi \left(\frac{sin\phi \cos\phi}{r^2} + \frac{\cos^2\phi \cos\phi \sin\phi}{r^2 \sin^2\theta} + \frac{\sin\phi \cos\phi}{r^2 \sin^2\theta} \right) \tag{32}$$

$$u_{yy} = (u_y)_r r_y + (u_y)_\theta \theta_y + (u_y)_\phi \phi_y$$

$$= \left(u_r \sin\theta \sin\phi + u_\theta \frac{\cos\theta\, sin\phi}{r} + u_\phi \frac{\cos\phi}{r\sin\theta} \right),$$

$$+\left(u_r \sin\theta \sin\phi + u_\theta \frac{\cos\theta \sin\phi}{r} + u_\phi \frac{\cos\phi}{\sin\theta}\right)_\theta \frac{\cos\theta \sin\phi}{r}$$

$$+\left(u_r \sin\theta \sin\phi + u_\theta \frac{\cos\theta \sin\phi}{r} + u_\phi \frac{\cos\phi}{\sin\theta}\right)_\phi \frac{\cos\phi}{r\sin\theta}$$

$$= \left(\sin^2\theta \sin^2\phi\right)u_{rr} + \frac{\cos^2\theta \sin^2\phi}{r^2} u_{\theta\theta} + \frac{\cos^2\phi}{r^2 \sin^2\theta} u_{\phi\phi}$$

$$= +u_{r\theta}\left(\frac{2\sin\theta\cos\theta\sin^2\phi}{r}\right) + u_{r\phi}\left(\frac{2\cos\phi\sin\phi}{r}\right)$$

$$+u_{\theta\phi}\left(\frac{2\cos\theta cos\phi\sin\phi}{r^2\sin\theta}\right) + u_r\left(\frac{\cos^2\theta\sin^2\phi}{r} + \frac{\cos^2\phi}{r}\right)$$

$$+u_\theta\left(-\frac{\sin\phi\cos\phi}{r^2} - \frac{\sin\phi\cos\phi}{r^2\sin^2\theta} - \frac{\cos^2\theta\sin\phi\cos\phi}{r^2\sin\theta}\right)$$

Similarly,

$$u_{zz} = \left(u_z\right)_r r_z + \left(u_z\right)_\theta \theta_z + \left(uz\right)_\phi \phi_z$$

$$= \left(u_r \cos\theta - u_\theta \frac{\sin\theta}{r}\right)_r (\cos\theta) + \left(u_r \cos\theta - u_\theta \frac{sin\theta}{r}\right)_\theta - \frac{\sin\theta}{r}$$

$$= u_{rr} \cos^2\theta - u_{r\theta} \frac{2\sin\theta \cos\theta}{r} + u_{\theta\theta} \frac{\sin^2\theta}{r^2} + u_r \frac{\sin^2\theta}{r} + u_\theta \frac{cos\theta \sin\theta}{r^2} \tag{34}$$

Adding Eq. (32) - (34), we obtain

$$\nabla^2 u = u_{rr} + \frac{1}{r^2} u_{\theta\theta} + \frac{1}{r^2 \sin^2\theta} u_{\phi\phi} + \frac{2}{r} u_r + \frac{\cos\theta}{r^2 \sin\theta} u_\theta = 0$$

which can be rewritten as

$$\nabla^2 u = \frac{\partial}{\partial r}\left(r^2 \frac{\partial u}{\partial r}\right) + \frac{1}{\sin\theta} \frac{\partial}{\partial\theta}\left(\sin\theta \frac{\partial u}{\partial\theta}\right) + \frac{1}{\sin^2\theta} \frac{\partial^2 u}{\partial\phi^2} = 0 \tag{35}$$

4.5. SEPARATION OF VARIABLES

The method of Separation of variables is applicable to a large number of classified linear homogeneous equations. The choice of the coordinate system in general depends on the shape of the boundary For example, consider a two –dimensional Laplace equation in Cartesian coordinates .

$$\nabla^2 u = u_{xx} + u_{yy} = 0 \qquad (36)$$

We assume the solution in the form

$$u(x,y) = X(x)Y(y) \qquad (37)$$

Subtracting in Eq. (36), we get

$$X''Y + Y''X = 0$$

i.e.

$$\frac{X''}{X} = \frac{Y''}{Y} = k$$

where k is separation constant. Three cases arise.

Case I Let $k = p^2, p$ is real. Then

$$\frac{d^2X}{dx^2} p^2 X = 0 \, and \, \frac{d^2Y}{dy^2} + p^2 Y = 0$$

whose solution is given by

$$X = c_1 e^{px} + c_2 e^{-px}$$

and

$$Y = c_3 \cos py + c_4 \sin py$$

Thus the solution is

$$u(x,y) = \left(c_1 e^{px} + c_2 e^{-px}\right)\left(c_3 \cos py + c_4 \sin py\right) \qquad (38)$$

Case II Let $k - 0$. Then

$$\frac{d^2X}{dx^2} = 0 \, and \, \frac{d^2Y}{dy^2} = 0$$

Integrating twice, we get

$$X = c_5 x + c_6$$

and

$$Y = c_7 y + c_8$$

The solution is therefore,

$$u(x,y)=(c_5x+c_6)(c_7y+c_8) \tag{39}$$

Case III let $k=-p^2$. proceeding as in Case I, we obtain

$$X=c_9\cos px+c_{10}\sin px$$

$$Y=c_{11}^{e^{py}}+c_{12}^{e^{-py}}$$

Hence, the solution in this case is

$$u(x,y)=(c_9\cos px+c_{10}\sin px)\left(c_{11}^{e^{py}}+c_{12}^{e^{-py}}\right) \tag{40}$$

In all these cases, $c_1\,(i=1,2,..,12)$ refer to integration constants, which are calculated by using the boundary conditions.

4.6. DIRICHLET PROBLEM FOR A RECTANGLE

The Dirichlet problem for a rectangle is defined as follows:

PDE: $\nabla^2 u=0, \quad 0\le x\le a, 0\le y\le b$

BCs: $u(x,b)=u(a,y)=0, u(0,y)=0, u(x,0)=f(x)$ (41)

This is an interior Dirichlet problem. The general solution of the governing PDE, using the method of variables separable, is discussed in previous section. The various solution of the Laplace equation are given by Eq. (38-40). Of three solutions, we have to choose that solution which is consistent with the physical nature of the problem and the given boundary conditions as despaired .

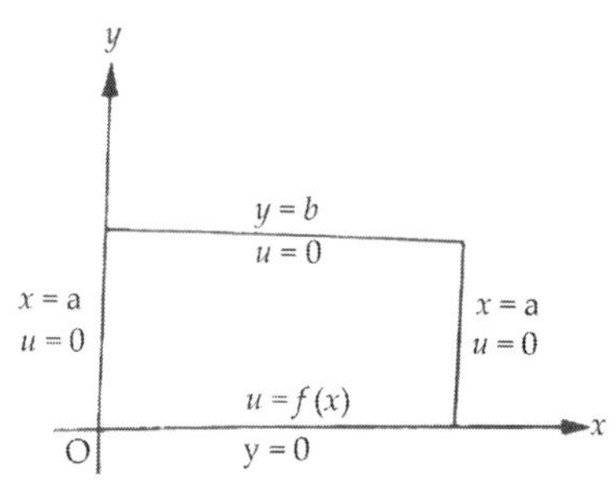

Figure 3: Dirichlet Boundary Conditions

Consider the solution given by Eq. (38)

$$u(x,y)=\left(c_1e^{px}+c2^{e\,px}\right)\left(c_3\cos py+c_4\sin py\right)$$

Using the boundary condition : $u(0\ ,y)=0, we\,get$

$$\left(c_1+c_2\right)\left(c_3\cos py+c_4\sin py\right)=0$$

which means that either

$$c_1+c_2=0\,or\,c_3\cos py+c_4\sin py=0.$$

But

$$c_3\ \cos py+c_4\ \sin\ py\neq 0;$$

Therefore

$$c_1+c_2=0 \tag{42}$$

Again, using the BC; $u(a\ ,y)=0\ ,Eq.(38)$ gives

$$\left(c_{1e^{ap}}+c_2^{e-ap}\right)\left(c_3\cos py+c_4\sin py\right)=0$$

Implying thereby

$$c_1e^{ap}+c_2e^{-ap}=0 \tag{43}$$

To determine the constants $c_1\ ,c_2,$ we have to solve Eq. (42) and (43), being homogeneous, the determinant

$$\begin{vmatrix} 1 & 1 \\ e^{ap} & e^{-ap} \end{vmatrix}=0$$

For the exercise of non-trivial solution, which is not the case. Hence only trivial solution $\pm(\ ,\ )=0$ is possible.

If we consider the solution given by Eq. (39) $u(x,y)=\left(c_5x+c_6\right)\left(c_7y+c_8\right),$ the boundary conditions: $u(0,y)-u(a,y)-0$ again yields a trivial solution. Hence, the possible solutions given by Eq. (38) and (39) are ruled out. Therefore, the only possible solutions given by Eq. (38) and (39) are rule out Therefore, the only possible solutions obtained from Eq.(40) is

$$u(x, y) = (c_9 \cos px + c_{10} \sin px)(c_{11}e^{py} + c_{12}e^{-py})$$

Using the BC : $u(0, y) = 0,$ we get $c_9 = 0.$ Also, the other BC: $u(a, y) = 0$ yeilds

$$c_{10} \sin pa(c_{11}e^{py} + c_{12}e^{-py}) = 0$$

For non trivial solution, c_{10} cannot be zero , implying sin $pa = 0$, which is possible if $pa = n\pi$ or $p = n\pi / a.,\ n = 1, 2, 3, \ldots$ therefore, the possible non-trivial solution after using the superposition principle is

$$u(x, y) = \sum_{n=1}^{\infty} \sin \frac{n\pi x}{a} \left[a_n \exp(n\pi y / a) + b_n \exp(-n\pi y / a) \right] \tag{44}$$

Now, using the BC : $u(x, b) = 0$, we get

$$\sin \frac{n\pi x}{a} \left[a_n \exp(n\pi b / a) + b_n \exp(-n\pi b / a) \right] = 0$$

Implying thereby

$$a_n \exp(n\pi b / a) + b_n \exp(-n\pi b / a) = 0$$

which gives

$$b_n = -a_n \frac{exp(n\pi b / a)}{\exp(-n\pi b / a)}, \quad n = 1, 2, \ldots, \infty$$

The solution (44) now becomes

$$u(x, y) = \sum_{n=1}^{\infty} \frac{2a_n \sin(n\pi x / a)}{\exp(-n\pi b / a)} \left[\frac{\exp\{n\pi(y-b)/a\} - \exp\{-n\pi(y-b)/a\}}{2} \right]$$

$$= \sum_{n=1}^{\infty} \frac{2a_n}{\exp(-n\pi b / a)} \sin(n\pi x / a) sinh\{n\pi(y-b)/a\}$$

Let $2a_n / \left[\exp\left(\frac{n\pi x}{a} \right) \right] = A_n.$ Then the solution can be written in the form

$$u(x,y)=\sum_{n=1}^{\infty} A_n \sin(n\pi x/a) sinh\{-n\pi(y-b)/a\}$$

Finally , using the non-homogenous boundary condition: $u(x,y)=f(x)$, we get

$$\sum_{n=1}^{\infty} A_n \sin(n\pi x/a) sinh(-n\pi x/b)=f(x)$$

which is half-range Fourier series. Therefore ,

$$A_n \sin h(-n\pi x/b)=\frac{2}{a}\int_0^a f(x)\sin(n\pi x/b)\,dx$$

Thus, required solution for the given Dirichlet problem is

$$u(x,y)=\sum_{n=1}^{\infty} A_n\, sin(n\pi x/b)\ \sin h\{n\pi(y-b)/a\} \tag{47}$$

where

$$A_n=\frac{2}{a}\frac{1}{\sin h(-n\pi x/b)}\int_0^a f(x)\sin(n\pi x/b)\,dx$$

4.7. THE NEUMANN PROBLEM FOR A RECTANGLE

The Neumann problem for a rectangle is defined as follows

$$PDE : \nabla^2 u = 0,\ 0\le x\le a, 0\le y\le b$$

$$BCs : u_x(0,y)=u_x(a,y)=0, u_y(x,y)=0\ , u_y(x,b)=f(x) \tag{48}$$

The general solution of the Laplace equation using the method of variables separated is found to be

$$u(x,y)=(c_1\ \cos px+c_2 \sin px)(c_3e^{py}+c_4e^{-py})$$

The BC : $u_x(a,y)=0$ gives

$$0=-c_2\, p(c_3e^{py}+c_4e^{-py})$$

Implying $c_2 = 0.$ Therefore,

$$u(x, y) = c_1 \cos px\left(c_3 e^{py} + c_4 e^{-py}\right) \tag{49}$$

The BC: $u_x(a, y) = 0$ gives

$$0 = c_1 p \sin pa\left(c_3 e^{py} + c_4 e^{-py}\right)$$

For non-trivial solution, $c_1 \neq 0,$ implying

$$\sin pa = 0,\ pa = n\pi,\ = \frac{n\pi}{a}\ (\pi = 0, 1, 2, \ldots)$$

Thus the possible solution is

$$u = (x, y) = \cos\frac{n\pi x}{a}\left(Ae^{n\pi y/a} + Be^{-n\pi y/a}\right) \tag{50}$$

Now, using the BC : $u_y(x, 0) = 0,$ we get

$$0 = \cos\frac{n\pi x}{a}\left(A\frac{n\pi}{a} - B\frac{n\pi}{a}\right)$$

Implying $B = A.$ Thus, the solution is

$$u(x, y) = A\cos\frac{n\pi x}{a}\left[\exp\left(\frac{n\pi y}{a}\right) + \exp\left(-\frac{n\pi y}{a}\right)\right]$$

$$2A\cos\frac{n\pi x}{a}\cosh\frac{n\pi y}{a}$$

Using the superposition principle and defining $2A = A_n, we\,get$

$$u = \sum_{n=0}^{\infty} A_n \cos\frac{n\pi x}{a}\cosh\frac{n\pi y}{a}$$

Finally, using the BC: $u_y(x, b) = f(x),$ we get

$$f(x) = \sum_{n=1}^{\infty} A_n \cos\frac{n\pi x}{a}\frac{n\pi}{a}\sin h\ \frac{n\pi b}{a}$$

which is the half-range Fourier cosine series. Therefore,

$$A_n \frac{n\pi}{a} \sinh\frac{n\pi b}{a} = \frac{2}{a}\int_0^a f(x)\cos\frac{n\pi x}{a}dx$$

Hence, the required solution is

$$u = A_0 + \sum_{n=1}^{\infty} A_n \cos\frac{n\pi x}{a} \cosh\frac{n\pi y}{a} \tag{52}$$

where

$$A_n = \frac{2}{n\pi}\frac{1}{\sinh(n\pi b/a)}\int_0^a f(x)\cos\frac{n\pi x}{a}dx$$

4.8. INTERIOR DIRICHLET PROBLEM FOR A CIRCLE

The Dirichlet problem for the circle is defined as follows:

PDE: $\nabla^2 u = 0,\ 0 \le r \le a, 0 \le 0 \le 2\pi$

BC: $u(a,\theta) = f(\theta), 0 \le \theta \le 2\pi$

where $f(\theta)$ is a continuous function on ∂IR. The task is to find the value of u at any point in the interior of the circle IR in terms of its value on ∂IR such that u is single valued and continuous on $\overline{IR}$.

In view of circular geometry , it us natural to choose polar coordinates to solve this problem and then use the variables separable method. The requirement of single-valuedness of $u\, in\, \overline{IR}$ implies the periodicity condition, i.e.,

$$u(r,\theta+2\pi) = u(r,\theta),\ 0 \le r \le a, \tag{54}$$

From Eq. (27) , $\nabla^2 u = 0$ which in polar coordinates can be written as

$$u_{rr} + \frac{1}{r}u_r + \frac{1}{r^2}u_{\theta\theta} = 0$$

If $u(r,\theta)\quad R(r)H(\theta),$ the above equations reduces to

$$R''H + \frac{1}{r}R'H + \frac{1}{r^2}RH'' = 0$$

This equation can be rewritten as

$$\frac{r^2R'' + rR'}{R} = -\frac{H''}{H} = k \tag{55}$$

which means that a function of r is equal to a function of θ and, therefore, each must be equal to a constant k (a separation constant).

Case I Let $k = \lambda^2$. Then

$$r^2 R'' + rR' - \lambda^2 R = 0 \tag{56}$$

which is Euler type of equation and can be solved by setting $r = e^z$. Its solution is

$$R = c_1 e^{\lambda z} + c_2 e^{-\lambda z} = c_1 r^{\lambda} + c_2 r^{-\lambda}$$

Also,

$$H'' + \lambda^2 H = 0$$

whose solution is

$$H = c_3 \cos\lambda\theta + c_4 \sin\lambda\theta$$

Therefore,

$$u(r,\theta) = \left(c_1 r^{\lambda} + c_2 r^{-\lambda}\right)\left(c_3\cos\lambda\theta + c_4\sin\lambda\theta\right) \tag{57}$$

Case II Let $k = -\lambda^2$. Then

$$r^2R'' + rR' + \lambda^2 R = 0\ ,\ H'' - \lambda^2 H = 0$$

Their respective solutions are

$$R = c_1\cos(\lambda in r) + c_2\sin(\lambda in r)$$

$$H = c_3 e^{\lambda\theta} + c_4 e^{-\lambda\theta}$$

Thus

$$u(r,\theta) = \left[c_1\cos(\lambda in r) c_2\ \sin(\lambda in r)\right]\left(c_3 e^{\lambda\theta} + c_4 e^{-\lambda\theta}\right) \tag{58}$$

Case III Let $k = 0$, then we have

$$rR'' + R' = 0$$

Setting $R'(r) = V(r)$, we obtain

$$r\frac{dV}{dr} + V = 0 \text{ , i.e. , } \frac{dV}{V} + \frac{dr}{r} = 0$$

Integrating , we get in $Vr \quad inc$. Therefore,

$$V = \frac{c_1}{r} = \frac{dR}{dr}$$

On integration ,

$$R = c_1\, In\, r + c_2$$

Also,

$$H'' = 0$$

After integrating twice , we get

$$H = c_3\theta + c_4$$

Thus,

$$u(r,\theta) = (c_1\, in\, r + c_2)(c_3\theta + c_4) \tag{59}$$

Now, for the interior problem, $r=o$ is a point in the domain IR and since in $r = 0$ is a point in the domain IR and since In r is not defined at $r = 0$, the solutions (58) and (59) are not acceptable. Thus the required solution is obtained from Eq. (57). The periodicity condition in θ implies

$$c_3 \cos\lambda + c_4 \sin(\lambda(\theta + 2\pi)) + c_4 \sin(\lambda(\theta + 2\pi))$$

i.e.,

$$c_3\left[\cos\lambda\theta - \cos(\lambda\theta + 2\lambda\pi)\right] + c_4[\sin(\lambda(\theta + 2\pi))] = 0$$

or

$$2\sin\lambda\pi\left[c_3 \sin(\lambda\theta + \lambda\pi) - c_4\cos(\lambda\theta + \lambda\pi)\right] = 0$$

Implying sin $\lambda\pi = 0, \lambda\pi = n\pi, \lambda = n(n = 0, 1, 2, ,...)$. Using the principle of superposition and renaming the constants, the acceptable general solution can be written as

$$u(r,\theta)=\sum_{n=0}^{\infty}\left(c_n r^n+d_n r^{-9}\right)\left(a_n\cos n\theta+b_n\sin n\theta\right)$$
(60)

At $r=o,$ the solution should be finite, which requires $d_n=0.$ Thus the appropriate solution assumes the form

$$u(r,\theta)=\sum_{n=0}^{\infty}r^n\left(A_n\cos n\theta+B_n\sin n\theta\right)$$

$For\, n-0,$ let the constant $A_0\ be\ A_0/2.$ Then the solution is

$$u(r,\theta)=\frac{A_0}{2}+\sum_{n=1}^{\infty}r^n\left(a_n\cos n\theta+B_n\sin n\theta\right)$$
(61)

which is a full-range Fourier series. Now we have to determine $A_n\ and\ B_n$ so that the BC: $u(a,\theta)=f(\theta)$ is satisfied.

i.e.,

$$f(\theta)=\sum_{n=1}^{\infty}a^n\ (A_n\cos n\theta+B_n\sin n\theta\ )$$

Hence,

$$A_0=\frac{1}{\pi}\int_0^{2\pi}f(\theta)d\theta$$

$$a^nA_n=1\pi\int_0^{2\pi}f(\theta)\cos\theta\,d\theta$$
(62)

$$a^nB_n=\frac{1}{\pi}\int_0^{2\pi}f(\theta)\sin n\theta\,d\theta,n=1,2,\ldots$$

In Eq. we replace the dummy variables $\theta\,by\,\phi$ to distinguish this variable from the current variables θ in Eq. Substituting Eq. into Eq., we obtain the relation

$$u(r,\theta)=\frac{1}{2\pi}\int_0^{2\pi}f(\phi)d\phi+\sum_{n=1}^{\infty}\left[\frac{r^n}{a^n}\frac{\cos n\theta}{\pi}\int_0^{2\pi}\cos(n\phi)f(\phi)d\phi\right.$$

$$+\frac{r^n}{a^n}\frac{\sin n\theta}{\pi}\int_0^{2\pi}\sin(n\phi)f(\phi)d\phi\Bigg]$$

Interchanging the order of summation and integration, we get

$$u(r,\theta)=\frac{1}{2\pi}\int_0^{2\pi}f(\phi)d\phi+\frac{1}{\pi}\int_0^{2\pi}f(\phi)\sum_{n=1}^{\infty}\left(\frac{r}{a}\right)^n\{\cos n\phi\cos n\theta+\sin n\phi\sin n\theta\}d\phi$$

$$=\frac{1}{\pi}\int_0^{2\pi}f(\phi)\left[\frac{1}{2}+\sum_{n=1}^{\infty}\left(\frac{r}{a}\right)^n\cos n(\phi-\theta)\right]d\phi \qquad (63)$$

$$c=\sum_{n=1}^{\infty}\left(\frac{r}{a}\right)^2\cos n(\phi-\theta)$$

$$s=\sum_{n=1}^{\infty}\left(\frac{r}{a}\right)^n\cos n(\phi-\theta)$$

So that

$$c+is=\sum_{n=1}^{\infty}\left[\frac{r}{a}e^{i(\phi-\theta)}\right]^n$$

Since $r<a,\left(\frac{r}{a}\right)<1 and\left|e^{1(\phi-\theta}\right|\le 1,$

$$c+is=\sum_{n=1}^{\infty}\left[\left(\frac{r}{a}\right)e^{i(\phi-\theta)}\right]^n=\frac{(r/a)r^{i(\phi-\theta)}}{\left[1-\left(\frac{r}{a}\right)e^{i(\phi-\theta)}\right]}$$

$$=\frac{\left(\frac{r}{a}\right)\left\{e^{i(\phi-\theta)}-\left(r^2/a^2\right)\right\}}{\left[1-\left(\frac{r}{a}\right)e^{i(\phi-\theta)}\right]\left[1-\left(\frac{r}{a}\right)e^{i(\phi-\theta)}\right]}$$

Equating the real part on both sides, we get

$$c=\frac{\left[\left(\frac{r}{a}\right)\cos(\phi-\theta)-\left(\frac{r^2}{a^2}\right)\right]}{\left[1-\left(\frac{2r}{a}\right)\cos(\phi-\theta)+r^2/a^2)\right]}=\frac{a^2-r^2}{2\left[a^2-2ar\cos(\phi-\theta)+r^2\right]}$$

Thus , the required solution takes the form

$$u(r,\theta)=\frac{1}{2\pi}\int_0^{2\pi}\frac{\left(a^2-r^2\right)f(\phi)}{\left[a^2-2ar\cos(\phi-\theta)+r^2\right]}d\phi \tag{64}$$

This is known as Poisson's integral formula for a circle, which gives a unique solution for the Dirichlet problem. The solution can be interpreted physically in many ways: it can be thought of as finding the potential $u(r,\theta) as$ a weighted average of the boundary potential $f(\phi)$ weighted by the Poisson kernel p, given by

$$p=\frac{a^2-r^2}{\left[a^2-2ar\cos(\phi-\theta)+r^2\right]}$$

It can also be thought of as a steady temperature distribution $u(r,\theta)$ in a circular disc, when the temperature u on its boundary ∂IR is given by $u=f(\phi)$ which is independent of time.

4.9. EXTERIOR DIRICHLET PROBLEM FOR A CIRCLE

The exterior Dirichlet problem is described by

$PDE:\nabla^2 u=0$

$$BC:u(a,\theta)=f(\theta) \tag{65}$$

u must be bounded as $r\to\infty$

By the method of separation of variables, the general solution of $\nabla^2 u=0$ in polar coordinates can be written as

$$u(r,\theta)=\sum_{n=0}^{\infty}\left(c_n r^n+d_n r^{-n}\right)\left(a_n\cos n\theta+b_n\sin n\theta\right)$$

Now as , $r \to \infty$, we require u tobe bounded, and therefore, $c_n = 0.$ After adjusting the constants, the general solution now reads

$$u(r,\theta) = \sum_{n=0}^{\infty}(A_n \cos n\theta + B_n \sin n\theta)$$

With no loss of generality, it can also be written as

$$u(r,\theta) = \frac{A_0}{2} + \sum_{n=1}^{\infty} r^{-n}\left(A_n \cos n\theta + B_n \sin n\theta\right) \qquad (66)$$

Using the BC: $u(a,\theta) = f(\theta)$, we obtain

$$f(0) = \frac{A_0}{2} + \sum_{n=1}^{\infty} a^{-n}\left(A_n \cos n\theta + B_n \sin n\theta\right)$$

This is a full range Fourier series in $f(\theta)$, where

$$A_0 = \frac{1}{\pi}\int_0^{2\pi} f(\theta)\,d\theta$$

$$a^{-n} A_n = \frac{1}{\pi}\int_0^{2\pi} f(\theta)\cos n\theta \, d\theta \qquad (67)$$

$$a^{-n} B_n = \frac{1}{\pi}\int_0^{2\pi} f(\theta)\sin n\theta \, d\theta$$

In Eq. (67) we replace the dummy variables θ by ϕ so as to distinguish it from the current variable θ. We then introduced the changed variables into solution (66) which becomes

$$u(r,\theta) = \frac{1}{2\pi}\int_0^{2\pi} f(\phi)\,d\phi + \sum_{n=1}^{\infty}\left[\frac{r^{-n} a^n}{\pi} \cos n\theta \int_0^{2\pi} \cos(n\phi) f(\phi)\,d\phi\right.$$

$$\left. + \frac{r^{-n} a^n}{\pi} \sin n\theta \int_0^{2\pi} \sin(n\theta) f(\phi)\,d\phi\right] \text{ or}$$

$$u(r,\theta)=\frac{1}{\pi}\int_0^{2\pi} f(\phi)\left[\frac{1}{2}+\sum_{n=1}^{\infty}\left(\frac{a}{r}\right)^n \cos n(\phi-\theta)\right]d\phi \tag{68}$$

Let

$$C=\sum_{n=1}^{\infty}\left(\frac{a}{r}\right)^n \cos n(\phi-\theta)$$

$$S=\sum_{n=1}^{\infty}\left(\frac{a}{n}\right)^n \sin\ n(\phi-\theta)$$

Then,

$$C+iS=\sum_{n=1}^{\infty}\left[\left(\frac{a}{r}\right)e^{i(\phi-\theta)}\right]^n$$

Since

$$\frac{a}{r}<1\ ,\left|e^{i(\phi-\theta)}\right|\leq 1$$

we have

$$C+iS=\frac{a}{r}\frac{e^{i(\phi-\theta)}}{\left[1\left(\frac{a}{r}\right)e^{i(\phi-\theta)}\right]}=\frac{\left(\frac{a}{r}\right)\left[e^{i(\phi-\theta)}\right]}{\left[1-\left(\frac{a}{r}\right)e^{i(\phi-\theta)}\right]\left[1-\left(\frac{a}{r}\right)e^{-i(\phi-\theta)}\right]}$$

Hence,

$$C=\frac{\left[\left(\frac{a}{r}\right)\cos(\phi-\theta)-\left(\frac{a^2}{r^2}\right)\right]}{\left[1-\left(\frac{2a}{r}\right)\cos(\phi-\theta)+\left(\frac{a^2}{r^2}\right)\right]}$$

Thus the quantity in the square bracket on the right-hand side of Eq. (68) becomes

$$\frac{1}{2}+\frac{\left[\left(\frac{a}{r}\right)\cos(\phi-\theta)\left(\frac{a^2}{r^2}\right)\right]}{2\left[1-\left(\frac{2a}{r}\right)\cos(\phi-\theta)+\left(a^2/r^2\right)\right]}=\frac{r^2+a^2}{2\left[r^2-2ar\cos(\phi-\theta)+a^2\right]}$$

Therefore, the solution of the exterior Dirichlet problem reduce to that of an integral equation of the form

$$u(r,\theta)=\frac{1}{2\pi}\int_0^{2\pi}\frac{\left(r^2-a^2\right)f(\phi)}{\left[r^2-2ar\cos(\phi-\theta)+a^2\right]}d\phi \qquad (69)$$

Example. Find the steady state temperature distribution in a semi-circular plate of radius a. insulated on both the faces with its curved boundary kept at a constant temperature U_0 and its bounding diameter kept at zero temperature .

Solution The Governing heat flow equation

$$u_t=\nabla^2 u$$

In the steady state, the temperature is independent of time ; hence 0, and the temperature satisfies the Laplace equation . The problem can now be stated as follows : To solve

$$PDE:\nabla^2 u(u,\theta)=u_{rr}+\frac{1}{u_r}+\frac{1}{r^2}u_{\theta\theta}=0$$

BCs: $u(a,\theta)=U_0,\ u(r,0)=0,\ u(r,\pi)=0$

The acceptable general solution is

$$u(r,\theta)=\left(cr^{\lambda}+Dr^{-\lambda}\right)\left(A\cos\lambda\theta+B\sin\lambda\theta\right) \qquad (70)$$

From the BC : $u(r,0)=0,$ we get $A=0;$ however, the BC : $u(r,\pi)=0$ also gives

$$B\sin\lambda\pi\left(cr^{\lambda}+Dr^{\lambda-}\right)=0$$

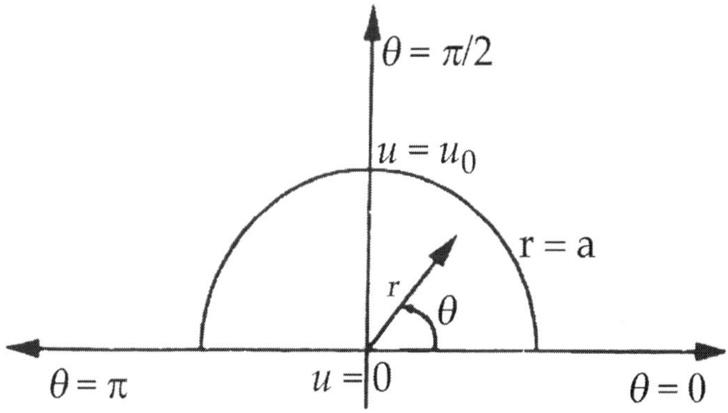

Figure 4

Implying either $B=0$ or sin $\lambda\pi=0$. $B=0$ gives a trivial solution. For a non-trivial solution. We $\lambda=n$. Hence the possible solution is

$$u(r,\theta)=B\sin n\theta\left(Cr^{\lambda}+Dr^{-\lambda}\right) \tag{71}$$

In eq. (71), we observe that as $r\to 0,$ the term $r^{-\lambda}\to\infty.$ But the solution should be finite at $r=0\, and\, so\, D=0$. Then after adjusting the constants, it follows form the superposition principle that

$$u(r,\theta)=\sum_{n=1}^{\infty}B_n\, a^n\ \sin n\theta$$

which is a half –range Fourier sine series . Therefore,

$$B_n a^n=\frac{2}{\pi}\int_0^{\pi}U_0\ \sin n\theta\, d\theta=\begin{cases}\dfrac{4U_0}{n\pi}, & for\, n=1,3,\ldots\\ 0, & for\, n=2,4.,,,,,\end{cases}$$

Hence,

$$B_n=\frac{4U_0}{n\pi a^n},\ n=1,3,\ldots$$

With these values of $B_n,$ the required solution is

$$u(r,\theta)=\frac{4U_0}{\pi}\sum_{\pi}^{\infty}\frac{1}{n}\left(\frac{r}{a}\right)^n\ \sin\, n\theta$$

4.10. INTERIOR NEUMANN PROBLEM FOR A CIRCLE

The interior Neumann problem for a circle is described by

$$PDE : \nabla^2 u = 0, \quad 0 \le r < a \; ; 0 \le \theta 2\pi \tag{72}$$

$$BC : \frac{\partial u}{\partial n} = \frac{\partial u(a, r)}{\partial r} = g(\theta), r = a$$

Following the method of separation of variables, the general solution (60) of equation $\nabla^2 u = 0$ in polar coordinates is given by

$$u(r, \theta) = \sum_{n=0}^{\infty} \left(c_n r^n + d_n r^{-n} \right) \left(a_n \cos n\theta + b_n \sin n\theta \right)$$

At $r = 0,$ the solution should be finite and , therefore, $d_n = 0.$ Hence , after adjusting the constant the general solution becomes

$$u(r, \theta) = \sum_{n=0}^{\infty} r^n \left(A_n \cos n\theta \right)$$

With no loss of generally , this equation can be written as

$$u(r, \theta) = \sum_{n=0}^{\infty} r^n \left(A_n \cos n\theta + B_n \sin n\theta \right) \tag{73}$$

$$\frac{\partial u}{\partial r} = \sum_{n=1}^{0} n r^{n-1} \left(A_n \cos n\theta + B_n \sin n\theta \right)$$

Using the BC:

$$\frac{\partial u}{\partial r}(a, \theta) = g(\theta)$$

we get

$$g(\theta) = \sum_{n=1}^{\infty} n a^{n-1} \left(A_n \cos n\theta + B_n \sin \theta \right)$$

which is a full-range. Fourier series in $g(\theta)$, where

$$n a^{n-1} A_n = \frac{1}{\pi} \int_0^{2\pi} g(\theta) \cos n\theta \, d\theta$$

$$na^{n-1} B_n = \frac{1}{\pi}\int_0^{2\pi} g(\theta)\sin n\theta \, d\theta$$

Here, we replace the dummy variables $\theta \, by \, \phi$ to distinguish form the current variables $\theta \, in \, Eq(74).$ now introducing Eq. (75) into Eq.(73), we obtain

$$u(r,\theta) = \frac{A_0}{2} + \sum_{n=1}^{\infty} \frac{r^n}{n\pi a^{n-1}} \int_0^{2\pi} g(\phi)(\cos n\phi \cos n\theta \sin n\phi \sin n\theta)d\phi$$

or

$$u(r,\theta) = \frac{A_0}{2} + \int_0^{2\pi} g(\phi) \sum_{n=1}^{\infty} \left(\frac{r}{a}\right)^n 1/n\pi \cos n(\phi - \theta) d\phi \tag{76}$$

or this solution can also be expressed in an alternative integral form as follows: Let

$$C = \Sigma\left(\frac{r}{a}\right)^n \frac{a}{n\pi} \cos n(\phi - \theta)$$

$$S = \Sigma\left(\frac{r}{a}\right)^n \frac{a}{n\pi} \sin n(\phi - \theta)$$

Therefore,

$$C + iS = \Sigma\left(\frac{r}{a}\right)^n \frac{a}{n\pi} e^{in(\phi-\theta)} = \frac{a}{\pi} \sum_{n=1}^{\infty} \left[\frac{r}{a} e^{i(\phi-\theta)}\right]^n \frac{1}{n}$$

$$= \frac{a}{\pi}\left[\frac{\left\{\frac{r}{a} e^{i(\phi-\theta)}\right\}}{1} + \frac{\left\{\frac{r}{a} e^{i(\phi-\theta)}\right\}}{2} + \frac{\left\{\frac{r}{a} e^{i(\phi-\theta)}\right\}}{3} + \ldots\right]$$

or

$$C + iS = \frac{a}{\pi} In\left[1 - \frac{r}{a} e^{i(\phi-\theta)}\right] = -\frac{a}{\pi} In\left[1 - \frac{r}{a}\cos(\phi - \theta) - i\frac{r}{a}\sin(\phi - \theta)\right] \tag{77}$$

To get the real part of $In\,z$, we may note that

$$W = In\,z \; or \; z = e^{w}$$

i.e.,

$$x + iy = e^{u+iv} = e^{u}\cos v + ie^{u}\sin v.$$

Therefore,

$$x = e^{u}\cos v,\; y = e^{u}\sin v$$

$$e^{2u} = x^2 + y^2 = |z|^2$$

i.c.,

$$u = In|z|.$$

Therefore ,

$$C = -\frac{a}{\pi} In\sqrt{\left(1 - \frac{r}{a}\cos(\phi - \theta)\right)^2 + \left(\frac{r}{a}\sin(\phi - \theta)\right)^2}$$

$$= -\frac{a}{\pi} In\sqrt{\frac{a^2 - 2ar\cos(\phi - \theta) + r^2}{a^2}}$$

Thus the required solution is

$$u(r,\theta) = \frac{A_0}{2} - \frac{a}{\pi}\int_0^{2\pi} In\sqrt{\frac{a^2 - 2ar\cos(\phi - \theta) + r^2}{a^2}}\, g(\phi)\,d\phi \tag{78}$$

which is again an integral equation.

4.11. SOLUTION OF LAPLACE EQUATION IN CYLINDRICAL COORDINATES

The Laplace equation in cylindrical coordinates assumes the following form:

$$\nabla^2 u = u_{rr} + \frac{1}{r}u_r + \frac{1}{r^2}u_{\theta\theta} + u_{zz} = 0 \tag{79}$$

We now seek a separable solution of the form

$$u(r,\theta,z) = F(r,\theta)Z(z) \tag{80}$$

Subtracting Eq. (80) into Eq. (79), we get

$$\frac{\partial^2 F}{\partial r^2}Z+\frac{1}{r}\frac{\partial F}{\partial r}Z+\frac{1}{r^2}\frac{d^2 F}{d\theta}Z+F\frac{d^2 Z}{dz^2}=0$$

or $$\left(\frac{\partial^2 F}{\partial r^2}+\frac{1}{r}\frac{\partial F}{\partial r}+\frac{1}{r^2}\frac{d^2 F}{d\theta^2}\right)\frac{1}{F}=-\frac{d^2 Z}{dz^2}\frac{1}{Z}=k(say)$$

where k is a separation constant. Therefore , either

$$\frac{d^2 Z}{dz^2}+kZ=0 \tag{81}$$

or

$$\frac{\partial^2 F}{\partial r^2}+\frac{1}{r}\partial F \backslash \partial r+\frac{1}{r^2}\frac{\partial^2 F}{\partial \theta^2}-KF=0 \tag{82}$$

If k is real and positive , the solution of Eq. (81) is

$$Z=c_1 \cos\sqrt{kz}+c_2 e^{-\sqrt{kz}}$$

If k is equal to zero, the solution of Eq. (81) is

$$Z=c_1 z+c_2$$

From physical considerations, one would except a solution which decays with increasing z and, therefore, the solution corresponding to negative k is acceptable. Let $k=-\lambda^2$. Then

$$Z=c_1 e^{\lambda z}+c_2 e^{-\lambda z} \tag{83}$$

Equation (82) now become

$$\frac{\partial^2 F}{\partial r^2}+\frac{1}{r}\frac{\partial F}{\partial r}+\frac{1}{r^2}\frac{\partial^2 F}{\partial \theta^2}+\lambda^2 F=0$$

Let $F(r,\theta)=f(r)H(\theta)$ .Substituting into the above equations we get

$$f''H+\frac{1}{r}f'H+\frac{1}{r^2}fH''+\lambda^2 fH=0$$

or

$$\left(r^2 f''+rf'+\lambda^2 r^2\right)\frac{1}{f}=-\frac{H''}{H}=K''(say)$$

Form physical consideration, we except the solution to be periodic in θ, which can be obtained when k' is positive and $k' = n^2$. therefore , the acceptable solution will be

$$H = c_3 \cos n\theta + c_4 \sin n\theta \quad (84)$$

When $k' = n^2$, we will also have

$$r^2 \frac{d^2 f}{dr^2} + r\frac{df}{dr} + \left(\lambda^2 r^2 - n^2\right) f = 0 \quad (85)$$

which is a Bessel 's equation whose general solution is given by

$$f = Aj_n(\lambda r) + BY_n(\lambda r) \quad (86)$$

Here , $J_n(\lambda r)$ and $Y_n(\lambda r)$ are the nth order Bessel function of first and second kind, respectively. Since $Y_n(\lambda r) \to -\infty$ as $r \to 0$,$Y_n(\lambda r)$ becomes unbounded at $r = 0$. Continuity of the solution demands $B = 0$. Hence the most general and acceptable solution of $\nabla^2 u = 0$ is

$$u(r, \theta, r) = J_n(\lambda r)\left(c_1 e^{\lambda z} + c_2 e^{-\lambda z}\right)\left(c_3 \cos n\theta + c_4 \sin n\theta\right) \quad (87)$$

Example. A homogeneous thermally conducting cylinder occupies the region $0 \le r \le a, 0 \le \theta \le 2\pi, 0 \le z \le h$, where r, θ, z are cylindrical coordinates . The top $z = h$ and the lateral surface $r = a$ are held at 0°, while the base $z = 0$ is held at 100^0. Assuiming that there are no sources of heat generation within the cylinder, find the steady –temperature distribution within the cylinder.

Solution. The temperature u must be a single valued continuous function. The steady state temperature satisfies the Laplace equation □E inside the cylinder. To compute the temperature distribution inside the cylinder , we have to solve the following BVP:

PDE: $\nabla^2 u = 0$

BCs : $u = 0^o \ on\, z = h$,

$u = 0^o \ on\, r = a$,

$u = 100^o \ on\, z = 0$

The general solution of the Laplace equation in cylindrical coordinates is

$$r(r,\theta,z)=J_n(\lambda r)(c_1\ \cos n\theta+c_2\ \sin n\theta)(c_3e^{\lambda z}+c_4e^{-\lambda z})$$

Since the face $z=0$ is maintained at 100^0 and since the other and lateral surface of the cylinder are maintained at 0^0, the temperature at any point inside the cylinder is obviously independent of θ. This is possible only when $n=0$ in the general solution .Thus,

$$u(r,z)=j_0(\lambda r)(Ae^{\lambda z}+Be^{-\lambda z})$$

Using the BC : $u=0\, on\, z=h,$ we get

$$0=J_0(\lambda r)(Ae^{\lambda h}+Be^{-\lambda h})$$

Implying thereby $Ae^{\lambda}+Be^{-\lambda h}=0$, from which

$$B=-\frac{Ae^{\lambda h}}{e^{-\lambda h}}$$

Therefore, the solution is

$$u(r,z)=\frac{J_0(\lambda r)A}{e^{-\lambda h}}\left[e^{\lambda(z-h)}-e^{-\lambda(z-h)}\right]$$

or

$$=u(r,z)=J_0(\lambda r)\ A_1\ \sinh\lambda(z-h)$$

where $A_1=2A\,/e^{-\lambda h}$, Now using the BC : $u=0$ on $r=a,$ we have $0=A_1J_0(\lambda a)\sinh\lambda(z-h)$

Implying $j_0(\lambda a)=0$, which has infinitely many positive roots. Denoting them by ξ_n, we have $\xi_n=\lambda a$, and therefore,

$$\lambda=\frac{\xi_n}{a}$$

Thus the solution is

$$u(r,z)=A_1 j_0\left(\frac{\xi_n r}{a}\right)\sinh\left[\frac{\xi_n}{a}(z-h)\right],\ n=1,2,\ldots$$

Using the principle of superposition , we have

$$u(r,z)=\sum_{n=1}^{\infty}A_n\ j_0\left(\frac{\xi_n r}{a}\right)\sinh\left[\frac{\xi_n}{a}(z-h)\right]$$

The $BC:u=100^0\ on\,z=0$ gives

$$100=\sum_{n=1}^{\infty}A_n\ j_0\left(\frac{\xi_n r}{a}\right)\sin h\left[\frac{\xi_n}{a}(z-h)\right]$$

The $BC:u=100\,on\,z=0$ gives

$$100=\sum_{n=1}^{\infty}A_n\ j_0\left(-\frac{\xi_n h}{a}\right)j_0\left(\frac{\xi_n r}{a}\right)$$

which is a Fourier-Bessel series. Multiplying both sides with $rJ_0\ (\xi_m r\,/a)$ and integrating, we get

$$100\int_0^a r\ j_0\left(\frac{\xi_{m'}}{a}\right)dr=\sum_{n=1}^{\infty}A_n\ \sin h\left(-\frac{\xi_n h}{a}\right)\int_0^a rj_o\left(\frac{\xi_n r}{a}\right)j_0\left(\frac{\xi_{mr}}{a}\right)dr$$

Using the orthogonality property of Bessel' s function , namely,

$$\int_0^a xJ_n(a_i x)J_n(a_j x)\,dx=\begin{cases}0, & if\ i\neq j\\ \dfrac{a^2}{2}\ j_{n+1}^2(\alpha_i), & if\ i=j\end{cases}$$

Where a_i, a_j are the zeros at $j_n(x)-0$, we have

$$100\int_0^1 r\ j_0\left(\frac{\xi_n r}{a}\right)dr=\sum_{n=1}^{\omega}A_n\ \sinh\left(-\frac{\xi_n h}{a}\right)\frac{a^2}{2}j_1^2(\xi_n)$$

Therefore,

$$A_n=\frac{200}{a^2\ \sin h\left(-\dfrac{\xi_n h}{a}\right)\ j_1^2(\xi_n)}\int_0^a rJ_0\left(\frac{\xi_n r}{a}\right)dr$$

Setting

$$\frac{\xi_n r}{a}=x\ ,\ \ dr=\frac{a}{\xi_n}dx$$

The relation for $A_n\, can$ also be written as

$$A_n=\frac{200}{\xi_n^2\ \sinh\left(-\frac{\xi_n h}{a}\right)j_1^2(\xi_n)}\int_0^{\xi_n} x\, j_0(x)\,dx$$

Using the recurrence relation

$$x^n\ j_{n-1}(x)=\frac{d}{dx}\left[x^n\ j_n(x)\right],$$

For $n=1$, we get

$$\int x\ j_0(x)dx=x\,j_1(x)$$

Now, A_n can be written as

$$A_n=\left[\frac{200\,x\,j_1(x)}{\xi_n^2\sinh\left(-\frac{\xi_n h}{a}\right)\ j_1^2(\xi_n)}\right]_0^{\xi}=\frac{200}{\xi_n \sin h\backslash a)\,j_1(\xi)}$$

Hence, the required temperature distribution inside the cylinder is

$$u(r,z)=200\sum_{n=1}^{\infty}\frac{j_0\left(\xi_n\frac{r}{a}\right)\ \sin h\left[\left(\frac{\xi_n}{a}\right)(z-h)\right]}{\xi_n\ \sinh(-\xi_n h/a)\,j_1(\xi_n)}$$

where x_n are the positive zeros of $j_0(\xi)$.

Example . Find the potential u inside the cylinder $0\le r\le a, 0\le\theta\le 2\pi, 0\le z\le h, if\ the\ potential$ on the top $z=h$ and on the lateral surface $r=a$ is held at zero, while on the base $z=0,$ the potential is given by $u(r,\theta,0)=V_0\left(1-r^2/a^2\right),$ where V_0 is a constant r,θ,z are cylindrical polar coordinates .

Solution. The potential u must be a single-valued continuous function and satisfy the Laplace equation inside the cylinder. To compute the potential inside the cylinder, we have to solve the following BVP:

PDE: $\nabla^2 u = 0$

BCs: $u = 0 \, on \, z = h,$

$u = 0 \;\; on \, r = a,$

$$u = V_0\left(1 - \frac{r^2}{a^2}\right) on \, z = 0$$

In cylindrical coordinates , the general solution of the Laplace equation\ /\ is

$$u(r,\theta,z) = J_n(\lambda r)(c_1 \cos n\theta + c_2 \sin n\theta)(c_3 e^{\lambda z} + c_4 e^{-\lambda z})$$

Since the face $z = 0$ has potential $V_0\left(1 - \frac{r^2}{a^2}\right)$ which is purely, a function of r and is independent of θ. This is possible only when $n = 0$ in the general solution. Thus ,

$$u(r,z) = j_0(\lambda r)(Ae^{\lambda z} + Be^{-\lambda z})$$

Using the BC : $u = 0 \, on \, z = h$, we obtain

$$0 = j_0(\lambda r)(A_e^{\lambda h} + Be^{-\lambda h})$$

Implying $Ae^{\lambda h} + Be^{-\lambda h} = 0,$ which yields

$$B = -\frac{Ae^{\lambda h}}{e^{-\lambda h}}$$

Hence , the solution is

$$u(r,z) = \frac{A}{e^{-\lambda h}} j_0(\lambda r)\left[e^{\lambda(z-h)} - e^{-\lambda(z-h)}\right]$$

or

$$u(r,z) = \frac{A}{e^{-\lambda h}} j_0(\lambda r) \sinh \lambda(z-h)$$

Where $A_1 = A/e^{\lambda h}$, Now, using the BC : $u = 0$ on the lateral surface , i.e., on $r = a,$ we get

$0 = A_1 J_0(\lambda a)\sinh\lambda(z-h)$

Implying $j_0(\lambda a)=0$. This has infinitely many positive roots; demoting them by ξ_n we shall have

$\xi_n = \lambda_a \ or \ \lambda=\xi_n/a$

The solution now takes the form

$$u(r,z) = A_1 j_0\left(\frac{\xi_n r}{a}\right)\sinh\left[\frac{\xi_n}{a}(z-h)\right], \quad n=1,2,,....$$

The principle of superposition gives

$$u(r,z)=\sum_{n=1}^{\infty} A_n J_0\left(\frac{\xi_n r}{a}\right)\sin h\left[\frac{\xi_n h}{a}(z-h)\right]$$

The last BC : $u = V_0\left(1-\dfrac{r^2}{a^2}\right) on\, z = 0$ yeilds

$$V_0\ (1-\frac{r^2}{a^2}=\sum_{n=1}^{\infty} A_n\ Sinh\left(-\frac{\xi_n h}{a}\right)J_0\left(\frac{\xi_n r}{a}\right)$$

This is a Fourier-Bessel's series . Multiplying both sides by $rj_0(\xi_m r/a)$ and integrating, we get

$$V_0\int_0^a\left(1-\frac{r^2}{a^2}\right)rj_0\left(\frac{\xi_m r}{a}\right)dr=\sum_{n=1}^{\infty}A_n\sinh\left(-\frac{\xi_n h}{a}\right)j_0\left(\frac{\xi_n r}{a}\right)$$

Using the orthogonality property of Bessel functions

$$\int_0^a xj_n(\alpha_i x)\ j_n(a_j x)dx=\begin{cases}0,\ if\, i=j\\ \dfrac{a^2}{2}j_{n+1}^2(\alpha_i),\ if\, i\neq j\end{cases}$$

Where α_i, α_j are the zeros of $j_n(x)=0$, we get

$$V_0 \int_0^a \left(1-\frac{r^2}{a^2}\right) r\, j_0\left(\frac{\xi_n r}{a}\right) dr = \sum_{n=1}^{\infty} A_n \sinh\left(-\frac{\xi_n h}{a}\right) \frac{a^2}{2} j_1^2(\xi_n)$$

which gives

$$A_n = \frac{2V_0}{a^2 \sinh\left(-\frac{\xi_n h}{a}\right) j_1^2(\xi_n)} \int_0^a \left(1-\frac{r^2}{a^2}\right) r\, j_0\left(\frac{\xi_n r}{a}\right) dr$$

Be letting $\xi r a = x$, this equation can be modified to

$$A_n = \frac{2V_0}{\xi_n^4 \sinh\left(-\frac{\xi_n h}{a}\right) j_1^2(\xi_n)} \int_0^{\xi_n} (\xi_n^2 - x^2)x\, j_0(x)dx$$

Using the well-known recurrence relation

$$x^{\alpha} j_{\alpha-1}(x) = \frac{d}{dx}\left[x^{\alpha} j_{\alpha}(x)\right] for = \alpha = 1, 2, \ldots$$

We get

$$\int x\, j_0(x) = x j_1(x), \int x^2 j_1(x) = x^2 j_2(x)$$

From these relations, we obtain

$$A_n = \frac{2V_0}{\xi_n^4 \sinh\left(-\frac{\xi_n h}{a}\right) J_1^2(\xi_n)} \int_0^{\xi n} x^2 j_1(x)dx$$

$$= \frac{2V_0}{\xi_n^4 \sinh\left(-\frac{\xi_n h}{a}\right) J_1^2(\xi_n)} \left[x^2 j_2(x)\right]_0^{\xi n}$$

Thus,

$$A_n = \frac{4V_0\, j_2(\xi_n)}{\xi_n^4 \sinh\left(-\frac{\xi_n h}{a}\right) J_1^2(\xi_n)}$$

The recurrence relation

$$J_{n-1}(x)+j_{n+1}(x)=\frac{2n}{x}\, j_n(x)$$

For $n=1$ gives

$$j_0(\xi_n)+j_2(\xi_n)=\frac{2}{\xi_n}\, j_1(\xi_n)$$

Hence,

$$j_2(\xi_n)=\frac{2}{\xi_n}\, j_1(\xi_n)$$

Since $j_0(\xi_n)=0$. Therefore,

$$A_n=\frac{8V_0\, j_1(\xi_n)}{\xi_n^3 \sinh\left(-\xi_n \dfrac{h}{a}\right) j_1^2(\xi_n)}$$

Thus, the required potential inside the cylinder is

$$u(r,z)=\sum_{n=1}^{\infty}\frac{8V_0\, j_0\left(\dfrac{\xi_n r}{a}\right)\sinh\left[\dfrac{\xi_n}{a}(z-h)\right]}{\xi_n^3\, j_1(\xi_n)\,\sinh\left(-\dfrac{\xi_n h}{a}\right)}$$

4.12. SOLUTION OF LAPLACE EQUATION IN SPHERICAL COORDINATES

In example 3 the Laplace equation is expressed in spherical and has the following form:

$$\nabla^2 u=\frac{\partial}{\partial r}\left(r^2\frac{\partial u}{\partial r}\right)+\frac{1}{\sin\theta}\frac{\partial}{\partial\theta}\left(\sin\theta\frac{\partial u}{\partial\theta}\right)+\frac{1}{\sin^2\theta}\frac{\partial^2 u}{\partial\phi^2}=0 \qquad (88)$$

Let us assume the separable solution in the form

$$u(r,\theta,\phi)=R(r)F(\theta,\phi) \qquad (89)$$

Subtracting of variables gives

$$\frac{\frac{d}{dr}\left(r^2\frac{dR}{dr}\right)}{R}=\frac{-\frac{1}{\sin\theta}\left\{\frac{\partial}{\partial\theta}\left(\sin\theta\frac{\partial F}{\partial\theta}\right)+\frac{1}{\sin\theta}\frac{\partial^2 F}{\partial\phi^2}\right\}}{F}=-\mu$$

where μ is separation constant. Therefore,

$$\frac{1}{R}\frac{d}{dr}\left(r^2\frac{dR}{dr}\right)=-\mu \qquad (90)$$

$$\frac{1}{F\sin\theta}\left[\frac{\partial}{\partial\theta}\left(\sin\theta\frac{\partial u}{\partial\theta}\right)+\frac{1}{\sin\theta}\frac{\partial^2 F}{\partial\phi^2}\right]=\mu \qquad (91)$$

Equation (90) gives

$$r^2\frac{d^2R}{dr^2}+2r\frac{dR}{dr}+\mu R=0$$

which is a Euler's equation. Hence, using the transformation $r=e^z$, the auxiliary equation can be written as

$$D(D-1)+2D+\mu=D^2+\mu=0$$

where $D=\frac{d}{dz}$. Its roots are given by

$$D=\frac{-1\pm\sqrt{1-4\mu}}{2}$$

Let $\mu=-\alpha(\alpha+1)$; then we get

$$D=\frac{\left\{-1\pm2\sqrt{\left(a+\frac{1}{2}\right)^2}\right\}}{2}=-\frac{1}{2}\pm\left(a+\frac{1}{2}\right)$$

Hence, $D-\alpha$ and $-(a+1)$ Therefore, the solution of Euler's equation is

$$R=c_1e^a+c_2r^{-(a+1)} \qquad (92)$$

Taking $-\mu=a(a+1), Eq.$ becomes

$$\frac{\partial}{\partial\theta}\left(\sin\theta\frac{\partial F}{\partial\theta}\right)+\frac{1}{\sin\theta}\frac{\partial^2 F}{\partial\phi^2}+a(a+1)F\sin\theta=0$$

Inserting

$$F=H(\theta)\,\Phi(\phi)$$

Into the above equation and separating the variables, we obtain

$$\frac{\sin\theta}{H}\left[\frac{d}{d\theta}\left(\sin\theta\frac{dH}{d\theta}\right)+\alpha(a+1)\sin\theta H\right]=-\frac{1}{\Phi}\frac{d^2\Phi}{d\phi}=v^2$$

where v^2 is another separation constant. Then

$$\frac{d^2\Phi}{d\phi^2}+v^2\Phi=0 \qquad (93)$$

$$\frac{\sin\theta}{H}\left[\frac{d}{d\theta}\left(\sin\theta\frac{dH}{d\theta}\right)+a(a+1)\sin(\theta)H\right]=v^2 \qquad (94)$$

The general solution of Eq. (93) is

$$\Phi=c_3\cos v\phi+c_4\sin v\phi$$

Provided $v\neq 0, if\ v=0$, the solution is independent of ϕ which corresponds to the ax symmetric case. Equation (94) becomes , for the ax symmetric , case,

$$\frac{d}{d\theta}\left(\sin\theta\frac{dH}{dx}\frac{dx}{d\theta}\right)+a(a+1)\sin(\theta)H=0$$

i.e.,

$$\frac{d}{dx}\left[\left(1-\cos^2\theta\right)\frac{dH}{dx}\right]+a(a+1)H=0$$

or

$$\frac{d}{dx}\left[\left(1-x^2\right)\frac{dH}{dx}\right]+a(a+1)H=0$$

This is the well-known Legendre equation. Its general solution is given by

$$H = c_5\, p_a(x) + c_6\, Q_a(x), -1 \le x \le 1$$

where p_a, Q_a are Legendre function of the first and second kind respectively. For connivance let α be a positive integer, say $a = n$. Then

$$H = c_5\, p_n(\cos\theta) + c_6\, Q_n(\cos\theta) \qquad (98)$$

Continuity of $H(\theta)\, at\, \theta = 0, \pi$ implies the continuity of $H(x)\, at\, x = \pm 1$. Since $Q_n(x)$ has a singularity at $x = 1,$ we choose $c_6 = 0.$ Therefore, in ax symmetric case the solution of Laplace equation in spherical coordinates is given by

$$u(r,\theta,\phi) = \left\{ c_1 r^a + c_2 r^{-(a+1)} \right\}(c_3)\left[c_5\, p_n(\cos\theta) \right]$$

After renaming the constant s and using the principle of superposition , we find the solution to be

$$u(r,\theta) = \sum_{n=0}^{\infty} \left[A_n r^n + B_n r^{-(n+1)} \right] p_n(\cos\theta)$$

Example. in a solid sphere of radius $'a',$ the surface is maintained at the temperature given by

$$f(\theta) = \begin{cases} k\cos\theta, \le \theta \dfrac{\pi}{2} \\ 0, \quad \dfrac{\pi}{2} < \theta < \pi \end{cases}$$

Prove that the steady state temperature within the solid is

$$u(r,\theta) = k\left[\frac{1}{4} p_0(\cos\theta) + \frac{1}{2}\left(\frac{r}{u}\right) p_1(\cos\theta) + \frac{5}{16}\left(\frac{r}{a}\right)^2 p_2(\cos\theta) - \frac{3}{32}\left(\frac{r}{a}\right)^4 p_4(\cos\theta) + \ldots \right]$$

Solution It is known that the steady state temperature distribution is governed by the Laplace equation. In spherical polar coordinates, the ax symmetric solution of the Laplace equation in general with the assumption that the temperature should be finite at the origin is given by Eq.(99)

In the form

$$u(r,\theta)=\sum_{n=0}^{\infty} A_n r^n \ p_n(\cos\theta) \tag{100}$$

Using the given BC: $u(a,\theta)=f(\theta)$, we have

$$u(a,\theta)=f(\theta)=\sum_{n=0}^{\infty} A_n a^n \ p_n(\cos\theta)=\sum_{n=0}^{\infty} b_n \ p_n(cos\theta)$$

where $A_n a^n = b_n$ This is a Fourier-Legendre series, where

$$b_n=\frac{2n+1}{2}\int_{-1}^{1} f(\theta)\,p_n(\cos\theta)\,d\theta$$

In the present problem,

$$b_n=\frac{2n+1}{2}\int_{-1}^{1} f(\theta)P_n(\cos\theta)\,d\theta$$

Let $\cos\theta=x$,

$$b_0=\frac{1}{2}\int_0^1 kx\,.p_0(x)\,dx=\frac{1}{2}\int_0^1 kx\,.1\,.dx=\frac{k}{4}$$

Hence, we get

$$A_0=\frac{k}{4}$$

Also, $$b_1=\frac{3}{2}\int_0^1 k_x\,..x\,.dx=\frac{k}{2}=A_1 a$$

Therefore,

$$A_1=\frac{k}{2}\left(\frac{1}{a}\right)$$

$$b_2=\frac{5}{2}\int_0^1 kxp_2(x)\,dx=\frac{5}{2}\int_0^1 kx\frac{3x^2-1}{2}\,dx=\frac{5}{16}k$$

Thus,

$$A_2=\frac{5}{16}k\,.\frac{1}{a^2}$$

Further,

$$b_3 = \frac{7}{2}\int_0^1 kxp_3(x)\,dx = \frac{7}{2}\int_0^1 kx\frac{5x^3-3x}{2}\,dx = 0$$

Similarly, noting that $p_4(x) = \frac{1}{8}\left(35x^4\right) - 30x^2 + 3\,)$, we get

$$b_4 = -\frac{3}{32}k = A_4 a^4$$

Hence,

$$A_4 = -\frac{3}{32}k\,.\frac{1}{a^4},\ldots$$

Subtracting these values of $A_0, A_1, A_2, \ldots$ into Eq. (100), we obtain , finally, the required temperature as

$$u(r,\theta) = k\left[\frac{1}{4}\,p_0(cos\theta) + \frac{1}{2}\left(\frac{r}{a}\right)p_1(cos\theta)\right]$$

$$\left. + \frac{5}{16}\left(\frac{r}{a}\right)^2 p_2(\cos\theta) + \left(-\frac{3}{32}\right)\left(\frac{r}{a}\right)^4 p_4(\cos\theta) + \ldots\right]$$

Example. find the potential at al point of space inside and outside of a sphere of radius $R = 1$ which is maintained at a constant distribution of electric potential $u(R,\theta) = f(\theta) = \cos 2\theta$.

Solution. It is known that the potential on the surface of a sphere is governed by the Laplace equation. The Laplace equation in spherical polar coordinates is

$$\frac{\partial^2 u}{\partial r^2} + \frac{2}{r}\frac{\partial u}{\partial r} + \frac{1}{r^2}\frac{\partial^2 u}{\partial \theta} + \frac{\cot\theta}{r^2}\frac{\partial u}{\partial \theta} + \frac{1}{r^2\sin^2\theta}\frac{\partial^2 u}{\partial \phi^2} = 0$$

The possible general solution by variables separable method, after using superposition principle , is given by Eq. (99). Thus have two possible solution:

$$u_1(r,\theta) = \sum_{n=0}^{\infty} A_n r^n p_n(\cos\theta) \tag{101}$$

$$u_2(r,\theta)=\sum_{n=0}^{\infty}\frac{B_n}{r^{n+1}}\,p_n(cos\,\theta) \tag{102}$$

For point inside the sphere, we take the series (101). Why is this so? Applying the BC : $u(R,\theta)=f(\theta)=\cos 2\theta$ we obtain

$$f(\theta)=\sum_{n=0}^{\infty}A_n\,R^n\,p_n(\cos\,\theta)$$

which is generalized Fourier series $f(\theta)$ in terms of the Legendre polynomials. Using the orthogonality property, we get

$$A_n R^n=\frac{2n+1}{2}\int_{-1}^{1}f(\theta)p_n(x)dx$$

Let $x=cos\,\theta$. Then we have

$$A_n=\frac{2n+1}{2}\int_{-1}^{1}f(\theta)p_n(\cos\,\theta)\,\sin\theta\,d\theta$$

For points outside the sphere, we takes the series (102). Why is this so ? Using the BC: $u(R,\theta)=f(\theta)$, we get

$$f(\theta)=\sum_{n=0}^{\infty}\frac{B_n}{R^{n+1}}\,p_n(\,\cos\theta)$$

Again using the orthogonality property of Legendre polynomials, we have

$$B_n=\frac{2n+1}{2}R^{n+1}\int_{0}^{\pi}f(\theta)p_n(\cos\,\theta)\,\sin\,\theta\,d\theta$$

In the present problem, it is assumed that at $R=1\,,f(\theta)=\,\cos 2\theta=2\,\cos^2\theta-1=2x^2-1$. Hence

$$A_n=\frac{2n+1}{2}\int_{-1}^{1}(2x^2-1)P_n(x)dx$$

However,

$$p_2(x)=\frac{1}{2}(3x^2-1)$$

Therefore,

$$2x^2 - 1 = \frac{4}{3} p_2(x) - \frac{1}{3}$$

Thus,

$$A_n = \frac{2n+1}{2} \int_{-1}^{1} \left[\frac{4}{3} \int_{-1}^{1} p_0^2(x)\,dx - \frac{1}{3} \int_{-1}^{1} p_0(x) p_n(x)\,dx \right]$$

Using the orthogonality property of Legendre polynomials, all integrals vanish except those corresponding to $n = 0\, and\, n = 2$. We obtain therefore,

$$A_0 = -\frac{1}{2} \cdot \frac{1}{3} \int_{-1}^{1} P_0^2(x)\,dx = -\frac{1}{3}$$

$$A_2 = \frac{5}{2} \cdot \frac{4}{3} \int_{-1}^{1} P_2^2(x)\,dx = \frac{4}{3}$$

Also ,

$$B_n = \frac{2n+1}{2} \int_{-1}^{1} (2x^2 - 1) p_n(x)\,dx$$

$$= \frac{2n+1}{2} \left[\frac{4}{3} \int_{-1}^{1} P_2(x) p_n(x)\,dx - \frac{1}{3} \int_{-1}^{1} P_0(x) p_n(x)\,dx \right]$$

which, on using the orthogonality property , gives the non-vanishing coefficients as

$$B_0 = -\frac{1}{3},\ B_2 = \frac{4}{3}$$

Subtracting these values of A_0 and A_2 into $Eq\,.(101)$, we obtain

$$u_1(r, 0) = -\frac{1}{3} + \frac{4}{3} r^2 p_2(\cos\theta)$$

Which gives the potential everywhere inside the sphere . Similarly. Subtracting the values of B_0 and B_2 into $Eq\,.(102)$, we get

$$u_2(r,\theta) = -\frac{1}{3r} + \frac{4}{3r^2} P_2(\cos\theta)$$

which gives the potential outside the sphere.

Example. Find a general spherically symmetric solution of the following Helmholtz equation : $\left(\nabla^2 - k^2\right)u = 0$

Solution In spherical polar coordinates , the Helmholtz equation can be written as

$$\frac{\partial^2 u}{\partial r^2} + \frac{2}{r}\frac{\partial u}{\partial \theta^2} + \frac{1}{r^2}\frac{\partial^2 u}{\partial \theta^2} + \frac{\cot\theta}{r^2}\frac{\partial u}{\partial \theta} + \frac{1}{r^2 \sin^2\theta}\frac{\partial^2 u}{\partial \phi^2} - k^2 u = 0 \qquad (103)$$

In view of spherical symmetry , we look for u to be a function of r alone. Hence . Eq. (103) becomes

$$\frac{\partial^2 u}{\partial r^2} + \frac{2}{r}\frac{\partial u}{\partial r} - k^2 u = 0$$

Therefore , we have to solve

$$r^2 \frac{\partial^2 u}{\partial r^2} + 2r\frac{\partial u}{\partial r} - k^2 r^2 u = 0$$

Let

$$u = \frac{1}{\sqrt{r}} F(r)$$

Differentiating twice with respect to r and rearranging, we obtain

$$2r\frac{\partial u}{\partial r^2} = \frac{F(r)}{\sqrt{r}} + 2\sqrt{r}\,F'(r)$$

$$r^2 \frac{\partial^2 u}{\partial r^2} = -\frac{3}{4} r^{-1/2} F(r) - r^{-1/2} F'(r) + r^{\frac{3}{2}} F''(r)$$

Substituting the above relations, Eq. (104) becomes

$$r^2 F''(r) + r F'(r) - \left(k^2 r^2 + \frac{1}{4}\right) F(r) = 0$$

or

$$r^2 F''(r) + rF'(r) + \left[(ik)^2 r^2 - \left(\frac{1}{2}\right)^2 \right] F(r) = 0$$

This is the Bessel equation whose solution is

$$F(r) = AJ_{\frac{1}{2}}(ikr) + BY_{1/2}(ikr)$$

where $J_{\frac{1}{2}}, Y_{\frac{1}{2}}$ are Bessel functions with imaginary arguments, and is rewritten as

$$F(r) = AI_{\frac{1}{2}}(kr) + BK_{\frac{1}{2}}(kr)$$

Therefore,

$$u(r) = r^{-1/2}\left[AI_{\frac{1}{2}}(kr) + BK_{\frac{1}{2}}(kr) \right]$$

But as $r \to \infty$, the solution should be finite, which is possible only if $A = 0$, . it is also known that for large z,

$$K_{\frac{1}{2}}(z) = \sqrt{\frac{\pi}{2z}}\, e^{-z}$$

Thus the acceptable spherically symmetric solution of the Helmholtz equation is given by

$$u(r) = Br^{-1/2}\sqrt{\frac{\pi}{2kr}}\, e^{-kz} = \frac{c}{r} e^{-kz}$$

where

$$c = B\sqrt{\frac{\pi}{2k}}$$

4.13. MISCELLANEOUS EXAMPLE

Example. Show that the velocity potential for an irrational flow of an incompressible fluids satisfies the Laplace solution.

Solution. Let us consider a closed surface ecnlosing a fixed volume in the region occupied by a moving fluids .

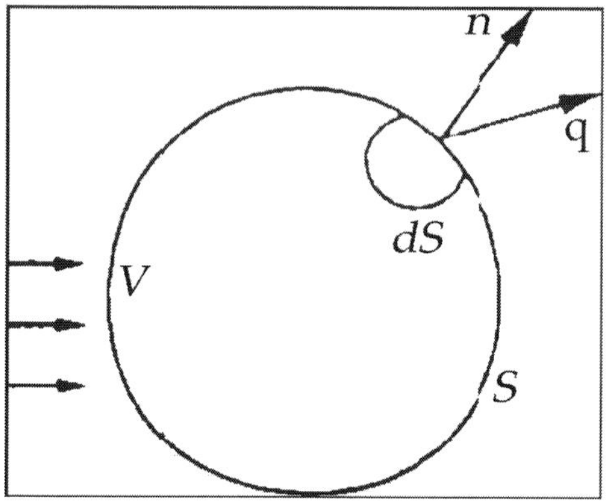

Figure 5

Let p be the density of the fluid. If n is a unit vector in the direction of the normal to the element dS and q the velocity of the fluid at that point , then the inward normal velocity is $(-q\ .\hat{n})$.. Hence the mass of the fluid entering per unit time through the element dS is $(-q\ .\hat{n})dS$. it follows therefore that the mass of the fluid entering the surface S is unit time is

$$\iint_S p(q\ .\hat{n})dS$$

Also , the mass of the fluid within S is

$$\iiint_V P\,dV$$

So the rate at which the mass goes on increasing is given by

$$\frac{\partial}{\partial t}\iiint_V P\,dV = \iiint_V \frac{\partial P}{\partial t}dV$$

By conservation of mass, the rate of generation of mass within a given volume under the assumption that no internal sources are present is equal to the net inflow of mass through the surface enclosing the given volume. Thus,

$$\iiint_V \frac{\partial P}{\partial t}dV = -\iint_S p(q.\hat{n})dS$$

$$= -\iiint_V div(pq)\,dV \left[using\, the\, divergence\, theorem \right]$$

Therefore, $$\iiint_V \left[\frac{\partial p}{\partial t} + div(pq) \right] dV = 0$$

Since the integrand is a continuous function and since this result is true for any arbitrary volume element dV · it follows that the integrand is zero . Therefore,

$$\frac{\partial p}{\partial t} + \nabla . pq = 0$$

which is called the equation of continuity. For an incompressible fluids, $p =$ constant and therefore,

$$\nabla . q = 0$$

Further, if the flow is irrotational, i.e., there exist a velocity potential ϕ such that

$$q = -\nabla \phi$$

Hence, $$\nabla . q = \nabla . \nabla \phi = \nabla^2 \phi = 0$$

Thus, an incompressible irrotational fluid satisfies the Laplace equation.

Example. A thin rectangular homogeneous thermally conducting plate lies in the $xy-$ plane defined by $0 \geq x \leq a, 0 \leq y \leq b$. The edge $y = 0$ is held at the temperature $Tx(x-a)$, where T is a constant , while the remaining edges are held at 0^0 . The other faces are insulated and no internal sources and sinks are present. Find the steady state temperature inside the plate.

Solution. Since no heat sources and sinks are present in the plate, the steady state temperature u must satisfy $\nabla^2 u = 0$ · hence the problem is to solve

$$PDE : \nabla^2 u = 0$$

BCs: $u(0, y) = 0, u(a, y) = 0,\ u(x, b) = 0, u(x, 0) = Tx(x-a)$

This is a typical Dirichlet problem . The general solution satisfying the first three BCs is given by Eq. (47). Therefore,

$$u(x,y)=\sum_{n=1}^{\infty} A_n \sin\left(\frac{n\pi}{a}x\right) \sinh\left[\frac{n\pi}{s}(y-b)\right]$$

$$A_n \sinh\frac{-n\pi}{a}=\frac{2}{a}\int_0^a f(x) \sin\left(\frac{n\pi}{a}x\right)dx$$

Using the last BC: $u(x,0)=Tx(x-a)=f(x)$, get

$$A_n \sinh\frac{-n\pi b}{a}=\frac{2}{a}\int_0^a Tx(x-a) \sin\left(\frac{n\pi}{a}x\right)dx$$

$$=\frac{2T}{a}\int_0^a x(x-a)d\left\{\frac{n\pi}{a}x\right\}dx$$

$$=-\frac{a}{n\pi}\cdot\frac{2T}{a}\left[\int_o^a x(x-a)d\left\{\cos\left(\frac{n\pi}{a}x\right)\right\}\right]$$

$$=-\frac{2T}{n\pi}\left[(2x-a)\sin\left(\frac{n\pi}{a}x\right)\right]_0^a-\int_o^a(2x-a)d\left[\sin\left(\frac{n\pi}{a}x\right)\right]$$

$$=\frac{2aT}{n^2\pi^2}\left[(2x-a)\sin\left(\frac{n\pi}{a}x\right)\right]_0^a-\int_o^a 2\sin\left(\frac{n\pi}{a}x\right)dx$$

$$=\frac{2aT}{n^2\pi^2}\left\{a\sin n\pi+\frac{2a}{n\pi}\left[\cos\left(\frac{n\pi}{a}x\right)\right]_0^a\right\}$$

$$=\frac{2aT}{n^2\pi^2}\frac{2a}{n\pi}(\cos n\pi-1)=\frac{4a^2T}{n^3\pi^3}\left[(-1)^n-1\right]$$

Thus the required temperature distribution is given by

$$u(x,y)=\sum_{n=1}^{\infty} cosech\left(-\frac{n\pi}{a}b\right)\frac{4Ta^2}{n^3\pi^3}[-1]\sin\left(\frac{n\pi}{a}x\right)\sinh\left[\frac{n\pi}{a}(y-b)\right]$$

Example. Solve

$\nabla^2 u=0,\ 0\le x\le a,\ 0\le y\le b$

Satisfying the BCs:

$$u(0,y)=0,\ u(x,0)=0,\ u(x,b)=0$$

$$\frac{\partial u}{\partial x}(a,y)=T\sin^3\frac{\pi y}{a}$$

Solution · Using the variables separable method, one of the acceptable general solutions is given by Eq. (38). Hence

$$u(x,y)=\left(c_1e^{px}+c_2e^{-px}\right)(c_3\cos py+c_4\sin py)$$

Using the BC: $u(x,0)=0$, we get

$$0=c_3\left(c_1e^{px}+c_2e^{-px}\right)$$

Implying $c_3=0$. Therefore,

$$u(x,y)=c_4\sin py\ (c_1e^{px}+c_2e^{-px})$$

Now, using the BC: $u(x,b)-0,$ we obtain

$$0=c_4\sin pb\left(c_1e^{px}+c_2e^{px}\right)$$

$c_4=0$, implying $\sin pb=0$ which gives

$ob=n\pi$ or $p=\frac{n\pi}{b}$, $n=1,2,3,\dots$.

Thus $u(x,y)=c_4\sin\left(\frac{n\pi}{b}x\right)\left(c_1e^{px}+c_2e^{-px}\right)$

Renaming the constants, we have

$$u(x,y)=\sin\left(\frac{n\pi}{b}y\right)\left[A\exp\left(\frac{n\pi}{b}x\right)+B\exp\left(-\frac{n\pi}{b}x\right)\right],n=1,2,\dots.$$

If we use the $BC:u(0,y)=0$, we get

$$0=\sin\left(\frac{n\pi}{b}y\right)(A+B)$$

Giving $A+B=0$;*therefore*, $A=-B$. Thus,

$$u(x,y)=A\sin\left(\frac{n\pi}{b}y\right)\left[\exp\left(\frac{n\pi}{b}x\right)-\exp\left(-\frac{n\pi}{b}x\right)\right]$$

$$=2A\sin\left(\frac{n\pi}{b}y\right)\sinh\left(\frac{n\pi}{b}x\right),\ n=1,2,.$$

Differentiating with respect to x, we obtain

$$\frac{\partial u}{\partial x}=2A\frac{n\pi}{b}\sin\left(\frac{n\pi}{b}y\right)\cosh\left(\frac{n\pi}{b}x\right)$$

The Last BC yields

$$T\,sin^3\frac{\pi y}{a}=2A\frac{n\pi}{b}\sin\left(\frac{n\pi}{b}y\right)\cosh\left(\frac{n\pi}{b}x\right)$$

The last BC yields

$$T\sin^3\frac{\pi y}{a}=2A\frac{n\pi}{b}\sin\left(\frac{n\pi}{b}y\right)\cosh\left(\frac{n\pi}{b}a\right)$$

Form which we can determine $2A$. Hence, the required solution is

$$u(x,y)=\frac{bT}{n\pi}\sin^3\frac{\pi y}{a}\operatorname{sech}\frac{n\pi}{b}a\sinh\left(\frac{n\pi}{b}x\right)$$

The principle of superposition gives the required solution as

$$u(x,y)=\sum_{n=1}^{\infty}\frac{bT}{n\pi}\sin^3\frac{\pi y}{a}\operatorname{sech}\left(\frac{n\pi}{b}a\right)\sinh\left(\frac{n\pi}{b}x\right)$$

Example. Find the potential function $u(x,y,z)$ in a rectangular box defined by $0\le x\le a,0\le y\le b,0\le z\le c$, if the potential is zero on all sides and the bottom, while $u=f(x,y)$ on the top of the box .

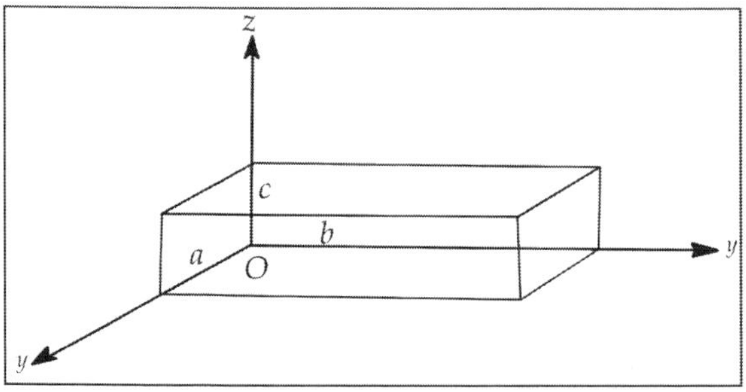

Figure 6

Solution. The potential distribution in the rectangular box satisfies the Laplace equation. Thus, the problem is to solve.

$$\nabla^2 u = u_{xx} + u_{yy} + u_{zz} = 0$$

Subject to the BCs:

$$u(0,y,z) = u(a,y,z) = 0$$

$$u(y,0,z) = u(x,b,z) = 0$$

$$u(x,y,0) = 0$$

$$u(x,y,z) = f(x,y)$$

Following the variables separable method, let us assume the solution in the form

$$u(x,y,z) = X(x)Y(y)Z(z)$$

Substituting into the Laplace equation, we get

$$X''(x)Y(y)Z(z) + X(x)Y''(y)Z(z) + X(x)Y(y)Z''(z) = 0$$

which can also be written as

$$\frac{Y''(y)}{Y(y)} + \frac{Z''(z)}{Z(z)} = -\frac{X''(x)}{X(x)} = \lambda_1^2$$

where λ_1 is a separation constant. Thus, we have

$$X''(x) + \lambda_1^2 X(x) = 0 \qquad (105)$$

After the second separation, we also have

$$\frac{Z''(z)}{Z(z)} - \lambda_1^2 = -\frac{Y''(y)}{Y(y)} = \lambda_2^2$$

$$Y''(y) + \lambda_2^2 Y(y) = 0 \qquad (106)$$

$$Z''(z) - \lambda_3^2 Z(z) = 0 \qquad (107)$$

where $\lambda_3^2 = \lambda_1^2 + \lambda_2^2$. The general solution of Eq. $(105)-(107)$ are

$$X(x) = c_1 \cos\lambda_1 x + c_2 \sin\lambda_1 x$$

$$Y(y) = c_3 \cos\lambda_2 x + c_4 \sin\lambda_2 y$$

$$Z(z) = c_5 \cosh\lambda_3 z + c_6 \sinh\lambda_3 z$$

From the BCs,

$$X(0) = X(a) = 0$$

$$Y(0) = Y(b) = 0$$

$$Z(0) = 0$$

$X(0) = 0$ gives $c_1 = 0$

$X(a) = 0$ gives $\lambda_1 a = m\pi$.

Therefore,

$$\lambda_1 = \frac{m\pi}{a}, m = 1, 2, \ldots.$$

Similarly,

$Y(0) = 0$ gives $c_3 = 0$

$Y(b) = 0$ gives $\lambda_2 b = n\pi$

Therefore,

$$\lambda_2 = \frac{n\pi}{b}, n = 1, 2, \ldots$$

Also, $Z(0)=0$ gives $c_5=0$ Further, we note that

$$\lambda_3^2=\lambda_1^2+\lambda_2^2=\pi^2\left(\frac{m^2}{a^2}+\frac{n^2}{b^2}\right)=\lambda_{mn}^2\,(say)$$

Then

$$\lambda_3=\pi\sqrt{\frac{m^2}{a^2}+\frac{n^2}{b^2}}=\lambda_{mn}$$

The solution now take the form

$$X(x)=c_{2m}\sin\frac{m\pi x}{a},\ m=1,2,\ldots$$

$$Y(y)=c_{4\pi}\sin\frac{n\pi y}{b},\ n=1,2,\ldots$$

$$Z(z)=c_{6mn}\sinh\lambda_{mn}z$$

Let $c_{mn}=c_{2m}c_{4n}c_{6mn}$; then, after using the principle of superposition, the required solution is

$$u(x,y,z)=X(x)Y(y)Z(z)=\sum_{m=1}^{\infty}\sum_{n=1}^{\infty}c_{mn}\sin\frac{m\pi x}{a}\sin\frac{n\pi y}{b}\sinh\lambda_{mn}z \tag{108}$$

Using the final BC: $f(x,y)=u(x,y,c)$, we get

$$f(x,y)=\sum\sum c_{mn}\sinh\lambda_{mn}c\sin\frac{m\pi x}{a}\sin\frac{n\pi y}{b}$$

which is a double Fourier sine series. Thus, we have

$$c_{mn}\sinh\lambda_{mn}c=\frac{4}{ab}\int_0^a\int_0^b f(x,y)\sin\frac{m\pi x}{a}\sin\frac{n\pi y}{b}\,dx\,dy \tag{109}$$

Therefore, $Eq.(108)\,and\,(109)$ constitute the required potential.

Example. Find the electrostatic potential u in the annular region bounded by the concentric sphere $r=a,r=b,0<a<b$, if the inner outer surfaces are kept at constant potential u_1 and u_2, $u_1\neq u_2$.

Solution. The electrostatic potential satisfies the Laplace equation $\nabla^2 u = 0$

It is natural that we choose spherical polar coordinates. from the problem. It is evident that we are looking for a solution with spherical symmetry which is independent of $\theta \, and \, \phi$. Hence. $u = u(r)$.

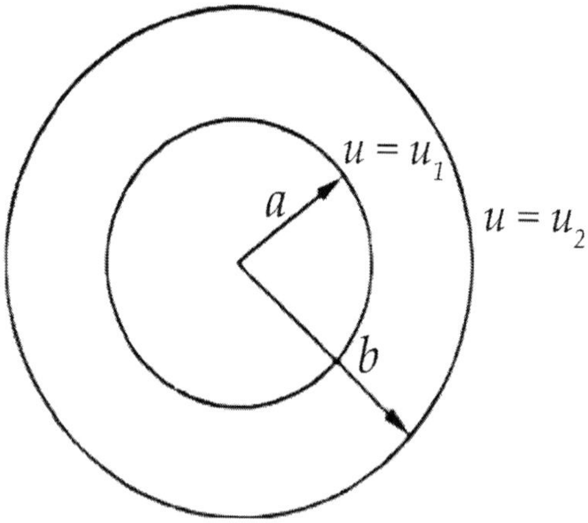

Figure 7

Thus, we have to solve

$$PDE: \frac{\partial}{\partial r}\left(r^2 \frac{\partial u}{\partial r} \right) = 0 \tag{110}$$

Subject to $BCs \; ; u = u_1 \; at \, r = a \; u = u_2 \; at \, r = b$

Integrating $Eq.(110)$ with respect to r, we obtain

$r^2 \frac{\partial u}{\partial r} = A$ (a constant of integration)

Again, integrating, we get

$$u = -\frac{A}{r} + B$$

Now, using the BCs : we have

$$u_1 = -\frac{A}{a} + B \; , \; u_2 = -\frac{A}{b} + B$$

Solving these equations, we get

$$A=u_2-\frac{u_1}{(1/a)-(1/b)}\left[u_1\left(\frac{1}{r}-\frac{1}{b}\right)+u_2\left(\frac{1}{a}-\frac{1}{r}\right)\right]$$

Example. A thermally conducting solid bounded by two concentric spheres of radius $a\,and\,b$, $a<b$, is such that the internal boundary is kept at $f_1(\theta)$ and the outer boundary at $f_2(\theta)$. Find the steady state temperature in the solid .

Solution. It is known that the steady temperature T satisfies the Laplace equationU?H. In the present problem,

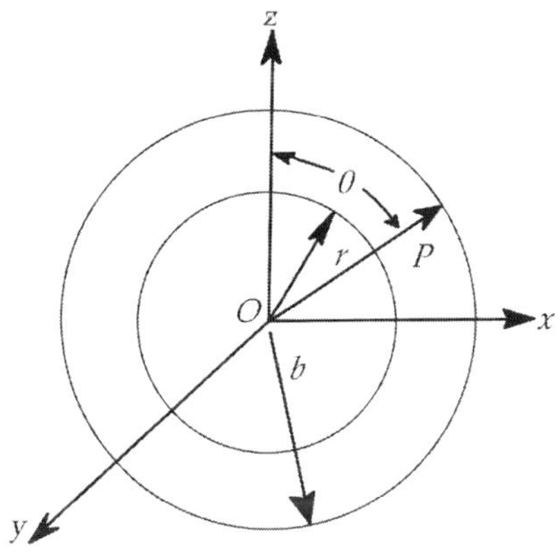

Figure 8

$T=T(r,\theta)$

Thus, we have to solve

PDE: $\nabla^2 T=0$

Subject to the boundary conditions

$T=f_1(\theta)\,at\,r=a$

$T=f_2(\theta)\,at\,r=b$

In spherical polar coordinates, for axially symmetric case, the solution of the Laplace equation is given by $Eq.(99)$ as follows :

$$T(r,\theta)=\sum_{n=0}^{\infty}\left(A_n r^n+\frac{B_n}{r^{n+1}}\right)p_n(\cos\theta)$$

Using the BCs:

$$f_1(\theta)=\sum_{n=0}^{\infty}\left(A_n a^n+\frac{B_n}{a^{n+1}}\right)p_n(\cos\theta) \tag{111}$$

$$f_2(\theta)=\sum_{n=0}^{\infty}\left(A_n a^n+\frac{B_n}{b^{n+1}}\right)p_n(\cos\theta) \tag{112}$$

In order to find the coefficient $A_n\ and\ B_n$, we have to express $f_1(\theta) and\ f_2(\theta)$ in terms of Legendre polynomials and compare the coefficient. In this process, the following orthogonality relation is useful:

$$\int_0^{\pi} p_m(cos\theta)P_n(\cos\theta)\sin\theta\ d\theta=\begin{cases}0, & if\ m\neq n\\ \dfrac{2}{2n+1}, & if\ m=n\end{cases}$$

Thus, multiplying both sides of $Eq.(111)$ by $p_m(\cos\theta)\sin\theta$ and integrating, we obtain

$$\int_0^{\pi} f_1(\theta)p_m(cos\theta)\sin\theta\ d\theta=\sum_{n=0}^{\infty}\left(A_n a^n+\frac{B_n}{a^{n+1}}\right)\int_0^{\pi} p_n(\cos\theta)p_m(cos\theta)\sin\theta\, d\theta$$

$$=\left(A_n a^n+\frac{B_n}{a^{n+1}}\right)\frac{2}{2m+1} \tag{113}$$

Similarly, $Eq.(112)$ gives

$$\int_0^{\pi} f_2(\theta)p_m(\cos\theta)\sin\theta\, d\theta=\sum_{n=0}^{\infty}\left(A_n b^n+\frac{B_n}{b^{m+1}}\right)\int_0^{\pi} p_n(\cos\theta)p_m(cos\theta)\sin\theta\, d\theta$$

$$=\left(A_m b^m+\frac{m}{a^{m+1}}\right)\frac{2}{2m+1} \tag{114}$$

Let

$$\frac{2m+1}{2}\int_0^{\pi} f_1(\theta)p_m(\cos\theta)\sin\theta\, d\theta=C_m$$

$$\frac{2m+1}{2}\int_0^{\pi} f_2(\theta)\, p_m(\cos\theta)\, \sin\theta\, d\theta = D_m$$

Then $Eq.(113) and (114)$ reduce to

$$A_m a^m + \frac{B}{a^{m+1}} = C_m$$

$$A_m a^n + \frac{B}{b^{m+1}} = D_m$$

Solving this pair of equations, we obtain

$$A_m = \frac{C_m a^{m+1} - D_m b^{m+1}}{a^{2m+1} - b^{2m+1}} \tag{115}$$

$$B_m = \frac{a^{m+1} b^{m+1}\left(C_m b^m - D_m a^m\right)}{b^{2m+1} - a^{2m+1}} \tag{116}$$

Hence, the require steady temperature is

$$T(r,\theta) = \sum_{m=0}^{\infty}\left(A_m r^m + \frac{B_m}{r^{m+1}}\right) p_m(\cos\theta)$$

where A_m and B_m are given by $Eq.(115) and (116)$.

Example. A thin annulus occupies the region $0 < a \le r \le b, 0 < \theta \le 2\pi$. The faces insulated. Along the inner edge the temperature is maintained at 0^0, while along the outer edge the temperature is held at $T = K\cos\left(\frac{\theta}{2}\right)$, where K is a constant. Determine the temperature distribution in the annulus.

Solution. Mathematically, the problem is to solve

PDE: $\nabla^2 T = 0,\ a \le r \le b, 0 \le \theta \le 2\pi$

BCs: $T(a,\theta) = 0$

$$T(b,\theta) = k\cos\frac{\theta}{2}$$

The required general solution is given by $Eq.(57)$ in the form

$$T(r,\theta)=\left(c_1 r^n + c_2 r^{-n}\right)\left(c_3 \cos n\theta + c_4 \sin n\theta\right)$$

Using the first BC. We get

$$0=\left(c_1 a^n + c_2 a^{-n}\right)\left(c_3 \cos n\theta + c_4 \sin n\theta\right)$$

Implying thereby $c_1 a^n + c_2 a^{-n} = 0, or\, c_2 = -c_1 a^{2n}$. After adjusting the constants suitably, we have

$$T(r,\theta)=\left(r^n - \frac{a^{2n}}{r^n}\right)\left(A \cos n\theta + B \sin n\theta\right)$$

The principal of superposition gives

$$T(r,\theta)=\sum_{n=1}^{\infty}\left(r^n - \frac{a^{2n}}{r^n}\right)\left(A \cos n\theta + B \sin n\theta\right)$$

Now, using the second boundary condition, we obtain

$$T(b,\theta)= K\cos\frac{\theta}{2}=\sum_{n=1}^{\infty}\left(b^n - b^{-n} a^{2n}\right)\left(A_n \cos n\theta + B_n \sin n\theta\right)$$

which is a full-range Fourier series . Hence,

$$A_n\left(b^n - b^{-n} a^{2n}\right)=\frac{1}{\pi}\int_0^{2\pi} K \cos\frac{\theta}{2} \cos n\theta\, d\theta$$

$$=\frac{k}{2\pi}\int_0^{2\pi}\left[\cos\left(n+\frac{1}{2}\right)\theta + \cos\left(n-\frac{1}{2}\right)\theta\right]d\theta$$

$$=\frac{k}{2\pi}\int_0^{2\pi}\left[\cos\left(n+\frac{1}{2}\right)\theta + \cos\left(n-\frac{1}{2}\right)\theta\right]d\theta$$

$$=\frac{k}{2\pi}\left[\frac{\left(n+\frac{1}{2}\right)\theta}{n+\frac{1}{2}}+\frac{\left(n-\frac{1}{2}\right)\theta}{n-\frac{1}{2}}\right]_0^{2\pi}$$

$$=0$$

Implying $A_n = 0$. Also,

$$B_n\left(b^n - b^{-n}a^{2n}\right) = \frac{k}{\pi}\int_0^{2\pi}\cos\frac{\theta}{2}\sin n\theta\, d\theta$$

$$= \frac{k}{2\pi}\int_0^{2\pi}\left[\sin\left(n+\frac{1}{2}\right)\theta + \sin\left(n-\frac{1}{2}\right)\theta\right]d\theta$$

$$= -\frac{k}{2\pi}\left[\frac{cos\left(n+\frac{1}{2}\right)\theta}{n+\frac{1}{2}} + \frac{cos\left(n-\frac{1}{2}\right)\theta}{n-\frac{1}{2}}\right]_0^{2\pi}$$

$$= -\frac{k}{2\pi}\left(-\frac{1}{n+\frac{1}{2}} - \frac{1}{n+\frac{1}{2}} - \frac{1}{n-\frac{1}{2}} - \frac{1}{n-\frac{1}{2}}\right)$$

$$\frac{k}{\pi}\left(\frac{1}{n+\frac{1}{2}} + \frac{1}{n+\frac{1}{2}}\right) = \frac{k}{\pi}\frac{2n}{n^2 - \frac{1}{4}}$$

or

$$B_n\left(b^n - b^{-n}a^{2n}\right) = \frac{8kn}{\pi\left(4n^2 - 1\right)}$$

Thus the temperature distribution in the annulus is given by

$$T(r,\theta) = \frac{8k}{\pi}\sum_{n=1}^{\infty}\frac{n}{4n^2-1}\left[\frac{\left(\frac{r}{a}\right)^n - \left(\frac{a}{r}\right)^n}{\left(\frac{b}{a}\right)^n - \left(\frac{a}{b}\right)^{-n}}\right]\sin n\theta$$

Example. V is a function of r and θ satisfying the equation

$$\frac{\partial^2 V}{\partial r^2} + \frac{1}{r}\frac{\partial V}{\partial r} + \frac{1}{r^2}\frac{\partial^2 V}{\partial \theta^2} = 0$$

within the region of the plane bounded by $r=a\ ,r=b\ ,\theta=0\ ,\theta=\pi/2$. its value along the boundary $r=a\,is\,\theta\left(\frac{\pi}{2}-\theta\right)$, along the other boundaries is zero. Prove that

$$V=\frac{2}{\pi}\sum_{n=1}^{\infty}\frac{(r/b)^{4n-2}-(b/r)^{4n-2}}{(a/b)^{4n-2}-(b/a)^{4n-2}}\left[\frac{sin(4n-2)\theta}{(2n-1)^3}\right]$$

Solution. The task is to solve the PDE

$$\frac{\partial^2 V}{\partial r^2}+\frac{1}{r}\frac{\partial V}{\partial r}+\frac{1}{r^2}\frac{\partial^2 V}{\partial \theta^2}=0$$

Subject to the following boundary conditions:

$(i)\ V(b\ ,\theta)=0\ ,\quad 0<\theta<\pi\ /2$

$(ii)\ V(r\ ,\pi/2)=0\ ,\ a<r\le b$

$(iii)\ V(r\ ,0)=0\ ,\quad a<r\le b$

$(iv)\ V(a\ ,\theta)=\theta\ ,\left(\frac{\pi}{2}-\theta\right),0<\theta<\pi/2.$

The three possible solution are given as follows:

$$V=\left(c_1r^p+c_2r^{-p}\right)\left(c_3\ \cos p\theta+c_4\sin p\theta\right)$$

$$V=\left[c_1\ \cos(p\,In\,r)+c_2\sin(p\,In\,r)\right]\left(c_3e^{p\theta}+c_4e^{-p\theta}\right)$$

$$V=(c_1\,In\,r+c_2)(c_3\theta+c_4)$$

Since the problem is not defined for $r=0\ ,\infty$, the second and third solutions are not acceptable .Hence , the generally acceptable solution is the first one. The boundary condition (iii) gives

$$0=c_3\left(c_1r^p+c_2r^{-p}\right)$$

Implying $c_3=0$. The boundary condition (ii) implies

$$0=c_4\ \sin p\frac{\pi}{w}\left(c_1\,r^p+c_2r^{-p}\right)$$

Therefore,

$$\sin p\frac{\pi}{2}=0 \ or \ p=2n\ ,n=1\ ,2\ ,...$$

Thus, the possible solution of the given equation has the form

$$V(r\ ,\theta)=c_4\ \sin\ (2n\theta\ )\left(c_1r^{2n}+c_2r^{-2n}\right)$$

Now, applying the boundary condition (i), we get

$$0=c_4\ \sin\left(2n\theta\right)\left(c_1b^{2n}+c_2b^{-2n}\right)$$

which gives $c_2=-c_1\,b^{4n}$. Therefore,

$$V(r\ ,\theta)=c_1c_4\ \sin\left(2n\theta\right)\left[r^{2n}-r^{-2n}\,b^{4n}\right]$$

Superposing all the solution, we obtain

$$V(r\ ,\theta)=\sum_{n=1}^{\infty}c_n\ \sin\left(2n\theta\right)\left(r^{2n}-r^{-2n}b^{4n}\right)$$

Satisfying boundary conditions (iv), we get

$$\theta\left(\frac{\pi}{2}-\theta\right)=\Sigma c_n\ \sin\left(2n\theta\right)\left(\frac{a^{4n}-b^{4n}}{a^{2n}}\right)$$

which is a Fourier sine series. Thus, we have

$$\frac{2}{\pi/2}\int_0^{\frac{\pi}{2}}\theta\left(\frac{\pi}{2}-\theta\right)\sin\left(2n\theta\right)=c_n\left(\frac{a^{4n}-b^{4n}}{a^2n}\right)$$

Integrating by parts , we obtain

$$\left[\left(\frac{\pi}{2}\theta-\theta^2\right)\left\{-\frac{\cos\left(2n\theta\right)}{2n}\right\}-\left(\frac{\pi}{2}-2\theta\right)\left\{-\frac{\sin\left(2n\theta\right)}{4n^2}\right\}+\left(-2\right)\left\{\frac{\cos\left(2n\theta\right)}{8n^3}\right\}\right]_0^{\frac{\pi}{2}}$$

$$=c_n\ \frac{\pi}{4}\left(\frac{a^{4n}\quad b^{4n}}{a^{2n}}\right)$$

On simplification , we get

$$-\frac{1}{4n^3}\left\{(-1)^n-1\right\}=\frac{\pi}{4}c_n\left(\frac{a^{4n}-b^{4n}}{a^{2n}}\right)$$

Thus,

$$\frac{\pi}{4}c_n\left(\frac{a^{4n}-b^{4n}}{a^{2n}}\right)=\begin{cases}\frac{1}{2n^3}, & for\,n\,odd\\ 0, & for\,n\,even\end{cases}$$

Hence, the required solution is

$$V(r,\theta)=\sum_{1}^{\infty}\frac{2}{\pi}\frac{1}{(2n-1)^3}\left(\frac{a}{r}\right)^{4n-2}\sin(4n-2)\theta\left(\frac{r^{8n-4}-b^{8n-4}}{a^{8n-4}-b^{8n-4}}\right)$$

which can be recast in the form given in the problem .

Example. Determine the potential of grounded conducting sphere in a uniform field defined by

PDE: $\nabla^2 u=0,\ 0\leq r<a,0<\theta<\pi,0\leq\phi<2\pi$

BCs: $(i)\ u(a,\theta)=0$.

$(iii)\,u\rightarrow E_0 r\cos\theta\, as\, r\rightarrow\infty$,

Solution. In spherical polar coordinates , with axial symmetry, the solution of the Laplace equation is given by Eq. (99) in the form

$$u(r,\theta)=\sum_{n=0}^{\infty}\left(A_n r^n+\frac{B_n}{r^{n+1}}\right)P_n(\cos\theta)$$

Using the boundary conditions (ii), we have

$$u(r,\theta)=\sum_{n=0}^{\infty}A_n r^n P_n(\cos\theta)=-E_0\,r\cos\theta$$

which is true only for

$n=1$, *when* $p_1(\cos\theta)=\cos\theta$. *Also* , $A_n=0$ *for* $n\geq 2$. Therefore ,

$u(r,\theta)=A_1 r\cos\theta=-E_0\,r\cos\theta$

Implying $A_1=-E_0$. Hence,

$$u(r,\theta)=-E_0\cos\theta+\sum_{n=1}^{\infty}\frac{B_n}{r^{n+1}}p_n(\cos\theta)$$

Now, applying the boundary condition (i), we get

$$0-E_0\,a\cos\theta+\sum_{a^{n+1}}^{\infty}\frac{B_n}{a^{n+1}}p_n(\cos\theta)$$

Multiplying both sides of $p_m(\cos\theta)\sin\theta\, and$ integrating between the limits $0\, to\, \pi$, we have

$$E_0 a\int_0^{\pi}\cos(\theta)p_m(\cos\theta)\sin\theta\,d\theta=\sum_{n=1}^{\infty}\frac{B_n}{a^{n+1}}\int_0^{\pi}p_n(\cos\theta)p_m(\cos\theta)\sin\theta\,d\theta$$

Using the orthogonality property

$$\int_0^{\pi}p_n(\cos\theta)p_m(\cos\theta)\sin\theta\,d\theta=\begin{cases}0, & for\, m\neq n\\ \dfrac{2}{2m+1}, & for\, m=n\end{cases}$$

we obtain

$$\frac{B_m}{a^{m+1}}\frac{2}{2m+1}=E_0 a\int_0^{\pi}\cos(\theta)p_m(\cos\theta)\sin\theta\,d\theta$$

or

$$B_m=\frac{2m+1}{2}E_0\,a^{m+2}\int_0^{\pi}\cos(\theta)p_m(cos\theta)\sin\theta\,d\theta$$

It can verified that the integral on the right-hand side of the Eq. (117) vanishes for all m except when $m=1$, I which case

$$B_1=E_0\,a^3$$

Therefore, the required potential is given by

$$u(r,\theta)=-E_0\,r\cos\theta+E_0\,a^3 r^2\cos\theta$$

Example. The steady, two –dimensional, incompressible viscous fluid flow past a circular cylinder, when the inertial terms are neglected (stoke flow) , is governed by the inharmonic

PDE: $\nabla^2 \backslash \Psi = 0$ where ψ is the stream function. Find its solution subject to the BCs:

$$(i)\psi(r,\theta) = \frac{\partial \psi}{\partial r} = 0 \quad \text{on } r = 1$$

$$(ii)\psi(r,\theta) \to r\sin\theta \quad as\ r \to \infty$$

Solution. In view of the cylindrical geometry , we can write

$$\nabla^4 \psi = \nabla^2\left(\nabla^2 \psi\right) = 0$$

where

$$\nabla^2 \psi = \left(\frac{\partial^2}{\partial r^2} + \frac{1}{r}\frac{\partial}{\partial r} + \frac{1}{r^2}\frac{\partial^2}{\partial \theta^2}\right)\psi$$

Using the variables separable method , let us look for a solution of the form

$$\psi(r,\theta) = f(r)\sin\theta$$

Therefore,

$$\frac{\partial \psi}{\partial r} = f'(r)\sin\theta, \frac{\partial \psi}{\partial \theta} = f(r)\cos\theta$$

$$\frac{\partial^2 \psi}{\partial r^2} = f''(r)\sin\theta, \frac{\partial^2 \psi}{\partial \theta^2} = f(r)\sin\theta$$

Hence,

$$\nabla^2 \psi = \left[f''(r) + \frac{1}{r}f'(r) - \frac{1}{r^2}f(r)\right]\sin\theta$$

which can also be written in the form

$$\nabla^2 \psi = F(r)\sin\theta$$

where

$$F(r) = f''(r) + \frac{1}{r}f'(r) - \frac{1}{r^2}f(r)$$

Therefore,

$$\nabla^4 \psi = \nabla^2\left(\nabla^2 \psi\right) = \nabla^2\left[F(r)\sin\theta\right] = 0$$

i.e.,

$$\left[F''(r) + \frac{1}{r}F'(r) - \frac{1}{r^2}F(r)\right]\sin\theta = 0$$

Implying

$$F''(r) + \frac{1}{r}F'(r)\frac{1}{r^2}F(r) = 0$$

Introducing the transformable $r = e^z, D = \frac{d}{dz}$, the above equation becomes

$$\left[D(D-1) + D - 1\right]F(r) = 0$$

or

$$\left(D^2 - 1\right)F(r) = 0$$

Its complementary function is

$$F(r) = Ae^z + Be^{-z} = Ar + \frac{B}{r}$$

or

$$f''(r) + \frac{1}{r}f'(r) - \frac{1}{r^2}f(r) = Ar + \frac{B}{r}$$

i.e.,

$$r^2 f''(r) + rf'(r) - f(r) = Ar^3 + Br$$

which is a homogeneous ordinary differential equation. Again using the transformable $r -= e^z, D = \frac{d}{dz}$. We get

$$\left[D(D-1) + D - 1\right]f = A^{3z} + Be^z$$

or

$$\left(D^2-1\right)f=Ae^{3z}+Be^{z}$$

Its complementary function is

$$f\left(r\right)=Ce^{z}+De^{-z}$$

while its particular integral is

$$\frac{1}{D^2-1}\left(Ae^{3z}+Be^{z}\right)=\frac{Ae^{3z}}{8}+\frac{Bze^{z}}{2}$$

Therefore,

$$f\left(r\right)=Cr+\frac{D}{r}+\frac{A}{8}r^3+\frac{B}{2}r\,In\,r$$

Thus, we have

$$\psi=\left(\frac{A}{8}r^3+\frac{B}{2}r\,In\,r\,Cr+\frac{D}{r}\right)\sin\theta$$

Now to satisfy the BC: $\psi\to r\sin\theta\ ,as\,r\to\infty$ and form physical considerations, we choose $A=0$, Therefore,

$$\psi=\left(\frac{B}{2}r\,in\,r+Cr+\frac{D}{r}\right)sin\,\theta$$

The boundary condition

$\psi=0\,on\,r=1\,gives\left(C+D\right)\sin\theta=0\ ,implying\ C=-D$. Also , the boundary condition $\frac{\partial\psi}{\partial r}=0$ on $r=1$ gives

$$\frac{B}{2}+C-D=0=\frac{B}{2}-2D$$

Implying $B=4D.Hence$, the general solution is

$$\psi=2D\ (r\,in\,r-\frac{r}{2}+\frac{1}{2r}\ \sin\theta$$

Example. The problem of axisymmetric fluid flow in a semi-infinite or in a finite circular pipe of radius a is described as follows in cylindrical coordinates :

$PDE : \nabla^2 u = 0 \, , \, 0 < r < a$

BCs: $(i) \; \dfrac{\partial u}{\partial r} = 0 \; at \; r = 0$

$(ii) \dfrac{\partial u}{\partial z} = 0 \, , \, \dfrac{\partial u}{\partial r} = V(z) \; at \; r = a$

Show that the speed of suction is given by

$$V(z) = -\sum_{n=1}^{\infty} a_n \left(A_n \cosh a_n z + B_n \sinh a_n z \right) j_1 \left(a_n a \right)$$

Solution. In cylindrical coordinates (r, θ, z),

$$\nabla^2 u = \frac{\partial^2 u}{\partial r^2} + \frac{1}{r}\frac{\partial u}{\partial r} + \frac{1}{r^2}\frac{\partial^2 u}{\partial \theta^2} + \frac{\partial^2 u}{\partial z^2} = 0$$

In axisymmetric case , the above equation becomes

$$\frac{\partial^2 u}{\partial r^2} + \frac{1}{r}\frac{\partial u}{\partial r} + \frac{\partial^2 u}{\partial z^2} = 0 \qquad (118)$$

Let $u(r, z) = f(r)\phi(z) \, which$, when substituted into Eq. (118) , gives

$$\frac{f'' + \left(\frac{1}{r}\right) f'}{f} = -\frac{\phi''}{\phi} = -\, a^2 \, (say)$$

Then

$$\phi'' - a^2 \phi = 0 \qquad (119)$$

$$f'' + \frac{1}{r} f' + a^2 \phi = 0 \qquad (120)$$

The solution of $Eq .(119)$ is

$\phi = A \cosh \alpha z \;\; B \sinh az$

Equation (120) can be rewritten as

$r^2 f'' + rf' + \alpha^2 r^2 f = 0$

which is a Bessel's equation of Zeroth order whose general solution is

$$f = j_0(\alpha r) + DY_0(\alpha r)$$

Here, $j_0(ar)\, and\, Y_0(ar)$ are Zeroth order Bessel functions of the first and second kind respectively , Therefore , the typical solution is

$$u = (A\ \cosh az + B\ \sin \alpha z)[\ j_0(\alpha)]$$

Now, $Y_0(\alpha r)$ is infinite at $r=0$, and hence $D=0$, Therefore , the possible solution is

$$u = (A\ \cosh \alpha z + B\ \sinh \alpha z)\, j_0(ar)$$

The condition $\frac{\partial u}{\partial r} = 0$ at $r=0$ is automatically satisfies , since $j_0'(ar)=0$ at $r=0$. Now the boundary condition $\frac{\partial u}{\partial z} = 0\, at\, r = a$ gives $j_0(aa)=0$, implying that aa are the zeros of the Bessel function j_0. Let these zeros be $a_n a(n=0,1,2,\ldots)$. Thus the appropriate solution is

$$u(r,z) = \sum_{n=1}^{\infty} a_n\ (A_n\ \cosh a_n\ z + B - n\ \sinh a_n z\ )\, j_0(a_n r)$$

Using the fact that, $j_0' = -j_1$, the speed of suction is given by

$$V(z) = \left(\frac{\partial u}{\partial r}\right)_{r=a} = -\sum_{n=1}^{\infty} a_n (A_n\ \cosh a_n z + B_n\ \sinh a_n\ z\ )\ j_1(a_n a)$$

Example. Solve the following Poisson equation:

$$\frac{\partial^2 u}{\partial x^2} + \frac{\partial^2 u}{\partial y^2} = 2$$

Subject to the boundary conditions

$$u(0,y) = u(5,y) = u(x,0) = u(x,4) = 0$$

Solution. We assume the solution of the form

$$u = v + \omega \qquad (121)$$

where v is a particular solution of the Poisson equation and ω is the solution of the corresponding homogenous Laplace equation. That is,

$$\nabla^2 v = 2 \tag{122}$$

$$\nabla^2 \omega = 0 \tag{123}$$

It is customary to assume that υ has the form

$$v(x, y) = a + bx + cy + dx^2 + exy + fy^2$$

Substituting this into Eq. (122), we get

$$2d + 2f = 2$$

Let $f = 0$. Then $d = 1$ The remaining coefficients can be chosen arbitrarily. Thus, we take

$$v(x, y) = -5x + x^2 \tag{124}$$

So that v reduce to zero (satisfies the boundary conditions) on the sides $x = 0 \, and \, x = 5$.

Now, we shall find ω form

$$\nabla^2 \omega = 0, 0 < x < 5, 0 < y < 4 \tag{125}$$

Satisfying

$$\omega(0, y) = -v(0, y) = 0$$

$$\omega(5, y) = -v(5, y) = 0$$

$$\omega(x, 0) = -v(x, 0) = -(-5x + x^2)$$

$$\omega(x, 4) = -v(x, 4) = -(-5x + x^2)$$

The above conditions are obtained by using $Eq.(121),(124)$ and the given boundary conditions. By using the superposition principle the general solution of $Eq.(125)$ is found to be

$$\omega(x, y) = \sum_{n=1}^{\infty} \sin\left(\frac{n\pi x}{5}\right)\left[a_n \exp\left(\frac{n\pi y}{5}\right) + b_n \exp\left(-\frac{n\pi}{5}\right)\right] \tag{126}$$

Now, applying the non-homogeneous BC: $\omega(x,0)=-(-5x+x^2)$, we get, after renaming the constants, the equation

$$\omega(x,0)=-(-5x^2+x^2)=\sum A_N \sin\left(\frac{n\pi x}{5}\right)$$

Also, applying the non-homogeneous BC : $\omega(x,0)=-(-5x+x^2)$, we get , after renaming the constants , the equation

$$-(-5x+x^2)=\sum_{n=1}^{\infty}\left(a_n \cosh\frac{4n\pi}{5}+b_n \sinh\frac{4n\pi}{5}\right)\sin\frac{n\pi x}{5}$$

which gives

$$a_n=\frac{2}{5}\int_0^5 (5x-x^2)\sin\frac{n\pi x}{5}dx$$

Now. Integrating by parts, the right-hand side yields

$$a_n=\frac{2}{5}\left[(5x-x^2)\left(-\frac{5}{n\pi}\cos\frac{n\pi}{5}x\right)-(5-2x)\left(-\frac{5^3}{n^2\pi^2}\sin\frac{n\pi x}{5}\right)+(-2)\left(\frac{5^3}{n^3\pi^3}\cos\frac{n\pi}{5}x\right)\right]_0^5$$

$$=\frac{2}{5}\left[-\frac{2(5)^3}{n^3\pi^3}\cos n\pi+\frac{2}{n^3}\left(\frac{5^3}{\pi^3}\right)\right]$$

$$=\frac{4(5^2)}{\pi^3}\left[\frac{1}{n^3}-\frac{\cos n\pi}{n^3}\right]=\frac{4(5)^2}{\pi^3}\left[\frac{1}{n^3}-\frac{(-1)^n}{n^3}\right]$$

Hence,

$$a_n=\begin{cases}\dfrac{8(5^2)}{\pi^3 n^3}, & when\, n\, is\, odd\\ 0, & when\, n\, is\, even\end{cases} \tag{128}$$

Also, from $Eq.(127)$, we have

$$a_n\cosh\frac{4n\pi}{5}+b_n\sinh\frac{4n\pi}{5}=\frac{2}{5}\int_0^5(5x-x^2)\sin\left(\frac{n\pi}{5}x\right)dx=a_n$$

Therefore,

$$b_n = \frac{a_n \left[1 - \cosh\left(\frac{4}{5}n\pi\right)\right]}{\sinh\left(\frac{4}{5}n\pi\right)} \tag{129}$$

Substituting a_n, b_n from $Eq.(128)$ and (129) into $Eq.(126)$, we get

$$\omega(x, y)\sum_{1}^{\infty} a \left[\cosh\left(\frac{n\pi}{5}y\right)\sinh\left(\frac{4}{5}n\pi\right) - \cosh\left(\frac{4}{5}n\pi\right)\sinh\left(\frac{n\pi}{5}y\right) + \sinh\left(\frac{n\pi}{5}y\right)\sin\left(\frac{n\pi}{5}x\right)/\sinh\left(\frac{4}{5}n\pi\right)\right]$$

or

$$\omega(x, y) - \sum_{n=1}^{\infty}\left[\sinh\left\{\frac{n\pi}{5}(4 - y)\right\} + \sinh\left(\frac{n\pi}{5}y\right)\right]\sin\left(\frac{n\pi}{5}x\right)/\sinh\left(\frac{4}{5}n\pi\right)$$

Combining $Eq.(124)$, (128) and (130), the solution of the given Poisson equation is

$$u(x, y) = x(x - 5) + \frac{8\times 5^2}{\pi^3}\sum_{n=1}^{\infty}\left[\frac{\sinh(2n-1)\pi(4-y)/5 + \sinh\left[(2n-1)\frac{\pi y}{5}\right]}{\sinh\left[(2n-1)\frac{4}{5}\pi\right]}\right]$$

$$\times\left\{\frac{\sin\left[(2n-1)\frac{\pi x}{5}\right]}{(2n-1)^3}\right\}$$

Example. Let IR be a region bounded by ∂IR. Let $p(x, y, z)$ be any point in the interior of IR , as shown in Fig . Let ϕ be a harmonic function in $\overline{IR}$; also , let $\psi = \frac{1}{r}$ where r is the distance form p. Applying Green's second identity , show that

$$\phi(p) = \frac{1}{4\pi}\iint_{\partial IR}\left[\frac{1}{r}\frac{\partial\phi}{\partial n} - \phi\frac{\partial}{\partial n}\left(\frac{1}{r}\right)\right]ds$$

Solution. Since ψ processes a point of discontinuity in IR at $p(x, y, z)$, Green's second identity cannot be directly applied to $\phi \, and \, \psi$. However, $\psi = \frac{1}{r}$ is bounded in IR $-\sum_{\dot{o}}$ with the boundary $\partial IR \cup \partial \sum_{\dot{o}}$, where $\sum_{\dot{o}}$ is a sphere of radius ε with centre at p. Now applying Green's second identity (19) to functions $\phi \, and \, \psi \, in \, IR - \sum_{\dot{o}}$., we get

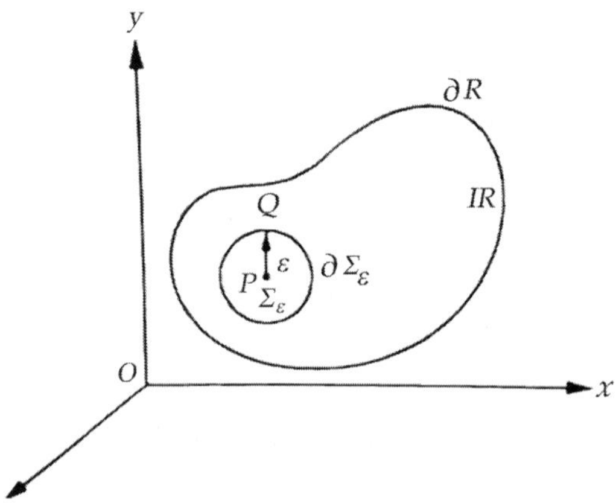

Figure 9

$$\iiint_{IR - \sum_{\dot{o}}} \left[\phi \nabla^2 \left(\frac{1}{r} \right) - \frac{1}{r} \nabla^2 \phi \right] dV = \iint_{\partial IR} \left[\phi \frac{\partial}{\partial n} \left(\frac{1}{r} \right) - \left(\frac{1}{r} \right) \frac{\partial \phi}{\partial n} \right] ds + \iint_{\partial \sum_{\dot{o}}} \phi \frac{\partial}{\partial n} \left(\frac{1}{r} \right) dS - \iint_{\partial \sum_{\dot{o}}} \left(\frac{1}{r} \right) \left(\frac{\partial \phi}{\partial n} \right) dS$$

(131)

From the right-hand side of $Eq.(131)$, we observe that the last two integrals depend only on ε. But in the direction of the exterior normal to $\partial \sum_{\dot{o}}$., we find that

$$\frac{\partial}{\partial n} \left(\frac{1}{r} \right) \bigg|_{\partial \sum_{\dot{o}}} = - \frac{\partial}{\partial r} \left(\frac{1}{r} \right) \bigg|_{r = \varepsilon} = \frac{1}{\varepsilon^2}$$

Therefore,

$$\iint\limits_{\partial\sum_{\dot{o}}} \phi \frac{\partial}{\partial n}\left(\frac{1}{r}\right) dA = \frac{1}{\varepsilon^2} \iint\limits_{\partial\sum_{\dot{o}}} \phi\, dS = \frac{4\pi\varepsilon^2}{\varepsilon^2} \phi(Q) = 4\pi\phi^*(Q)$$

where $\phi^*(Q)$ is the average value of $\phi(Q)\, on\, \partial\sum_{\dot{o}}$. Further, the third integral

$$-\iint\limits_{\partial\sum_{\dot{o}}} \left(\frac{1}{r}\right)\frac{\partial\phi}{\partial n} dS = -\frac{1}{\varepsilon} \iint\limits_{\partial\sum_{\dot{o}}} \left(\frac{\partial\phi}{\partial n}\right) dS = -4\pi\varepsilon \left(\frac{\partial\phi^*}{\partial n}\right)$$

Where $\left(\frac{\partial\phi}{\partial n}\right)^*$ is an average value of the normal derivative on $\partial\sum_{\dot{o}}$. Substituting these results and suing the fact that $\nabla^2\left(\frac{1}{r}\right) = 0\, in\, IR - \sum_{\dot{o}}$, we obtain

$$\iiint\limits_{IR-\sum_{\dot{o}}} \left(-\frac{1}{r}\right)\nabla^2\phi\, dV = \iint\limits_{\partial IR}\left[\phi\frac{\partial}{\partial n}\left(\frac{1}{r}\right) - \left(\frac{1}{r}\right)\frac{\partial\phi}{\partial n}\right] dS + 4\pi\phi^*(p) + 4\pi\varepsilon\left(\frac{\partial\phi}{\partial n}\right)^*$$

(132)

Now, taking the limit as $\varepsilon \to 0$, and using the fact that ϕ is harmonic in $IR - \sum_{\dot{o}}$ we arrive at the fundamental result

$$\phi(p) - \frac{1}{4\pi}\iint\limits_{\partial IR}\left[\left(\frac{1}{r}\right)\frac{\partial\phi}{\partial n} - \phi\frac{\partial}{\partial n}\left(\frac{1}{r}\right)\right] dS$$

Thus, the value of a harmonic function at any point of IR can be obtained in terms of the values of ϕ and $\partial\phi / \partial n$ on the boundary ∂IR of the region IR.

4.14. SUMMARY AND DISCUSSIONS

In this chapter we discussed various properties and technique for solving Laplace and Poisson equations which are elliptic in nature. Various physical phenomena are governed by the well-known Laplace and poison equations. A few of them, frequently encountered in applications are: study heat condition, magnetic potential , torsion of prismatic shaft, etc.

$\nabla^2 = div\nabla = \dfrac{\partial^2}{\partial x^2}+\dfrac{\partial^2}{\partial y^2}+\dfrac{\partial^2}{\partial z^2}$ is called the Laplace operator.

$\nabla^2 V = 0$ is called the Laplace equation

$\nabla^2 V = -4\pi Gp$ is known as *poisson's equation* .

4.14.1. Harmonic Functions

Solutions of Laplace equation1 are called harmonic functions which possess a number of interesting properties, and they are presented in the following theorems.

Theorems. If a harmonic function vanishes everywhere on the boundary , then it is identically zero everywhere.

If ϕ is a harmonic function in IR and $\partial\phi/\partial n = 0$ on ∂IR, then ϕ is a constant in $\overline{IR}$.

4.14.2. Separation Of Variables

The method of Separation of variables is applicable to a large number of classified linear homogeneous equations. The choice of the coordinate system in general depends on the shape of the boundary

4.14.3. Dirichlet Problem For A Rectangle

The Dirichlet problem for a rectangle is defined as follows:

PDE: $\nabla^2 u = 0 , \quad 0 \le x \le a , 0 \le y \le b$

BCs: $u(x,b) = u(a,y) = 0, u(0,y) = 0 , u(x,0) = f(x)$ (41)

This is an interior Dirichlet problem. The general solution of the governing PDE, using the method of variables separable

4.14.4. The Neumann Problem For A Rectangle

The Neumann problem for a rectangle is defined as follows

$PDE: \nabla^2 u = 0,\ 0 \le x \le a, 0 \le y \le b$

$BCs: u_x(0, y) = u_x(a, y) = 0, u_y(x, y) = 0\ , u_y(x, b) = f(x)$

4.14.5. Interior Dirichlet Problem For A Circle

The Dirichlet problem for the circle is defined as follows:

PDE: $\nabla\ \ u = 0,\ 0 \le r \le a, 0 \le 0 \le 2$

BC: $u(a, \theta) = f(\theta), 0 \le \theta \le 2\pi$

$$u(r,\theta) = \frac{1}{2\pi}\int_0^{2\pi} \frac{(a^2 - r^2) f(\phi)}{\left[a^2 - 2ar\cos(\phi - \theta) + r^2\right]} d\phi$$

This is known as Poisson's integral formula for a circle, which gives a unique solution for the Dirichlet problem. The solution can be interpreted physically in many ways: it can be thought of as finding the potential $u(r,\theta) as$ a weighted average of the boundary potential $f(\phi)$ weighted by the Poisson kernel p, given by

$$p = \frac{a^2 - r^2}{\left[a^2 - 2ar\cos(\phi - \theta) + r^2\right]}$$

It can also be thought of as a steady temperature distribution $u(r,\theta)$ in a circular disc, when the temperature u on its boundary ∂IR is given by $u = f'(\phi)$ which is independent of time.

4.14.6. Exterior Dirichlet Problem For A Circle

The exterior Dirichlet problem is described by

$PDE: \nabla^2 u = 0$

$BC: u(a, \theta) = f(\theta)$, u must be bounded as $r \to \infty$

4.14.7. Interior Neumann Problem For A Circle

The interior Neumann problem for a circle is described by

$$PDE: \nabla^2 u = 0, \quad 0 \le r < a \; ; 0 \le \theta 2\pi$$

$$BC: \frac{\partial u}{\partial n} = \frac{\partial u(a, r)}{\partial r} = g(\theta), r = a$$

4.14.8. Laplace Equation In Cylindrical Coordinates

The Laplace equation in cylindrical coordinates assumes the following form:

$$\nabla^2 u = u_{rr} + \frac{1}{r} u_r + \frac{1}{r^2} u_{\theta\theta} + u_{zz} = 0$$

4.14.9. Laplace Equation In Spherical Coordinates

In example 3 the Laplace equation is expressed in spherical and has the following form:

$$\nabla^2 u = \frac{\partial}{\partial r}\left(r^2 \frac{\partial u}{\partial r} \right) + \frac{1}{\sin\theta} \frac{\partial}{\partial \theta}\left(\sin\theta \frac{\partial u}{\partial \theta} \right) + \frac{1}{\sin^2\theta} \frac{\partial^2 u}{\partial \phi^2} = 0$$

CHAPTER 5

HYPERBOLIC DIFFERENTIAL EQUATION

CONTENTS

5.1 INTRODUCTION

One of the most important and typical homogenous hyperbolic differential equations is the wave equation. It is of the form

$$\frac{\partial^2 u}{\partial r^2} = c^2 \nabla^2 u \qquad (1)$$

where c is the wave speed. This differential equation is used in many branches of physics and longitudinal vibrations in a bar , propagation of sound waves, electromagnetic waves, sea waves, elastic waves in solids, and surface waves as in earthquakes. The solution of a waves equation is called a wave function.

An example for inhomogeneous wave equation is

$$\frac{\partial^2 u}{\partial t^2} - c^2 \nabla^2 u = F \qquad (2)$$

where F is a given function of spatial variables and time. In physical problem F represents an external driving force such as gravity force. Another related equation is

$$\frac{\partial^2 u}{\partial t^2} + 2\gamma \frac{\partial u}{\partial t} - c^2 \nabla^2 u = F \qquad (3)$$

where γ is a real positive constant. This equation is called a wave equation with damping term. The amplitude of which decreases exponentially as t increases . , we shall derive the partial differential equation describing the transverse vibration of a string .

5.1.1. Derivation of one-Dimensional wave equation

Suppose a flexible string is stretched under tension τ between two points at a distance L apart as shown in . We assumed the following

- The mention takes place in one planes only and in the plane each particle moves in a direction perpendicular to the equilibrium position of the string.
- The tension τ in the string is constant.
- The gravitational force is neglected as compared with tension τ of the string.
- The slope of the deflection curve is a small .

Let two fixed ends of the string be at the origin O and $A(L,0)$ which lies along the $x-axis$ in its equilibrium position. Consider an infinitesimal segment PQ of the string. Let ρ be the mass per unit length of the string. If the string is set vibrating in the $xy-plane$, the subsequent displacement, y from the equilibrium position of a point P of the string will be a function of and time t, while an element of lenfght dx is stretched into an element of length ds given by

$$ds=\sqrt{1+\left(\frac{\partial y}{\partial x}\right)^2}\,dx=\left[1+\frac{1}{2}\left(\frac{\partial y}{\partial x}\right)^2\right]dx$$

The elementary elongation is given by

$$dL=ds-dx=\frac{1}{2}\left(\frac{\partial y}{\partial x}\right)^2 dx$$

While the work done by this element against the tension τ is

$$\frac{1}{2}\tau\left(\frac{\partial y}{\partial x}\right)^2 dx$$

Therefore, the total work done, W, for the whole string is

$$W=\frac{1}{2}\int_0^L \tau\left(\frac{\partial y}{\partial x}\right)^2 dx$$

If U is the potential energy of the string, then

$$U=W=\frac{1}{2}\int_0^L \tau\left(\frac{\partial y}{\partial x}\right)^2 dx$$

Also, the total kinetic energy of the string is given by

$$T=\frac{1}{2}\int_0^L \rho\left(\frac{\partial y}{\partial t}\right)^2 dx$$

Using Hamilton's principle, we have

$$\delta\int_{t_0}^{t_1}(T-U)\,dt=0$$

i.e.,

$$\int_{t_0}^{t_1}(T-U)\,dt$$

Is stationary. In other words,

$$\frac{1}{2}\int_{t_0}^{t_1}\int_{0}^{L}\left[\rho\left(\frac{\partial y}{\partial t}\right)^2-\tau\left(\frac{\partial y}{\partial x}\right)^2\right]dx\,dt$$

$$\iint F(x,t,y,y_x,y_t)\,dx\,dt$$

Noting that x and t are independent variables, form the Euler-Ostrogradsky equation, we have

$$\frac{\partial F}{\partial y}-\frac{\partial}{\partial t}\left(\frac{\partial F}{\partial y_t}\right)-\frac{\partial}{\partial x}\left(\frac{\partial F}{\partial y_x}\right)=0$$

which gives

$$\frac{\partial}{\partial t}\left(\rho\frac{\partial y}{\partial t}\right)-\frac{\partial}{\partial x}\left(\tau\frac{\partial y}{\partial x}\right)=0$$

If the string is homogenous, then ρ and τ are constants, in which case the governing equation representing the transverse vibration of a string is given by

$$\frac{\partial^2 y}{\partial t^2}=c^2\frac{\partial^2 y}{\partial x^2} \tag{4}$$

where

$$c^2=\frac{\tau}{\rho} \tag{5}$$

Example. Consider Maxwell's equations of electromagnetic theory given by

$$\nabla.E=4\pi\rho$$

$$\nabla . H = 0$$

$$\nabla \times E = -\frac{1}{c}\frac{\partial H}{\partial t}$$

$$\nabla \times H = \frac{4\pi i}{c} + \frac{1}{c}\frac{\partial E}{\partial t}$$

where E is an electric field, ρ is electric density, H is the magnetic field, i is the current density, and c is the velocity of light . Show that in the absence of charges, i.e., when $\rho = i = 0,\ E\, and\, H$ satisfy the wave equations.

Solution. Given

$$curl\, E = \nabla \times E = -\frac{1}{c}\frac{\partial H}{\partial t}$$

Taking its curl again, we get

$$\nabla \times (\nabla \times E) = \nabla \times \left(-\frac{1}{c}\frac{\partial H}{\partial t} \right) = -\frac{\partial}{\partial t}\left(\frac{1}{c}\nabla \times H \right)$$

$$= -\frac{1}{c}\frac{\partial}{\partial t}\left(-\frac{1}{c}\frac{\partial H}{\partial t} \right) = -\frac{\partial}{\partial t}\left(\frac{1}{c}\nabla \times H \right)$$

$$= -\frac{1}{c}\frac{\partial}{\partial t}\left(\frac{1}{c}\frac{\partial E}{\partial t} \right) = -\frac{1}{c^2}\frac{\partial^2 E}{\partial t^2}$$

Moreover, using the identity

$$\nabla \times (\nabla \times E) = \nabla\, (\nabla . E) - \nabla^2 E = -\nabla^2 E$$

It follows directly that

$$\nabla^2 E - \frac{1}{c^2}\frac{\partial^2 E}{\partial t^2}$$

Similarly, we can observe that the magnetic field H also satisfies

$$\nabla^2 H = \frac{1}{c^2}\frac{\partial^2 H}{\partial t^2}$$

which is also a wave equation.

5.2. SOLUTION OF ONE-DIMENSIONAL WAVE EQUATION BY CANONICAL REDUCTION

The one-dimensional wave equation is

$$u_{tt} - c^2 u_{xx} = 0 \tag{6}$$

Choosing the characteristic lines

$$\xi = x - ct \ , \ \eta = x + ct \tag{7}$$

The chain rule of partial differential gives

$$u_x = u_\xi \xi_x + u_\eta \eta_x = u_\xi + u_\eta$$

$$u_t = u_\xi \xi_t + u_\eta \eta_t = c\left(u_\eta - u_\xi\right)$$

In the operator notation we have

$$\frac{\partial}{\partial x} = \frac{\partial}{\partial \xi} + \frac{\partial}{\partial \eta} \ , \ \frac{\partial}{\partial t} = c\left(\frac{\partial}{\partial \eta} - \frac{\partial}{\partial \xi}\right)$$

Thus, we get

$$\frac{\partial^2 u}{\partial x^2} = \left(\frac{\partial}{\partial \xi} + \frac{\partial}{\partial \eta}\right)^2 u = u_{\xi\xi} + 2u_{\xi\eta} + u_{\eta\eta} \tag{8}$$

$$\frac{\partial^2 u}{\partial t^2} = c^2\left(u_{\xi\xi} - 2u_{\xi\eta} - u_{\eta\eta}\right) \tag{9}$$

Substituting Eq. (8) and (9) into Eq. (6) , we obtain

$$4u_{\xi\eta} = 0 \tag{10}$$

Integrating, we get

$$u(\xi, \eta) = \phi(\xi) + \psi(\eta),$$

where ϕ and ψ are arbitrary functions. Replacing ξ *and* η as defined in Eq., we have the general solution of the wave equation in the form

$$u(x, t) = \phi(x - ct) + \psi(x + ct) \tag{11}$$

The two terms a wave traveling to the right with speed c whose shapes does not change as it travels, the initial shape being given by a known function $\phi(x)$. In fact, by setting $t=0$ in the form argument of ϕ, it can be observed that the initial wave profile is given by

$$u_1(x,0)=\phi(x)$$

At time $t=\frac{1}{c}$,

$$u_1\left(x,\frac{1}{c}\right)=\phi(x-1)$$

Let $x'=x-1$. Then $\phi(x-1)=\phi(x')$. That is, the same shape is retained even if the origin is shifted by one unit along the $x-axis$. in other words, the graph of $u_1\left(x,\frac{1}{c}\right)$ is the same as the graph of the original wave profile translated one unit to the right . At $t=\frac{2}{c}$, the graph of $u_1\left(x,\frac{2}{c}\right)$ is the graph of the wave profile translated two units to the right. Thus, in particular, at $t=1, we have u_1(x,1)=\phi(x-c)$. Hence in one unit of time, the profile has moved c units to the right. Therefore, c is the waves speed or speed of propagation Using similar argument, we can conclude that the equation $u_2(x,t)=\psi(x+ct)$ is also a wave profile travelling to the left with speed c along the $x-axis$. Hence the general solution (11) of one-dimensional wave equation represents the superposition of two arbitrary wave profiles, both of which are travelling with a common speed but in the opposite directions along the $x-axis$, while their forms remain unaltered as they travel. This situation us described in

Let k be an arbitrary real parameter. Consider then

$$u(x,t)=\phi[k(x-ct)]+\psi[k(x+ct)] \tag{12}$$

This is also a solution of the one-dimensional wave equation. Further, let $\omega=kc$. Then

$$u(x,t)=\phi[(kx-\omega t)]+\psi[k+(kx+\omega t)] \tag{13}$$

A function of the type given in Eq. (13) is a solution of one-dimensional wave equation if $\omega = kc.$ Therefore, waves travelling with speeds which are not the same as c cannot be described by the solution of the wave equation (6). Here, $(kx+\omega t)$ is called the phase for the left travelling wave. We have already noted that $x \pm ct$ are the characteristics of the one-dimensional wave equation.

Example. Obtain the periodic solution of the wave equation in the form $u(x,t) = Ae^{i(kx \pm \omega t)}$

where $i = \sqrt{-1}, k = \pm\frac{\omega}{c}, A$ is constant; and hence define various term involved in wave propagation.

Solution. Let $u(x,t) = f(x)\, e^{\pm i\omega t}$ be a solution of the wave equation

$$u_{tt} = c^2 u_{xx}$$

then

$$u_{xx} = f''(x)\, e^{\pm i\omega t},\ u_{tt} = f(x)\omega^2\, e^{\pm i\omega t}$$

Substituting into the wave equation , we get

$$f''(x) + \frac{\omega^2}{c^2} f(x) = 0$$

Its general solution is found to be

$$f(x) = c_1 \exp\left[i\left(\frac{\omega}{c}\right)x\right] + c_2 \exp\left[-i\left(\frac{\omega}{c}\right)x\right]$$

Therefore, the required solution of the wave equation is

$$u(x,t) = \left[c_1 \exp\left\{i\left(\frac{\omega}{c}\right)x\right\} + c_2 \exp\left\{-i\left(\frac{\omega}{c}\right)x\right\}\right] e^{\pm i\omega t}$$

Since $k = \pm\frac{\omega}{c},$ the time- dependent wave function are of the form

$$u(x,t) = Ae^{i(kx \pm \omega t)}$$

Hence, $u(x,t)=Ae^{i(kx\pm\omega t)}$ is a solution of the wave equation, and is called a wave function. It is also called a plane harmonic wave or monochromatic wave. Here, A is called the amplitude, ω the angular or circular frequency, and k is the wave number, defined as the number of waves per unit distance. By Taking the real and imaginary parts of the solution. We find the linear combination of term of the form

$$A\cos(kx\pm\omega t),\quad A\sin(kx\pm\omega t)$$

Representing periodic plane waves. For instance, consider the function $u(x,t)=A\sin(kx-\omega t)$. This is a sinusoidal wave profile moving towards the right along the $x-axis$ with speed c. Defining the wave length λ as the length over which one full cycle is completed . we have $\lambda=2\pi/k$. Thereby implying that $k=2\pi/\lambda$.

Suppose an observe is stationed at a fixed point x_0; then ,

$$u\left(x_0\ ,t+\frac{\lambda}{c}\right)=A\ \sin\left(kx_0\ -\omega t-\omega\frac{\lambda}{c}\right)$$

$$=A\sin(kx_0-\omega t-2\pi)=A\sin(kx_0-\omega t)$$

Thus, we have

$$u\left(x_0,t+\frac{\lambda}{c}\right)=u(x_0\ ,t)$$

Hence, exactly one computer wave passes the observer in time $T=\frac{\lambda}{c}$, which is called the period of the wave. The reciprocal of the period is called frequency and is denoted by

$$f=\frac{1}{T}$$

The function, $u\ A\cos(kx-\omega t)=A\sin\left(\frac{\pi}{2}+kx-\omega t\right)$, also represents a wave train except that it differ in phase by $\frac{\pi}{2}$ form the sinusoidal wave.

Now consider the superposition of the sinusoidal waves having the same amplitude, speed , frequency , but moving in opposite directions. Thus, we have

$$u(x,t)=A\sin[k(x-ct)]+A\sin[k(x+ct)]$$

$$2A\sin kx\cos(kct)=2A\cos(kct)\sin kx$$

Its amplitude factor $[2A\cos(kct)]$ varies sinusoidal with frequency ω. This situation is described as standing wave. The point $x_n=\frac{n\pi}{k}, n=0,\pm 1,\pm 2,\ldots$ are called nodes . No displacement takes place at a node . Therefore,

$$u(x_n,t)=0 \ for\ all\ t$$

The nth standing wave profile will have $(n-1)$ equally spaced nodes in a given interval as shown in

5.3. THE INITIAL VALUE PROBLEM; D'ALEMBERT'S SOLUTION

Consider the initial value problem of Cauchy type described as

$$PDE: u_{tt}-c^2u_{xx}=0, \quad -\infty<x<\infty, t\geq 0 \tag{14}$$

$$\text{ICs: } u(x,0)=\eta(x), u_t(x,0)=v(x) \tag{15}$$

where the curve on which the initial data $\eta(x) and\, v(x)$ are prescribed is the x $axis.$ (x) and $v(x)$ are assumed to be twice continuously differentiable . Here, the string considered is of an infinite extent. Let $u(x,t)$ denote the displacement for any x and t . At $t=0$, let the displacement and velocity of the string be prescribed. We have already noted that the general solution of the wave equation is given by

$$u(x,t)=\phi(x+ct)+\psi(x-ct) \tag{16}$$

where $\phi\, and\, \psi$ are arbitrary functions. Substituting the ICs (15) into Eq. (16), we obtain

$$\phi(x)+\psi(x)=\eta(x) \tag{17}$$

Integrating the second equation of (17) , we have

$$\phi(x)-\psi(x)=\frac{1}{c}\int_0^x v(\xi)d\xi$$

Addition and substituting of this equation with the first relation of Eq. (17) yield

$$\phi(x)=\frac{\eta(x)}{2}+\frac{1}{2c}\int_0^x v(\xi)\,d\xi$$

$$\psi(x)=\frac{\eta(x)}{2}-\frac{1}{2c}\int_0^x v(\xi)d\xi$$

Respectively, substituting these relations for $\phi(x)$ and x $\psi(x)$ *into Eq.*(16), we at once have

$$u(x,t)=\frac{1}{2}\left[\eta(x+ct)+\eta(x-ct)\right]+\frac{1}{2c}\int_{x-ct}^{x+ct} v(\xi)d\xi \tag{18}$$

Thus is known as the D'ALEMBERT's Solution of the one-dimensional wave equation. If $v=0$,*i.e.*, if the string is released form rest, the required solution is

$$u(x,t)=\frac{1}{2}\left[\eta(x+ct)+\eta(x-ct)\right] \tag{19}$$

The D'ALEMBERT's solution has an interesting interpretation as given in figure

Consider the $xt-plane$ and a point $p(x_0,t_0)$. Draw two characteristics through p backwards, until they intersect the initial line, i.e., the $x-axis$ at A and B. The equation of these two characteristics are

$$x\pm ct=x_0\pm ct_0$$

Equation (18) reveals that the solution $u(x,t)\,at\,P(x_0,t_0)$ can be obtained by averaging the value of η at A and B and integrating $v\,along$ the $x-axis$ between $A\,and\,B.$ Thus, to find the solution of the eave equation at a given point p in the $xt-plane$ we should know the initial data on the segment $AB\,of\,the$ initial line which is obtained by drawing the characteristics backward from p to the initial line. Here the segment AB of the initial line, on which the value of $u(x,t)\,at\,p$ depend is called the domain of dependence of p, and the triangle PAB is called the characteristics triangle , which is also called the domain of determinacy of the interval.

Example. A stretched string of finite length L is held fixed at its ends and is subjected to an initial displacement $u(x,0)=u_0\sin\left(\frac{\pi x}{L}\right).$ The string is released from this position with zero initial velocity. Find the resultant time dependent motion of the string.

Solution. One of the practical application of the theory of wave motion is the vibration of a stretched string, say, that of a musical instrument. In the present problem , let us consider a stretched string of finite L, which is subjected to an initial disturbance. The governing equation of motion is

PDE: $u_{tt}-c^2u_{xx}=0\,,0\le x\le L\,,\,t>0$ (1)

BCs: $u(0,t)=u(L,t)=0$ (2)

ICs: $u(x,0)=u_0\sin\left(\frac{\pi x}{l}\right),$ (3)

$\frac{\partial u}{\partial t}(x,0)=0$ (4)

We have shown the solution of the one-dimensional wave equation by canonical reduction as

$$u(x,t)=\phi(x-ct)+\psi(x+ct) \quad (5)$$

One of the known method for solving this problem is based on trial function approach. Let us choose a trial function of the form

$$u(x,t)=A\left[\sin\frac{\pi}{L}(x+ct)+\sin\frac{\pi}{L}(x-ct)\right] \tag{6}$$

where A is an arbitrary constant. Now, we rewrite Eq. (6) as

$$u(x,t)=2A\sin\left(\frac{\pi x}{L}\right)\cos\left(\frac{c\pi t}{L}\right) \tag{7}$$

Obviously, (7) satisfies the initial condition (3) with $A=\frac{u_0}{2}$, while the second initial condition (4) is satisfied identically. In fact Eq. (7) also satisfies the boundary condition (2). Therefore, the final solution is found to be

$$u(x,t)=u_0\sin\left(\frac{\pi x}{L}\right)\cos\left(\frac{cx\pi}{L}\right) \tag{8}$$

It may be noted that the trial function approach is easily adoptable if the initial condition is specified as a sin function . However, it is difficult if the initial conditions are specified as a general function such as $f(x)$. In such case, it is better to follow variable separable method as explained in previous sections .

Example. Solve the following initial value problem of the wave equation (Cauchy problem), described by the inhomogeneous wave equation

PDE: $u_{tt}-c^2u_{xx}=f(x,t)$

subject to the initial conditions

$u(x,0)=\eta(x)$, $u_t(x,0)=v(x)$

Solution. To make the task easy, we shall set $u=u_1+u_2$, so that u_1 is a solution of the homogenous wave equation subject to the general initial conditions given above. Then u_2 will be a solution of

$$\frac{\partial^2 u_2}{\partial t^2}-c^2\frac{\partial^2 u_2}{\partial x^2}=f(x,t) \tag{20}$$

subject to the homogenous ICs

$$u_2(x,0)=0,\ \frac{\partial u_2}{\partial t}(x,0)=0 \tag{21}$$

To obtain the value of u at $p(x_0,t_0)$, we integrate the partial differential equation (20) over the region IR as shown in , we obtain

$$\int_{IR}\int\left(\frac{\partial^2 u_2}{\partial t^2}-c^2\frac{\partial^2 u_2}{\partial x^2}\right)dx\,dt=\int_{IR}\int f(x,t)\,dx\,dt$$

Using Green's theorem in a plane to the left-hand side of the above equation to replace the surface integral over IR by a line integral around the boundary ∂IR of IR, the above equation reduces to

$$-\int_{IR}\int\left[\frac{\partial}{\partial x}\left(c^2\frac{\partial u_2}{\partial x}\right)-\frac{\partial}{\partial t}\left(\frac{\partial u_2}{\partial t}\right)\right]dx\,dt=\int_{IR}\int f(x,t)\,dx\,dt$$

And finally to

$$\int_{\partial IR}\left(\frac{\partial u_2}{\partial t}dx+c^2\frac{\partial u_2}{\partial x}\right)dt=\int_{IR}\int f(x,t)\,dx\,dt \tag{22}$$

Now, the boundary ∂IR comprises three segments $BP, PA\,and\,AB$. Along BP, $\frac{dx}{dt}=-c;$ along $PB.\frac{dx}{dt}=c.$ Using these results , Eq. (22)

$$\int_{BP}c\left(\frac{\partial u_2}{\partial t}dt+\frac{\partial u_2}{\partial x}dx\right)-\int_{PA}c\left(\frac{\partial u_2}{\partial t}dt+\frac{\partial u_2}{\partial x}dx\right)$$

$$-\int_{AB}\left(\frac{\partial u_2}{\partial t}dx+c^2\frac{\partial u_2}{\partial x}dt\right)=\int_{IR}\int f(x,t)\,dx\,dt$$

The integrands of the first two integrals are simply the total differentials, while in the third integral, the first term vanishes on AB in view of the second IC in Eq. (21), and the second term also vanishes because AB is directed along the $x-axis$ on which $\frac{dt}{dx}=0.$ Then we arrive at the result

$$\int_{BP}c\,du_2-\int_{IR}\int f(x,t)\,dx\,dt$$

which can be rewritten as

$$cu_2(p)-cu_2(B)+cu_2(p)-cu_2(A)=\int\limits_{IR}\int f(x,t)\,dxdt \tag{23}$$

Using the first IC of Eq. (21), we get $u_2(A)=u_2(B)=0,$ and hence Eq. (23) becomes

$$u_2(p)=\frac{1}{2c}\int\limits_{IR}\int f(x,t)\,dx\,dt$$

with the help of , we deduce

$$u_2(p)=\frac{1}{2c}\int_0^{t_0}\int_{x_0-ct_0+ct}^{x_0+ct_0-ct} f(x,t)\,dx\,dt \tag{24}$$

Now, using the fact that $u=u_1+u_2$, as also using Eq. (24) and D'Alembert's solution (18) the required solution of the inhomogeneous wave equation subject to the given ICs is given by

$$u(x,t)=\frac{1}{2}\{\eta(x+ct)+\eta(x-ct)+\frac{1}{c}\int_{x-ct}^{x+ct} v(\xi)\,d\xi+\frac{1}{2}\int_0^{t_0}\int_{x_0-ct_0+ct}^{x_0+ct_0-ct} f(x,t)\,dxdt \tag{25}$$

The solution is known as the Riemann-Volterra solution.

Following Tychonov and Samarski, it is known that transverse vibration of a string is normally generated in musical instruments. We distinguish the string instrument depending on whether the string is plucked as in the case of guitar or struck as in the case of harmonic or piano. In the case of string which are struck we give a fixed initial velocity but does not undergo any initial displacement. In the case of plucked instruments, the string vibrate form a fixed initial displacement without any initial velocity .

Let a thin homogeneous string which is perfectly flexible under uniform tension lie in it equilibrium position along the $x-axis$. The ends of the string are fixed at $x=0$ and $x=L$. The string is pulled aside a short distance and released. If no external forces are present which correspond to the case of free vibrations, the subsequent motion of the string is described by the solution $u(x,t)$ of the following problem:

PDE: $u_{tt}-c^2u_{tt}=0,\ \ 0\le x\le L, t>0$ (26)

BCs: $u(0,t)=0,\ \ t>0$

$$u(L,t) = 0,\ t > 0 \tag{27}$$

ICs: $u(x,0) = f(x),\quad u_t(x,0) - g(x)$ (28)

To obtain the variables separable solution, we assume

$$u(x,t) = X(x)T(t) \tag{29}$$

Substituting into Eq. (26), we obtain

$$X\frac{d^2T}{dt^2} = c^2\,T\frac{d^2X}{dx^2}$$

i.e.,

$$\frac{d^2X/dx^2}{X} = \frac{d^2/dt^2}{c^2T} = k$$ (a separation constant)

Case I. When $K > 0$, we have $k = \lambda^2$. Then

$$\frac{d^2X}{dx^2} - \lambda^2 X = 0$$

$$\frac{d^2T}{dt^2} - c^2\,\lambda^2 T = 0$$

Their solution can be put in the form

$$X = c_1e^{\lambda x} + c_2e^{-\lambda x} \tag{30}$$

$$T = c_3e^{c\lambda t} + c_4\,e^{-c\lambda t} \tag{31}$$

Therefore,

$$u(x,t) = \left(c_1e^{\lambda x} + c_2e^{-\lambda x}\right)\left(c_3e^{c\lambda t} + c_4e^{-\lambda t}\right) \tag{32}$$

Now, use the BCs:

$$u(0,t) = 0 = (c_1 + c_2)\left(c_3e^{\lambda t} + c_4e^{-\lambda t}\right) \tag{33}$$

which imply that $c_1 + c_2 = 0$. Also, $u(L,t) = 0$ gives

$$c_1e^{-\lambda L} + c_2e^{-\lambda L} = 0 \tag{34}$$

Equation (33) and (34) possess a non- trivial solution if

$$\begin{vmatrix} 1 & 1 \\ e^{\lambda L} & e^{-\lambda L} \end{vmatrix} = e^{-\lambda L} - e^{\lambda L} = 0$$

or

$$1 - e^{2\lambda L} = 0 \ \ implying \, e^{2\lambda L} = 1 \ or \ \lambda L = 0$$

This implies that $\lambda = 0, since \, L$ cannot be zero, which is against the assumption as in the Case I, Hence, the solution is not acceptable.

Case II. Let $K = 0$. Then we have

$$\frac{d^2X}{dx^2} = 0, \ \frac{d^2T}{dt^2} = 0$$

Their solutions are found to be

$$X = Ax + B, \quad T = ct + D$$

Therefore, the required solution of PDE (26) is

$$u(x,t) = (Ax + B)(ct + D)$$

Using the BCs, we have

$$u(0,t) = 0 = B(ct + D), \ implying \ B = 0$$

$$u(L,t) = 0 = AL(ct + D), \ implying \ A = 0$$

Hence, only a trivial solution is possible. Since we are looking for a non-trivial solution, consider the following case.

Case III. When $k = 0,$ say $K = -\lambda^2$, the differential equations are

$$\frac{d^2X}{dx^2} + \lambda^2 X - 0, \ \frac{d^2T}{dt^2} + c^2\lambda^2 T = 0$$

Their general solution gives

$$u(x,t) = (c_1 \cos \lambda x + c_2 \sin \lambda x)(c_3 \cos c\lambda t + c_4 \sin c\lambda t) \qquad (35)$$

Using the BC: $u(0,t) = 0$ we obtain $c_1 = 0.$ Also, using the BC: $u(L,t) = 0,$ we get $\sin \lambda L = 0$ implying that $\lambda = \frac{n\pi}{L}, \ n = 1, 2, \ldots.$ which are the eigenvalue. Hence the possible solution is

$$u_n(x,t) = \sin\frac{n\pi x}{L}\left(A\cos\frac{n\pi ct}{L} + B\sin\frac{n\pi ct}{L}\right),\ n = 1, 2, \ldots \tag{36}$$

Using the superposition principle, we have

$$u(x,t) = \sum_{n=1}^{\infty} \sin\frac{n\pi x}{L}\left(A_n\cos\frac{n\pi ct}{L} + B_n\sin\frac{n\pi ct}{L}\right) \tag{37}$$

The initial condition give

$$u(x,0) = f(x) = \sum_{n=1}^{\infty} A_n \sin\frac{n\pi x}{L}$$

which is half – range Fourier sine series, where

$$A_n = \frac{2}{L} f(x)\sin\frac{n\pi x}{L}dx \tag{38}$$

Also,

$$u_t(x,0) = g(x) = \sum_{n=1}^{\infty} B_n \sin\frac{n\pi x}{L}\left(\frac{n\pi}{L}c\right)$$

which is also a half range sine series , where

$$B_n = \frac{2}{n\pi c}\int_0^L g(x)\sin\frac{n\pi x}{L}dx$$

Hence the required physically meaningful solution is obtained from Eq. (37), where A_n and B_n are given by Eq. (38) and (39). $u_n(x,t)$ *given by Eq.* (36) are called normal modes vibrations and $\frac{n\pi c}{L} = \omega_n,\ n = 1, 2, \ldots$ are called frequencies .

Example. Obtain the solution of the wave equation

$$u_{tt} = c^2 u_{xx}$$

under the following conditions:

(i) $u(0,t) = u(2,t) = 0$

(ii) $u(x,0) = \sin^3\frac{\pi x}{2}$

(iii) $u_t(x,0)=0.$

Solution. We have noted in Example 4 that the physically acceptable solution of thc wave equation is given by Eq. (35), and is of the form

$$u(x\ ,t)=\left(c_1\cos\lambda x\, c_2\sin\lambda x\right)\left[c_3\cos\left(c\lambda t\right)+c_4\sin\left(c\lambda t\right)\right]$$

Using the condition $u(0\ ,t),$ we obtain $c_1=0.$ Also, condition (iii) implies $c_4=0.$ The condition $u(2,t)=0$ gives

$\sin 2\lambda=0,$

Implying that

$$\lambda=\frac{n\pi}{2},\quad n-1,2\ ,\ldots$$

Finally, using condition (ii), we obtain

$$\sum_{n=1}^{\infty}A_n\sin\frac{n\pi x}{2}\cos\frac{n\pi ct}{2} \qquad (40)$$

Finally, using the condition (ii), we obtain

$$\sum_{n=1}^{\infty}A_n\sin\frac{n\pi x}{2}=\sin^3\frac{\pi x}{2}=\frac{3}{4}\sin\frac{\pi x}{2}-\frac{1}{4}\sin\frac{3\pi x}{2}$$

which gives $A_1=\dfrac{3}{4}$.

5.4. SUMMARY AND DISCUSSION

One of the most important and typical homogenous hyperbolic differential equations is the wave equation. It is of the form

$$\frac{\partial^2 u}{\partial r^2}=c^2\nabla^2 u$$

where c is the wave speed. This differential equation is used in many branches of physics and longitudinal vibrations in a bar , propagation of sound waves, electromagnetic waves, sea waves, elastic waves in solids, and surface waves as in earthquakes. The solution of a waves equation is called a wave function.

An example for inhomogeneous wave equation is $\frac{\partial^2 u}{\partial t^2} - c^2 \nabla^2 u = F$,Where F is a given function of spatial variables and time. In physical problem F represents an external driving force such as gravity force.

Another related equation is $\frac{\partial^2 u}{\partial t^2} + 2\gamma \frac{\partial u}{\partial t} - c^2 \nabla^2 u = F$,Where γ is a real positive constant . This equation is called a wave equation with damping term. The amplitude of which decreases exponentially as t increases . , we shall derive the partial differential equation describing the transverse vibration of a string.

$$u(x,t) = \phi\left[k(x - ct)\right] + \psi\left[k(x + ct)\right]$$

This is also a solution of the one-dimensional wave equation. Further, let $\omega = kc.$ Then

$$u(x,t) = \phi\left[(kx - \omega t)\right] + \psi\left[k + (kx + \omega t)\right]$$

A function of the type given in above equation is a solution of one-dimensional wave equation if $\omega = kc.$ Therefore, waves travelling with speeds which are not the same as c cannot be described by the solution of the wave equation. Here, $(kx + \omega t)$ is called the phase for the left travelling wave. We have already noted that $x \pm ct$ are the characteristics of the one-dimensional wave equation.

Solution of the wave equation is

$$u(x,t) = \left[c_1 \exp\left\{ i\left(\frac{\omega}{c}\right)x \right\} + c_2 \exp\left\{ -i\left(\frac{\omega}{c}\right)x \right\} \right] e^{\pm i\omega t}$$

Since $k = \pm\frac{\omega}{c},$ the time- dependent wave function are of the form

$$u(x,t) = Ae^{i(kx \pm \omega t)}$$

Hence, $u(x,t) = Ae^{i(kx \pm \omega t)}$ is a solution of the wave equation, and is called a wave function . it is also called a plane harmonic wave or monochromatic wave. Here, A is called the amplitude , ω the angular or circular frequency, and k is the wave number, defined as the number of waves per unit distance

. By Taking the real and imaginary parts of the solution. We find the linear combination of term of the form

$A\cos(kx \pm \omega t), \ A\sin(kx \pm \omega t)$

Representing periodic plane waves. For instance , consider the function $u(x,t) = A\sin(kx - \omega t)$. This is a sinusoidal wave profile moving towards the right along the $x - axis$ with speed c. Defining the wave length λ as the length over which one full cycle is completed . we have $\lambda = \frac{2\pi}{k}$. Thereby implying that $k = \frac{2\pi}{\lambda}$.

$$u(x,t) = \frac{1}{2}\left[\eta(x+ct) + \eta(x-ct)\right] + \frac{1}{2c}\int_{x-ct}^{x+ct} v(\xi)\,d\xi$$

Thus is known as the D'ALEMBERT's Solution of the one-dimensional wave equation. If $v = 0$,*i.e.* , if the string is released form rest, the required solution is

$$u(x,t) = \frac{1}{2}\left[\eta(x+ct) + \eta(x-ct)\right]$$

The D'ALEMBERT's solution has an interesting interpretation.

$$u(x,t) = \frac{1}{2}\{\eta(x+ct) + \eta(x-ct) + \frac{1}{c}\int_{x-ct}^{x+ct} v(\xi)\,d\xi + \frac{1}{2}\int_{0}^{t_0}\int_{x_0-ct_0+ct}^{x_0+ct_0-ct} f(x,t)\,dx\,dt$$

The solution is known as the Riemann-Volterra solution

CHAPTER 6

PARABOLIC DIFFERENTIAL EQUATIONS

CONTENTS

6.1. INTRODUCTION

The diffusion phenomena such as conduction of heat in solids and diffusion of vorticity in the case of viscous flow past a body are governed by a partial differential equation of parabolic type. For example , the flow of heat in a conducting medium is governed by the parabolic equation

$$\rho C \frac{\partial T}{\partial t} = div(K \nabla T) + H(r.T, t) \tag{1}$$

where ρ is the density, C is the specific heat of the solid, T is the temperature at a point with position vector r. K is the thermal conductivity , t is the time ,and $H(r, T, t)$ is the amount of heat generated per unit time in the element dV situated at a point (x, y, z) whose position vector is r. This equation is known as diffusion equation or heat equation. We shall now derive the heat equation form the basic concepts.

Let V be an arbitrary domain bounded by a closed surface $S\, and\, let\, \bar{V} = V \cup S$. Let $T(x, y, z, t)$ be the temperature at a point (x, y, z) at time t. if the temperature is not constant , heat flows from a region of high temperature to a region of low temperature and flows the Fourier law which states that heat flux $q(r, t)$ across the surface element dS with normal $\hat{n}$ is proportional

$$q(r, t) = -k \nabla T(r, t) \tag{2}$$

where k is the thermal conductivity of the body. The negative sign indicates that the heat flux vector points in the direction of decreasing temperature . Let $\hat{n}$ be the outward unit normal vector and $\boldsymbol{q}$ be the heat flux at the surface element dS. Then the rate of heat flowing through the element surface dS in unit time as shown in

$$dQ = (q \,.\hat{n})\, dS \tag{3}$$

Heat can be generated due to nuclear reactions or movement of mechanical parts as in inertial measurement unit (IMU) , or due to chemical sources which may be a function of position, temperature and time and may be denoted by $H(r, T, t)$. We also may be function of position , temperature and time may be substance as the amount of heat needed to raise the temperature of a unit mass by a unit $dm = \rho\, dV$ the value T is given by $dQ = C\rho T\, dV$ therefore,

$$Q = \iiint_V C\rho T\, dV$$

$$\frac{dQ}{dt} = \iiint_V C\rho \frac{\partial T}{\partial t} dV$$

These energy balance equation for a small control volume V is: The rate of energy storage in V equate to the sum of rate of heat entering V theorugh its bounding surface and the rate of heat generation in V. Thus,

$$\iiint_V C\rho \frac{\partial T(r,t)}{\partial t} dV = -- \iint_S q\,.\hat{n}\, dS + \iiint_V H(r,T,t)\, dV \tag{4}$$

Using the divergence theorem, we get

$$\iiint_V \left[C\rho \frac{\partial T(r,t)}{\partial t} dV = -div\, q(r,t) + H(r,T,t) \right] dv = 0 \tag{5}$$

Since the volume is arbitrary , we have

$$\rho C \frac{\partial T(r,t)}{\partial t} = -div\, q(r,t) + H(r,T,t) \tag{6}$$

Substituting Eq. (2) into Eq. (6), we obtain

$$\rho C \frac{\partial T(r,t)}{\partial t} = div\, q(r,t) + H(r,T,t) \tag{7}$$

If we define thermal diffusivity of the medium as

$$\alpha = \frac{k}{\rho C}$$

Then the differential equation of heat conduction with heat source is

$$\frac{1}{\alpha} \frac{\partial T(r,)}{\partial t} = \nabla^2 T(r,t) + \frac{H(r,T,t)}{K} \tag{8}$$

In the absence of heat sources, Eq. (8) reduces to

$$\frac{\partial T(r,t)}{\partial t} = \alpha \nabla^2 T(r,t) \tag{9}$$

This is called Fourier heat conduction equation or diffusion equation. The fundamental problem of heat conduction is to obtain the solution of Eq. (8) subject to the initial and boundary conditions which are called initial boundary value problems, hereafter referred to as IBVPs.

6.2. BOUNDARY CONDITIONS

The heat conduction equation may have numerous solution unless a set of initial and boundary conditions are specified .The boundary conditions are mainly of three types which we now briefly explain.

Boundary Condition 1: The temperature is prescribed all over the boundary surface. That is, the temperature $T(r,t)$ is a function of both position and time. In other words, $T = G(r,t)$ which is some prescribed function on the boundary. This type of boundary condition is called the Dirichlet condition. Specification of boundary conditions depends on the problem under investigation. Sometimes the temperature on the boundary surface is a function of position only or is a function of time only or a constant. A special case includes $T(r,t) = 0$ on the surface of the boundary, which is called a homogeneous boundary condition Specification of boundary conditions depends on the problem under investigation. Sometimes the temperature on the boundary surface is a function of position only or is a function of time only or a constant . A special case includes $T(r,t) = 0$ on the surface of the boundary, which is called a *homogeneous boundary condition.*

Boundary Condition II: The flux of heat, i.e., the normal derivative of the temperature $\frac{\partial T}{\partial n}$, is prescribed on the surface of the boundary. It may be a function of both position and time, i.e.,

$$\frac{\partial T}{\partial n} = f(r,t)$$

This is called the Neumann condition. Sometimes, the normal derivatives of temperature may be a function of position only or a function of time only. A special case includes $\frac{\partial T}{\partial n} = 0 \, on \, the$ boundary. This homogenous boundary condition is also called insulated boundary condition which states the heat flow is zero.

Boundary condition III. A linear combination of the temperature and its normal derivatives is prescribed on the boundary , i.e.,

$$K\frac{\partial T}{\partial n} + hT = G(r\ ,t)$$

where $K \, and \, h$ are constants. This type of boundary condition is called Robin's condition. It means that the boundary surface dissipates heat by convection. Following Newton's law of the cooling, which states that the rate at which heat is transferred from the body to the surrounding is Proportional to the difference in temperature between the body and the surrounding , we have

$$-K\frac{\partial T}{\partial n} = h(T - T_a)$$

As a special case, we may also

$$K\frac{\partial T}{\partial n} + hT = 0$$

which is homogeneous boundary. This mean that heat is converted by dissipation from the boundary surface into a surrounding maintained at zero temperature.

The other boundary conditions such as the heat transfer due to radiation obeying the fourth power temperature law and those associated with change of phase .like melting, ablation, etc, give rise to non-linear boundary conditions.

6.3. ELEMENTARY SOLUTIONS OF THE DIFFUSION EQUATION

Consider the one-dimensional diffusion equation

$$\frac{\partial^2 T}{\partial x^2} = \frac{1}{\alpha}\frac{\partial T}{\partial t}, \ -\infty < x \langle \infty \ ,t \rangle 0 \qquad (10)$$

The function

$$T(x,t)=\frac{1}{\sqrt{4xat}}\exp\left[-(x-\xi)^2/(4at)\right] \tag{11}$$

where ξ is an arbitrary real constant , is a solution of Eq. (10) . It can be verified easily as follows:

$$\frac{\partial T}{\partial t}=\frac{1}{\sqrt{4xat}}\left[\frac{(x-\xi)^2}{4\alpha t}\right]\exp[-(x-\xi)^2/(4at)]$$

$$\frac{\partial T}{\partial t}=\frac{1}{\sqrt{4xat}}\left[\frac{-2(x-\xi)}{4\alpha t}\right]\exp[-(x-\xi)^2/(4at)]$$

Therefore

$$\frac{\partial^2 T}{\partial x^2}=\frac{1}{\sqrt{4xat}}\left[-\frac{1}{2\alpha t}+\frac{(x-\xi)^2}{4\alpha^2 t^2}\right]\exp[-(x-\xi)^2/(4at)]=\frac{1}{a}\frac{\partial T}{\partial t}$$

which show that the function (11) is a solution of Eq. (10). The function (11), known as Kernel, is the elementary solution or the fundamental solution of the heat equation for the infinite interval. For $t>0$ *,the Kernel* $T(x,t)$ is an analytic function of x and t and it can also be noted that $T(x,t)$ is positive for every x . Therefore, the region of influence for the diffusion equation includes the entire x-axis. It can be observed that as $|x|\to\infty$, the amount of heat transported decrease exponentially .

In order to have an idea about the nature of the solution to the heat equation, consider a one-dimensional infinite region which is initially at temperature $f(x)$. Thus, the problem is described by

$$PDE:\frac{\partial T}{\partial t}=\alpha\frac{\partial^2 T}{\partial x^2},\quad -\infty<x\langle\infty,t\rangle 0$$

IC: $T(x,0)=f(x),\quad -\infty<x<\infty,t=0$

Following the method of variables separable, we write

$$T(x,t) = X(x)\,\beta(t) \tag{14}$$

Substituting into $Eq.(12)$. We arrive at

$$\frac{X''}{X} = \frac{1}{\alpha}\frac{\beta'}{\beta} = \lambda \tag{15}$$

where λ is a separation constant. The separated solution for β gives

$$\beta = Ce^{a\lambda t} \tag{16}$$

If $\lambda > 0$, we have $\beta\, and$, therefore, T growing exponentially with timc. From rcalistic physical considcrations, it is reasonable to assume that $f(x) \to 0\, as\, |x| \to \infty$, while $|T(x,t)| < M\, as\, |x| \to \infty$. But for $T(x,t)$ to remain bounded, λ should be negative and thus we take $\lambda = -\mu^2$. Now from $Eq.(15)$. we have

$$X'' + \mu^2 X = 0$$

Its solution is found to be

$$X = c_1 \cos \mu x + c_2 \sin \mu x$$

Hence

$$T(x,t,\mu) = A \cos \mu x\ B \sin \mu x\ )e^{-a\mu^2 t} \tag{17}$$

is a solution of $Eq.(12)$, where $A\, and\, B\, are$ arbitrary constants . Since $f(x)$ is in general not periodic , it Is natural to use Fourier integral instead of Fourier series in the present case . Also, since A and $B\, are$ arbitrary, we may consider them as functions of μ and take $A = A(\mu), B = B(\mu)$. In this particular problem, since we do not have any boundary conditions which limit our choice of μ, we should consider all possible values. From the principles of superposition, this summation of all the product solution will give in the relation

$$T(x,t) = \int_0^{\infty} T(x,t,\mu)\,d\mu = \int_0^{\infty} \left[A(\mu)\cos \mu x + B(\mu)\sin \mu x\right] e^{-\alpha\mu^2 t}\,d\mu \tag{18}$$

which is the solution of $Eq\ .(12)$ From the initial condition (13) , we have

$$T(x,0) = f(x) = \int_0^{\infty} \left[A(\mu)\cos\mu x + B(\mu)\sin\mu x \right] d\mu \tag{19}$$

In addition, if we recall the Fourier integral theorem, we have

$$f(t) = \frac{1}{\pi} \int_0^{\infty} \left[\int_{-\infty}^{\infty} f(x)\ \cos\ \ (t-x) dx \right] d \tag{20}$$

Thus, we may write

$$f(x) = \frac{1}{\pi} \int_0^{\infty} \left[\int_{-\infty}^{\infty} f(y) \cos\mu(x-y) dy \right] d\mu$$

$$\frac{1}{\pi} \int_0^{\infty} \left[\int_{-\infty}^{\infty} f(y)(\cos\mu x \cos\mu y + \sin\mu x \sin\mu y) dy \right] d\mu$$

$$\frac{1}{\pi} \int_0^{\infty} \left[\cos\mu x \int_{-\infty} f(y)\cos\mu dy + \sin\mu x \int_{-\infty} f(y)\sin\mu y dy \right] d\mu \tag{21}$$

Let

$$A(\mu) = \frac{1}{\pi} \int_{-\infty}^{\infty} f(y)\ \cos\mu\, dy$$

$$B(\mu) = \frac{1}{\pi} \int_{-\infty}^{\infty} f(y)\sin\mu y\, dy$$

Then Eq. (21) can be written in the form

$$f(x) = \int_0^{\infty} \left[A(\mu)\ \cos\mu x + B(\mu)\ \sin\mu x \right] d\mu \tag{22}$$

Comparing $Eq\ .(19) and (22)$, we shall write relation (19) as

$$T(x,0) = f(x) = \frac{1}{\pi} \int_0^{\infty} \left[\int_{-\infty}^{\infty} f(y)\cos\mu(x-y) dy \right] d\mu \tag{23}$$

Assuming that the conditions for the formal interchange of orders of integration are satisfied, we get

$$T(x,t)=\frac{1}{\pi}\int_{0}^{\infty} f(y)\left[\int_{-\infty}^{\infty}\exp\left(-\alpha\mu^{2}t\right)\cos\mu(x-)\,d\mu\right]dy \tag{25}$$

Using $s=\mu\sqrt{\alpha t}$, and choosing

$$b=\frac{x-y}{2\sqrt{\alpha t}}$$

Equation (26) becomes

$$\int_{0}^{\infty} e^{-\alpha\mu^{2}t}\cos\mu(x-y)\,d\mu=\frac{\sqrt{\pi}}{\sqrt{4\alpha t}}\exp\left[-(x-y)^{2}/(4at)\right] \tag{27}$$

Substituting $Eq.(27)$ into $Eq.(25)$, we obtain

$$T(x,t)=\frac{1}{\sqrt{4\alpha\pi t}}\int_{-\infty}^{\infty} f(y)\exp\left[-(x-y)^{2}/(4at)\right]dy \tag{28}$$

Hence, if $f(y)$ is bounded for all real values of y, Eq. (28) is the solution of the problem described by Eq.(12) and (13).

Example. In a one-dimensional infinite solid, $-\infty<x<\infty$, the surface $a<x<b$ is initially maintained at temperature T_0 and at zero temperature everywhere outside the surface. Show that

$$T(x,t)=\frac{T_0}{2}\left[\operatorname{erf}\left(\frac{b-x}{\sqrt{4at}}\right)-\operatorname{erf}\left(\frac{a-x}{\sqrt{4at}}\right)\right]$$

where erf is an error function.

Solution. The problem is described as follows:

PDE: $T_t=\alpha T_{xx}$, $-\infty<x<\infty$

IC: $T=T_0$, $a<x<b$ .outside the above region

The general solution of PDE is found to be

$$T(x,t)=\frac{1}{\sqrt{4at}}\int_{-\infty}^{\infty} f(\xi)\exp\left[-(x-\xi)^{2}/(4at)\right]d\xi$$

Introducing the new independent variable η defined by

$$\eta = -\frac{x-\xi}{\sqrt{4\alpha t}}$$

And hence

$$d\xi = \sqrt{4\alpha t}\, d\eta$$

The above equation becomes

$$T(x,t) = \frac{T_0}{\sqrt{\pi}} \int_{(a-x)/(4at)}^{(b-x)/(4at)} e^{-\eta^2}\, d\eta$$

$$= \frac{T_0}{2}\left[\frac{2}{\sqrt{\pi}} \int_0^{(b-x)/(4at)} e^{-\eta^2}\, d\eta - \frac{2}{\sqrt{\pi}} \int_0^{(a-x)/(4at)} e^{-\eta^2}\, d\eta\right]$$

Now we introduce the error function defined by defined by

$$Erf(z) = \frac{2}{\sqrt{\pi}} \int_0^z \exp(-\eta^2)\, d\eta$$

Therefore the required solution is

$$T(x,t) = \frac{T_0}{2}\left[\operatorname{erf}\left(\frac{b-x}{\sqrt{4at}}\right) - \operatorname{erf}\left(\frac{a-x}{\sqrt{4\alpha t}}\right)\right]$$

6.4. DIRAC DELTA FUNCTION

According the notion mechanic, we come across a very large force (ideally infinite) acting for a short duration (ideally zero time) knows as impulsive force. Thus we have a function which is non-zero in a very short interval. The Dirac delta function may be thought of as a generalization of this concept. This Dirac delta function and its derivative play a useful role in the solution of initial boundary value problem (IBVP).

Consider the function having the following property:

$$\delta_\varepsilon(t) = \begin{cases} 12\varepsilon, |t| < \varepsilon \\ 0, \quad |t| < \varepsilon \end{cases} \tag{29}$$

Thus,

$$\int_{-\infty}^{\infty} \delta_\varepsilon(t)\,dt = \int_{-\varepsilon}^{\varepsilon} \frac{1}{2\varepsilon}\,dt = 1 \tag{30}$$

Let $f(t)$ be any function which is integrable in the interval $(-\varepsilon, \varepsilon)$. Then using the Mean –value theorem of integral calculus , we have

$$\int_{-\infty}^{\infty} f\infty(t)\delta_\varepsilon(t)\,dt = \frac{1}{2\varepsilon}\int_{-\varepsilon}^{\varepsilon} f(t)\,dt = f(\xi), -\varepsilon < \xi < \varepsilon \tag{31}$$

Thus, we may regard $\delta(t)$ as limiting function approached by $\delta_\varepsilon(t)$ as $\varepsilon \to 0$, i.e.,

$$\delta(t) - LT_{\varepsilon \to 0}\delta_\varepsilon(t) \tag{32}$$

As $\varepsilon \to 0$, we have , from $Eq.(29)\,and\,(30)$, the relations

$$\delta(t) = LT_{\varepsilon \to 0}\delta_\varepsilon(t) = \begin{cases} (in\,the\,same\,of\,being\,very\,large) \\ \infty, & if\; t = 0 \\ 0, & if\; t \neq 0 \end{cases} \tag{33}$$

$$\int_{-\infty}^{\infty} \delta(t)\,dt 1 \tag{34}$$

This limiting function $\delta(t)$ defined by $Eq.(33)$ and (34) is known as Dirac delta function or the unit impulse function. Its profile is depicted in Dirac originally called it an improper function as there is no proper function with these properties . In fact, we can observed that

$$1 = \int_{-\infty}^{\infty} \delta(t)\,dt = LT_{\varepsilon \to 0} \int_{|t| > \varepsilon} \delta_\varepsilon(t)\,dt = LT_{\varepsilon \to 0} 0 = 0$$

Obviously, this contradiction implies that $\delta(t)$ cannot be a function in the ordinary sense. Some important properties of Dirac delta function are presented now:

Property 1:

$$\int_{-\infty}^{\infty} \delta(t)\,dt = 1$$

Property II: For any continuous functions $f(t)$,

$$\int_{-\infty}^{\infty} f(t)\delta(t)\,dt = f(0)$$

Proof (of property 2) Consider the equation

$$LT_{\varepsilon \to 0} \int_{-\infty}^{\infty} f(t)\delta_{\varepsilon}(t)\,dt = LT_{\varepsilon \to 0}\, f(\xi) - \varepsilon < \xi < \varepsilon$$

As $\varepsilon \to 0$, we have $\xi \to 0$. Therefore,

$$\int_{-\infty}^{\infty} f(t)\delta(t)\,dt = f(0)$$

Property III: Let $f(t)$ be any continuous function. Then

$$\int_{-\infty}^{\infty} \delta(t-a) f(t)\,dt = f(a)$$

Proof

Consider the function

$$\delta_e(t-a) = \begin{cases} \dfrac{1}{\varepsilon}\,. & a < 1 < a+\varepsilon \\ 0\,, & elsewhere \end{cases}$$

Using the mean-value theorem of integral calculus, we have

$$\int_{-\infty}^{\infty} \delta_{\varepsilon}(t-a) f(t)\,dt = \frac{1}{\varepsilon}\int_{a}^{a+\varepsilon} f(t)\,dt = f(a+\theta\varepsilon), 0<\theta<1$$

Now, taking the limit as $\varepsilon \to 0$, we obtain

$$\int_{-\infty}^{\infty} \delta(t-a) f(t)\,dt = f(a)$$

Thus, the operation of multiplying $f(t) by \delta(t-a)$ and integrating over all t is equivalent to substituting a for t in the original function .

Property IV:

$$\delta(-t)=\delta(t)$$

Property V:

$$\delta(at)=\frac{1}{a}\delta(t),\ a>0$$

Property VI: If $\delta(t)$ is a continuously differentiable. Dirac delta function vanishing for large t, then

$$\int_{-\infty}^{\infty} f(t)\delta'(t)dt=\left[f(t)\delta(t)\right]_{-\infty}^{\infty}-\int_{-\infty}^{\infty} f'(t)\delta(t)dt$$

Since $\delta t \to 0\, as\, t \to \infty, -\infty$, we have

$$\int_{-\infty}^{\infty} f(t)\delta'(t)dt=-f'(0)$$

Property VII:

$$\int_{-\infty}^{\infty} \delta'(t-a)f(t)dt=-f'(a)$$

Having discussed the one-dimensional Dirac delta function , we can extend the definition to two dimension, Thus, for every f which is continuous over the region S containing the point (ξ,η), we define $\delta(x \quad \xi, y-\eta)$ in such a way that

$$\iint_S \delta(x-\xi y-\eta)f(x,y)d\sigma-f(\xi,\eta)$$

Note that $\delta(x-\xi, y-\eta)$ is a form a limit of a sequence of ordinary functions, i.e.,

$$\delta(x-\xi, y-\eta) = LT_{\varepsilon\to 0}\,\delta_\varepsilon(r) \tag{36}$$

where $r^2 = (x-\xi)^2 + (y-\eta)^2$. Also observe that

$$\iint \delta(x-\xi)\delta(y-\eta) f(x, y)\,dx\,dy = f(\xi, \eta) \tag{37}$$

Now, comparing $Eq.(35) and (37)$, we see that

$$\delta(x-\xi, y-\eta) = \delta(x-\xi)\delta(y-\eta) \tag{38}$$

Thus, a two –dimensional Dirac delta function can be expressed as the product of two one –dimensional delta functions. Similarly, the definition can be higher dimensions.

Example. A one-dimensional infinite region $-\infty<x<\infty$ is initially kept at zero temperature. A heat source of strength g_s units, situated at $x=\xi$ releases its heat instantaneously at time $t=\tau$. Determine the temperature in the region for $t>\tau$.

Solution. Initially, the region $-\infty<x<\infty$ is at zero temperature. Since the heat source is situated at $x=\xi$ and releases heat instantaneously at $t=\tau$, the released temperature at $x=\xi$ and $t=\tau is$ a $\delta-$ function type. Thus, the given problem is a boundary value problem described by

PDE: $\dfrac{\partial^2 T}{\partial x^2} + \dfrac{g(x,t)}{k} = \dfrac{1}{a}\dfrac{\partial T}{\partial t}, \quad -\infty x\langle\infty, t\rangle 0$

IC: $T(x,t) = F(x) = 0, \quad -\infty<x<\infty, t=0$

The general solution to this problem as given , after using the initial condition $F(x)=0$, is

$$T(x,t) = \frac{\alpha}{k}\int_{t'=0}^{t} \frac{dt'}{\sqrt{4\pi\alpha(t-t')}} \int_{x'}^{\infty} g(x',t') \exp\left[-(x-x')^2/\{4\alpha(t-t')\}\right] dx' \tag{39}$$

Since the heat source term is of the Dirac delta function type, substituting

$$g(x,t) = g_s\,\delta(x-\xi)\delta(t-\tau)$$

into Eq. (39), and integrating we get, with the help of properties of delta function, the relation

$$T(x,t)=\frac{\alpha}{k}\frac{g_s}{\sqrt{4\pi\alpha}}\int_0^t\frac{\exp\left[-(x-\xi)^2/\left\{4\alpha\left(t-t'\right)\right\}\right]}{\sqrt{4\pi\alpha(t-\tau)}}\delta(t-\tau)\,dt'$$

Therefore, the required temperature is

$$T(x,t)=\frac{\alpha g_s}{k}\frac{\exp\left[-(x-\xi)^2/\left\{4\alpha(t-\tau)\right\}\right]}{\sqrt{4\pi\alpha(t-\tau)}}\quad \text{for } t>\tau$$

Example. An infinite one-dimensional solid defined by $-\infty<x<\infty$ is maintained at zero temperature initially. There is a heat source of strength $g_s(t)$ units, situated at $x=\xi$, which releases costant heat continuously for $t>0$. Find an expression for the temperature distribution in the solid for $t>0$.

Solution. Thisproblemissimilarto,exceptthat $g(x,t)=g_s(t)\delta(x-\xi)$ is a Dirac delta function type. The solution to this IBVP is

$$T(x,t)=\frac{\alpha}{k}\int_{t'=0}^{t}\frac{g_s(t')}{\sqrt{4\pi\alpha(t-t')}}\exp\left[-\left(x-\xi^2\right)/\left\{4\alpha(t-t')\right\}\right]dt' \tag{40}$$

It is given as $g_s(t)=constant=g_x(say)$. Let us introduce a new variable η defined by

$$\eta=\frac{x-\xi}{\sqrt{4\alpha(t-t')}}\ or\ t-t'=\frac{1}{\eta^2}\frac{(x-\xi)^2}{4\alpha}$$

Therefore,

$$dt'=\frac{1}{\eta^3}\frac{(x-\xi)^2}{2\alpha}d\eta$$

Thus, $Eq.(40)$ becomes

$$T(x,t)=g_s\frac{x-\xi}{2K\sqrt{\pi}}\int_{(x-\xi)/\sqrt{4at}}^{\infty}\frac{\exp\left(-\eta^2\right)}{\eta^2}d\eta$$

However,

$$\frac{d}{d\eta}\left(-\frac{e^{-n^2}}{\eta}\right)=\frac{e^{-n^2}}{\eta^2}+2e^{-\eta^2}$$

Hence,

$$T(x,t)=g_s\frac{x-\xi}{2K\sqrt{\pi}}\left[\left(-\frac{e^{-\eta^2}}{\eta}\right)_{x-\xi/\sqrt{4at}}^{\infty}-2\int_{(x-\xi)/\sqrt{4at}}^{\infty}e^{-\eta^2}\,d\eta\right]$$

Recalling the definitions of error function and its complement

$$Erf(x)=\frac{2}{\sqrt{\pi}}\int_0^x e^{-\eta^2}\,d\eta\,,\qquad \text{erf}(\infty)=1$$

$$erfc(x)=1-erf(x)=2/\sqrt{\pi}\left(\int_0^\infty \exp(-\eta^2)d\eta-\int_0^x \exp(-\eta^2)d\eta\right)$$

$$\frac{2}{\sqrt{\pi}}\int_0^x \exp(-n^2)d\eta$$

The temperature distribution can be expressed as

$$T(x,t)=\frac{\alpha g_s}{K}\left[\sqrt{\frac{t}{2\pi}}\exp\left[-\frac{(x-\xi)^2}{4at}\right]-\frac{|x-\xi|}{2\alpha}\left(1-\text{erf}\frac{x-\xi}{\sqrt{4at}}\right)\right]$$

Alternatively, the required temperature is

$$T(x,t)=\frac{\alpha g_s}{K}\left[\sqrt{\frac{t}{2\pi}}\exp\left[-\frac{(x-\xi)^2}{4at}\right]-\frac{|x-\xi|}{2\alpha}erfc\frac{x-\xi}{\sqrt{4at}}\right]$$

6.5. SEPARATION OF VARIABLES METHOD

Consider the equation

$$\frac{\partial T}{\partial t}=\alpha\frac{\partial^2 T}{\partial x^2} \tag{41}$$

Among the many methods that are available for the solution of the above parabolic partial differential equation. The method of separation of variables

of $T(x,t)$ as follows. Let

$$T(x,t)=X(x)\beta(t) \tag{42}$$

Be a solution of the differential Eq.(41). Substituting Eq.(42) into (41), we obtain $\frac{X''}{X}=\frac{1}{a}\frac{\beta'}{\beta}=K\ ,a$ separation constant. Then we have

$$\frac{d^2X}{dx^2}-KX=0 \tag{43}$$

$$\frac{d\beta}{dt}-ak\beta=0 \tag{44}$$

In solving Eq. (43) and (44), three distinct case arise:

Case I when K is positive , say λ^2, the solution of Eq. (43) and (44) will have the form

$$X=c_1e^{\lambda x}+c_2e^{-\lambda x}\ .\ \ \beta=c_3e^{a\lambda^2t} \tag{45}$$

Case II when K is negative , say $-\lambda^2$, then the solution of Eq. (43) and (44) will have the form

$$X=c_1\cos\lambda a+c_2\sin\lambda x,\ \ \beta=c_3e^{-a\lambda^2t} \tag{46}$$

Case III when K is zero, the solution of $Eq.(43)$ and (44) can have the form

$$X=c_1c+c_2.\ \ \beta=c_3 \tag{47}$$

Thus, various possible solutions of the heat conduction equation (41) could be the following

$$T(x,t)=\left(c_1'e^{\lambda x}+c_2'e^{-\lambda x}\right)e^{a\lambda^2t}$$

$$T(x,t)-(c_1'\cos\lambda x+c_2'\sin\lambda x)e^{-a\lambda^2t} \tag{48}$$

$$T(x,t)=c_1'x+c_2'$$

where

$$c_1'=c_1c_3,\quad c_2'=c_2c_3$$

Solve the one-dimensional diffusion equation in the region $0 \le x \le \pi, t \ge 0$, subject to the conditions

(i) T *remains* finite as $t \to \infty$

(ii) $T = 0$, *if* $x = 0$ *and* π *for all* t

(iii) At $t = 0, T = \begin{cases} x, & 0 \le x \le \frac{\pi}{2} \\ \pi - x, & \frac{\pi}{2} \le x \le \pi. \end{cases}$

Solution. Since T should satisfy the diffusion equation, the three possible solutions are:

$$T(x,t) = \left(c_1 e^{\lambda x} + c_2 e^{-\lambda x}\right) e^{a\lambda^2 t}$$

$$T(x,t) = (c_1 \cos \lambda x + c_2 \sin \lambda x) e^{-\alpha \lambda^2 t}$$

$$T(x,t) = c_1 x + c_2)$$

The first condition demand that T should remain finite as $t \to \infty$, we therefore reject the first solution. In view of BC (ii), the third solution gives

$$0 = c_1 .0 + c_2, \quad 0 = c_1 .\pi + c_2$$

Implying thereby the both c_1 *and* c_2 are zero and hence $T = 0$ for all t. This is a trivial solutions. Since we are looking for a non-trivial solutions, we reject the third solution too. Thus, the only possible solution satisfying the first condition is

$$T(x,t) = (c_1 \cos \lambda x + c_2 \sin \lambda x) e^{-\alpha \lambda^2 t}$$

Using the BC (ii), we have

$$0 = c_1 \cos \lambda x + c_2 \sin \lambda x \Big|_{x=0}$$

Implying $c_1 = 0$. Therefore, the possible solution is

$$T(x,t) = c_2 e^{-\alpha\lambda^2 t} \sin \lambda x$$

Applying the BC: $T = 0\, when\, x =$, we get

$$\sin \lambda\pi = 0 \Rightarrow \lambda\pi = n\pi$$

where $n\, is$ an integer. Therefore,

$$\lambda = n$$

Hence the solution is found to be of the form

$$T(x,t) = ce^{an^2 t} \sin nx$$

Nothing that the heat conduction equation is linear, its most general solution is obtained by applying the principle of superposition. Thus,

$$T(x,t) = \sum_{n=1}^{\infty} c_n e^{an^2 t} \sin nx$$

Using the third condition, we get

$$T(x,0) = \sum_{n=1}^{\infty} c_n \sin nx$$

which is a half-range Fourier-sine series and, therefore,

$$c = \frac{2}{\pi} \int_0^{\pi} T(x,0) \sin nx dx = \frac{2}{\pi} \left[\int_0^{\pi/2} + \left\{ -(\pi - x) \frac{\cos nx}{n} + \frac{\sin nx}{n^2} \right\}_{\pi/2}^{\pi} \right]$$

or

$$c_n = \frac{4 \sin\left(\frac{n\pi}{2}\right)}{n^2 \pi}$$

Thus, the required solution is

$$T(x,t) = \frac{4}{\pi} \sum_{n=1}^{\infty} \frac{e^{-\alpha n^2 t} \sin\left(\frac{n\pi}{2}\right)}{n^2 \pi}$$

Example. A uniform rod of length L whose surface is thermally insulated is initially at temperature $\theta = \theta_0$. At time $t = 0$, one end is suddenly cooled to $\theta=0$ and subsequently maintained at this temperature ; the other end remains thermally insulated. Find the temperature distribution $\theta(x,t)$.

Solution. The initial boundary value problems IBVPs of heat conduction is given by

PDE: $\frac{\lambda\theta}{\partial t}=a\frac{\partial^2\theta}{\partial x^2}\quad 0\le x\le L\ ,t>0$

BCs: $\theta(0,t)=0,\ t\ge 0$

$\frac{\partial\theta}{\partial x}(L,t)=0\ ,\ t>0$

IC: $\theta(x,0)=\theta_0,\ 0\le x\le L$

$\theta(x,t)=e^{-\lambda^2 t}(A\cos\lambda x B\sin\lambda x)$

Using the first boundary condition, we obtain $A=0$. Thus the acceptable solution is

$$\theta=Br^{a\lambda^2 t}\sin\lambda x$$

$$\frac{\partial\theta}{\partial x}=\lambda Be^{-a\lambda^2 t}\cos\lambda x$$

Using the second boundary condition, we have

$$0=\lambda Be^{a\lambda^2 t}\cos\lambda L$$

Implying $\cos\lambda L=0$. Therefore, the eigen value and the corresponding eigen functions are

$$\lambda_n=\frac{(2n+1)\pi}{2I\ .}n=0,1,2,\ldots$$

Thus the acceptable solution is of the form

$$\theta=B\exp\left[-\alpha\left\{\frac{2n+1}{2L}\right\}^2\pi^2 t\right]\sin\left(\frac{2n+1}{2L}\right)\pi x$$

Using the principle of superposition, we obtain

$$\theta(x,t)=\sum_{n=0}^{\infty}B_n\exp\left[-a\left\{\frac{2n+1}{2l}\right\}^2\pi^2 t\right]\sin\left(\frac{2n+1}{2L}\pi x\right)$$

Finally, using the initial condition , we have

$$\theta_0 = \sum_{n=0}^{\infty} B_n \sin\left(\frac{2n+1}{2L}\pi x\right)$$

which is a half range Fourier-sine series and, this

$$B_n = \frac{2}{L}\int_0^L \theta_0 \sin\left(\frac{2n+1}{2L}\pi x\right)dx$$

$$= \frac{2}{L}\left[-\theta_0 \frac{2L}{(2n+1)\pi}\left\{\cos\left(\frac{2n+1}{2L}\pi x\right)\right\}_0^L\right]$$

$$= -\frac{4\theta_0}{(2n+1)\pi}\left[\cos\left\{\frac{(2n+1)\pi}{2}\right\} - \cos 0 = \frac{4\theta_0}{(2n+1)\pi}\right.$$

Thus, the required temperature distribution is

$$\theta(x,t) = \sum_{n=0}^{\infty} \frac{4\theta_0}{(2n+1)\pi} \exp\left[-\alpha\left\{\frac{(2n+1)}{2L}\right\}^2 \pi^2 t\right] \sin\left(\frac{2n+1}{2L}\pi x\right)$$

Example. A conducting bat of uniform cross-section lies among the $x-axis$ with ends at $x = 0 \, and \, x = L$. it kept initially at temperature 0^0 and its lateral surface is insulated. There are no heat sources in the bar. The end $x = 0$ is kept at 0^0 , and heat is suddenly applied at the end $x = L$, so that there is a constant flux q_0 at $x = L$. find the temperature distribution in the bar for $t > 0$.

Solution. The given initial boundary value problem can be described as follows:

PDE: $\frac{\partial T}{\partial t} = \alpha \frac{\partial^2 T}{\partial x^2}$

BCs: $T(0,t) = 0, \; t > 0$

$\frac{\partial T}{\partial x}(L,t) = q_0, \quad t > 0$

IC: $T(x,0) = 0, \quad 0 \le x \le L$

Prior to applying heat suddenly to the end $x=L$, when $t=0$, the heat flow in the bar is independent of time (steady state condition). Let

$$T(x,t)=T_s(x)+T_1(x,t)$$

where T_s is a steady part and T_1 *is* the transient part of the solution . Therefore,

$$\frac{\partial^2 T_s}{\partial x^2}=0$$

whose general solution is

$$T_s=Ax+B$$

when $x=0, T_s=0$, implying $B=0$. Therefore,

$$T_s=Ax$$

Using the other BC: $\frac{\partial T_s}{\partial x}=q_0$. w get $A=q_0$. hence , the steady state solution is

$$T_s=q_0 x$$

For the transient part, the BCs and IC are redefined as

$(i)\ T_1(0,t)=T(0,t)-T_s(0)=0-0=0$

$(ii)\ \partial T_1(L,t)/\partial x=\partial T/\partial x-\partial T_s(Lt)/\partial x=q_0-q_0=0$

$(iii)\ T_1(x,0)=T(x,0)-T_s(x)=-q_0 x, 0<x<L$.

Thus, for the transient part, we have to solve the given PDE subject to these conditions. The acceptable solution is given by $Eq.(48), i.e.$

$$T_1(x,t)=e^{-a\lambda^{2t}}(A\cos\lambda x+B\sin\lambda x)$$

Applying the BC (i), we get $A=0$. Therefore ,

$$T_1(x,t)=Be^{-a\lambda^2 t}\sin\lambda x)$$

Applying the BC (i), we get $A=0$. Therefore ,

$$T_1(x,t) = Be^{-a\lambda^2 t} \sin \lambda x$$

And using the BC (ii), we obtain

$$\left.\frac{\partial T_1}{\partial x}\right|_{x=L} = B\lambda^{-a\lambda^2 t} \cos \lambda L = 0$$

Implying $\lambda L = (2n-1)\frac{\pi}{2}, n = 1, 2, \ldots$ Using the superposition principle , we have

$$T_1(x,t) = \sum_{n=1}^{\infty} B_n \exp\left[-\alpha \left\{\frac{2n-1}{2L}\right\}^2 \pi^2 t\right] \sin\left(\frac{2n-1}{2L}\pi x\right)$$

Now, applying the IC (iii), we obtain

$$T_1(x,0) = -q_0 x = \sum_{n=1}^{\infty} B_n \sin\left(\frac{2n-1}{2L}\pi x\right)$$

Multiplying both sides by $\sin\left(\frac{2m-1}{2L}\pi x\right)$ and integrating between 0 to L and noting that

$$\int_0^L B_n \sin\left(\frac{2n-1}{2L}\right)\pi x \sin\left(\frac{2m-1}{2L}\pi x\right) dx = \begin{cases} 0, & n \neq m \\ \frac{B_m L}{2}, & n = m \end{cases}$$

We get at once , after integrating by parts, the equation

$$-q_0 \frac{4L^2}{(2m-1)^2 \pi^2}\left[\sin\left(\frac{2m-1}{2L}\right)\right] = B_m \frac{L}{2}$$

which gives

$$B_m = \frac{(-1)^m 8Lq_0}{(2m-1)^2 \pi^2}$$

Hence, the required temperature distribution is

$$T(x,t)=q_0x+\frac{8Lq_0}{\pi^2}\sum_{n=1}^{\infty}\left[\frac{(-1)^m}{(2m-1)^2}\exp\left[-a\left\{\frac{(2m-1)}{L}\right\}^2\pi^2t\right]\sin\left(\frac{2m-1}{2L}n\pi\right)\right]$$

Example. The end A and B of a rod, $10cm$ on length , are kept at temperature $0^0 C$ and $100^0 C$ until the steady state condition prevails. Suddenly the temperature at the end A is increased to $20^0 C$, and the end B is described to $60^0 C$. find the temperature distribution in the rod at time t.

Solution. The problem is described by

PDE: $\frac{\partial T}{\partial t}=\alpha\frac{\partial^2 T}{\partial x^2}, \quad 0<x<10$

BCs: $T(0,t)=0\ ,\ T(10,t)=100$

Prior to change in temperature at the ends of the rod, the heat flow in the rod is independent of time as steady condition prevails. For study state,

$$\frac{d^2T}{dx^2}=0$$

whose solution is

$$T_s=Ax+B$$

when $x=0\ ,T=0$ implying $B=0$. Therefore,

$$T_s=Ax$$

when $x=10\ ,T=100$, implying $A=10$. Thus, the initial steady temperature distribution in the rod is

$$T_s(x)=10x$$

Similarly, when the temperature at the ends $A\,and\,B$ are changed to $20\,and\,60$, the final steady temperature in the rod is

$$T_s(x)=10x$$

which will be attained after a long time, To get the temperature distribution $T(x,t)$ in the intermediate period. Counting time the moment the end temperature were changed, we assume that

$$T(x,t)=(4x+20)+e^{-\alpha\lambda^2 t}(B\cos\lambda x+c\sin\lambda x)$$

Using the BC: $T=20\, when\, x=0$, we obtain

$$20=20+Be^{-\alpha\lambda^2 t}$$

Implying $B=0$. Using the BC: $T=60$ when $x=10$, we get

$$\sin 10\lambda=0\ ,imlying\ \lambda=\frac{n\pi}{10},n=1,2,\ldots$$

The principal of superposition yields

$$T(x,t)=(4x+20)+\sum_{n=1}^{\infty}c_n\exp\left[-a\left(\frac{n\pi}{10}\right)^2 t\right]\sin\left(\frac{n\pi}{10}\right)x$$

Now using the IC: $T=10x\ ,when t=0$, we obtain

$$10x=4x+20+\sum c_n\sin\left(\frac{n\pi}{10}x\right)$$

or

$$6x-20=\sum c_n\sin\left(\frac{n\pi}{10}x\right)$$

where

$$c_n=\frac{2}{10}\int_0^{10}(6x-20)\sin\left(\frac{n\pi}{10}x\right)dx=-\frac{1}{5}\left[(-1)^n\frac{800}{n\pi}-\frac{200}{n\pi}\right]$$

Thus required solution is

$$T(x,t)=4x+20-\frac{1}{5}\sum_{n=1}^{\infty}\left[(-1)^n\frac{800}{n\pi}-\frac{200}{n\pi}\right]\exp\left[-\alpha\left(\frac{n\pi}{10}\right)^2 t\right]\sin\left(\frac{n\pi}{10}x\right)$$

Example. Assuming the surface of the earth to be flat , which is initially at zero temperature and for times $t>0$, the boundary surface is being subject is being subject to a periodic heat flux $g_0\cos\omega t$. investigate the penetration of these temperature variations into the earth's surface and show that at a depth $x,$ the temperature fluctuates and the amplitude of the steady temperature is given by

$$\frac{g_0}{\sqrt{2}}\sqrt{\frac{2\alpha}{\omega}}\exp\left[-\sqrt{\left(\frac{\omega}{2a}\right)}x\right]$$

Solution. The given IBVP is described by

PDE: $\dfrac{\partial T}{\partial t}=\alpha\dfrac{\partial^2 T}{\partial x}$ (49)

BC: $-\dfrac{\partial T}{\partial x}=g_0\cos\,\omega t\;at\;x=0\;,t>0$ (50)

IC: $T(x\;,0)=0$ (51)

We shall introduce an auxiliary function $\overline{T}$ satisfying $Eq\;.(49)-(51)$ and then define the complex function Z such that

$Z=T+i\overline{T}$

We can easily verify that Z satisfies

PDE: $\dfrac{\partial Z}{\partial t}=\alpha\dfrac{\partial^2 Z}{\partial x^2}$ (52)

BC:- $\dfrac{\partial Z}{\partial x}=g_0\;\;e^{\omega t}\;at\;x=0\;,t>0$

IC: $Z=0$ in the region. $t=0$

Let us assume the solution of $Eq.(52)$ in the form

$Z=f(x)e^{i\omega t}$

where $f(x)$ satisfies

$$\frac{d^2 f(x)}{dx^2}-i\frac{\omega}{\alpha}f(x)=0 \qquad (53)$$

$$-\frac{df(x)}{dx}=g_0\;\;at\;x=0 \qquad (54)$$

Also,

$f(x)$ is finite for large x. (55)

The solution of $Eq.(53)$, satisfying the $BC(55)$, is

$$f(x) = A\exp\left[-\sqrt{\left(\frac{i\omega}{\alpha}\right)}x\right]$$

The constant A can be determined by using the $BC(54)$. Therefore,

$$f(x) = \frac{1}{\sqrt{i}}\, g_0 \sqrt{\frac{\alpha}{\omega}} \exp\left[-\sqrt{\left(\frac{i\omega}{\alpha}\right)}x\right]$$

Thus,

$$Z = g_0 \sqrt{\frac{a}{\omega}} \frac{1}{\sqrt{i}} \exp\left[i\omega t - \sqrt{\left(\frac{i\omega}{\alpha}\right)}x\right] \tag{56}$$

It can be shown for convenience that

$$\sqrt{i} = \frac{1+i}{\sqrt{2}}, \quad \frac{1}{\sqrt{i}} = \frac{1-i}{\sqrt{2}}$$

Thus, $q.(56)$ can be written in the form

$$Z = \frac{g_0}{2}\sqrt{\frac{2a}{\omega}} \exp\left(-\sqrt{\frac{\omega}{2a}}x\right)(1-i)\exp\left[i(\omega t - \sqrt{\frac{\omega}{2a}}x\right]$$

$$= \frac{g_0}{2}\sqrt{\frac{2a}{\omega}} \exp\left(-\sqrt{\frac{2u}{\omega}}x\right)(1-i)\left[\cos\left(\omega t - \sqrt{\frac{\omega}{2a}}x\right) + i\sin\left(\omega t - \sqrt{\frac{\omega}{2a}}x\right)\right]$$

Its real part gives the fluctuation in temperature is

$$T(x,t) = \frac{g_0}{2}\sqrt{\frac{2a}{\omega}} \exp\left(-\sqrt{\frac{\omega}{2a}}x\right)\left[\cos(\omega t - \sqrt{\frac{\omega}{2a}}x + \sin\left(\omega - \sqrt{\frac{\omega}{2a}}x\right)\right]$$

$$\frac{g_0}{2}\sqrt{\frac{2a}{\omega}} \exp\left(-\sqrt{\frac{\omega}{2a}}x\right) \cos\left(\omega t - \sqrt{\frac{\omega}{2a}}x - \frac{\pi}{4}\right)$$

Hence, the amplitude of the steady temperature is given by the factor

$$\frac{g_0}{\sqrt{2}}\sqrt{\frac{2a}{\omega}}\exp\left(-\sqrt{\frac{\omega}{2a}}x\right)$$

Example. Find the solution of the non-dimensional diffusion equation satisfying the following BCs.

$(i)\ T$ is bounded as $t\to\infty$

$(ii)\left.\frac{\partial T}{\partial x}\right|_{x=0,}=0$, for all t

$(iii)\left.\frac{\partial T}{\partial x}\right|_{x=0}=0$, for all t

Solution. This is an example with insulated boundary conditions. It can be seen that a physically acceptable general solution of the diffusion equation is

$$T(x,t)=\exp\left(\alpha\lambda^2 t\right)\left(A\cos\lambda x+B\sin\lambda x\right)$$

Thus,

$$\frac{\partial T}{\partial x}=\exp\left(-\alpha\lambda^2 t\right)\left(-A\lambda\sin\lambda x+B\lambda\cos\lambda x\right) \tag{57}$$

Using the BC (ii) , Eq. (57) , gives $B=0$, Since we are looking for a non-trivial solution, the use of BC (iii) into Eq. (57) at once gives $\sin\lambda\alpha=0$ implying $\lambda\alpha=n\pi\ ,\ n=0,1,2,\ldots$

Using the principle of superposition, we get

$$T(x,t)=\Sigma A_n\exp\left(-\alpha\lambda^2 t\right)\cos\lambda x=\sum_{n=0}^{\infty}A_n\exp\left[-\alpha\left(\frac{n\pi}{a}\right)^2 t\right]\cos\left(\frac{n\pi}{a}\right)x.$$

The boundary condition (iv) gives

$$T(x,0)=x(a-x)=A_0+\sum_{n=1}^{\infty}A_n\exp\left[-a\left(\frac{n\pi}{a}\right)^2 t\right]\cos\left(\frac{n\pi}{a}\right)x$$

where

$$A_0=\frac{2}{a}\int_0^a\left(ax-x^2\right)dx=\frac{a^2}{6}$$

$$A_n=\frac{2}{a}\int_0^a\left(ax-x^2\right)dx=\frac{a^2}{6}$$

$$=\frac{2a^2}{n^2\pi^2}(1+\cos n\pi)=\frac{2a^2}{n^2\pi^2}\left[1+(-1)^n\right]$$

Therefore,

$$A_n=\begin{cases}-\dfrac{4a^2}{n^2\pi^2}, & for\, n\, even\\ 0, & for\, n\, odd\end{cases}$$

Hence, the required solution is

$$T(x,t)=\frac{a^2}{6}-\frac{4a^2}{\pi^2}\sum_{n=2,4,\ldots even}^{\infty}\frac{1}{n^2}\cos\left(\frac{n\pi}{a}\right)x\exp\left[-\alpha\left(\frac{n\pi}{a}\right)^2 t\right]$$

Example. The boundaries of the rectangle $0\le x\le a, 0\le y\le b$ are maintained at zero temperature. if at $t=0$ the temperature T has the prescribed value $F(x,y)$, show that for $t>0$, the temperature at a point within the rectangle is given

$$T(x,y,t)=\frac{4}{ab}\sum_{m=1}^{\infty}\sum_{n=1}^{\infty}f(m,n)\exp\left(-a\lambda_{mn}^2 t\right)\sin\frac{m\pi x}{a}\sin\frac{n\pi y}{b}$$

where

$$f(m,n)=\int_0^a\int_0^b f(x,y)\sin\frac{m\pi x}{a}\sin\frac{n\pi y}{b}dx\,dy$$

and

$$\lambda_{mn}^2=\pi^2\left(\frac{m^2}{a^2}+\frac{n^2}{b^2}\right)$$

Solution. The problem is to solve the diffusion equation described by

$$PDE : \frac{\partial T}{\partial t} = \alpha\left(\frac{\partial^2 T}{\partial x^2} + \frac{\partial^2 T}{\partial y^2}\right), \ 0<x<a \ , \ 0<y\langle b \ ,t\rangle 0$$

BCs : $T(0\ ,y\ ,t) = T(a\ ,y\ ,t) = 0\ ,\ 0<y\langle b\ ,t\rangle 0$

$T(x\ ,0\ ,t) = T(x\ ,b\ ,t) = 0,\ \ 0<x\langle a\ ,t\rangle 0$

IC: $T(x\ ,y\ ,0) = f(x\ ,y),\ \ 0<x<a\ ,0<y<b$

Let the separable solution be

$$T = X(x)Y(y)\beta(t)$$

Substituting into PDE , we get

$$\frac{X''}{X} + \frac{Y''}{Y} = \frac{1}{a}\frac{\beta'}{\beta} = -\lambda^2$$

Then

$$\beta' + a\lambda^2\beta = 0$$

$$\frac{X''}{X} = -(\lambda^2 + \frac{Y''}{Y} = -\ p^2(say)$$

Hence,

$$X'' + p^2 X = 0$$

$$\frac{Y''}{Y} = -\lambda^2 + p^2 = -q^2(say)$$

Therefore ,

$$Y'' + q^2 Y = 0$$

Thus, the general solution of the given PDE is

$$T(x,y,t) = (A\cos px + B\sin px)(c\ \cos qy\ + D\ \sin qy\ )e\text{^}{-a}\ ^2 t$$

where

$$\lambda^2 = p^2 + q^2$$

Using the BC: $T(0\ ,y\ ,t) = 0\ , we\ get\ A = 0$. Then, the solution is of the form

$$T(x,y,t)=B\sin px\,(c\cos qy+D\sin qy)e^{-a\lambda^2}$$

Applying the BC: $T(x,0,t)=0$, $we\ get\ c=0$ Thus , the solution is given by

$$T(x,y,t)=BD\sin px\ \sin qye^{-a\lambda^2 t}$$

Applying the BC : $T(a,y,t)=0$ gives $\sin pa=0$, implying $pa=n\pi$

or

$$p-\frac{n\pi}{a}\ ,\ n-1,2,\ldots$$

Using the principle of superposition the solution can be written in the form

$$(x,y,t)=\sum_{n=1}^{\infty}A_n\ \sin\left(\frac{n\pi}{a}x\right)\ \sin qye^{-a\lambda^2 t}$$

Using the solution is found to be

$$T(x,y,t)=\sum_{m=1}^{\infty}\sum_{n=1}^{\infty}A_{mn}\ \sin\left(\frac{n\pi}{a}x\right)\ \sin\left(\frac{m\pi}{b}\right)e^{-a\lambda^2 t}$$

$$\lambda^2=p^2+q^2=\pi^2\left(\frac{m^2}{b^2}+\frac{n^2}{a^2}\right)$$

Finally, using the IC , we get

$$T(x,y,0)-f(x,y)-\Sigma A_{mn}\sin\left(\frac{n\pi}{u}x\right)\sin\left(\frac{m\pi}{b}y\right)$$

which is a double Fourier series , where

$$A_{mn}=\frac{2}{a}\cdot\frac{2}{a}\int_0^u\int_0^b f(x,y)\left(\frac{m\pi}{a}x\right)\sin\left(\frac{n\pi}{b}y\right)dx\,dy$$

Hence, the required general solution is

$$T(x,y,t)=\sum_{m=1}^{\infty}\sum_{n=1}^{\infty}f(m,n)e^{-a\lambda^2 t}\sin\left(\frac{m\pi}{a}x\right)\sin\left(\frac{n\pi}{b}y\right)dxdy$$

and

$$\lambda^2 = \pi^2\left(\frac{m^2}{b^2} + \frac{n^2}{a^2}\right)$$

6.5.1. Solution of Diffusion Equation in Cylindrical Coordinates

Consider a three dimensional equation

$$\frac{\partial T}{\partial t} = a\nabla^2 T$$

In cylindrical coordinates (r, θ, z), it becomes

$$\frac{1}{a}\frac{\partial T}{\partial t} = \frac{\partial^2 T}{\partial r^2} + \frac{1}{r}\frac{\partial T}{\partial r} + \frac{1}{r^2}\frac{\partial^2 T}{\partial \theta^2} + \frac{\partial^2 T}{\partial z^2} \tag{58}$$

where $T = T(r, \theta, z, t)$.

Let us assume separation of variables in the form

$$T(r, \theta, z, t) = R(r)H(\theta)Z(z)\beta(t)$$

Substituting into Eq. (58) , it becomes

$$R'' HZ\beta + \frac{1}{r}R' HZ\beta + \frac{1}{r^2}H'' RZ\beta + Z'' RH\beta = \frac{\beta'}{\alpha}RHZ$$

or

$$\frac{R''}{R} + \frac{1}{r}\frac{R'}{R} + \frac{1}{r^2}\frac{H''}{H} + \frac{Z''}{Z} = \frac{1}{a}\frac{\beta'}{\beta} = -\lambda^2$$

where $-\lambda^2$ is separation constant . Then

$$\beta' + a\lambda^2\beta = 0 \tag{59}$$

$$\frac{R''}{R} + \frac{1}{R}\frac{R'}{R} + \frac{1}{r^2}\frac{H''}{H} + \lambda^2 = -\frac{Z''}{Z} = -\mu^2(say)$$

Thus, the equation determining $Z, R \, and \, H$ become

$$Z'' - \mu^2 Z = 0 \tag{60}$$

$$\frac{R''}{R}+\frac{1}{r}\frac{R'}{R}+\frac{1}{r^2}\frac{H''}{H}+\lambda^2+\mu^2=0$$

or

$$r^2\frac{R''}{R}+r\frac{R'}{R}+\left(\lambda^2+\mu^2\right)r^2=-\frac{H''}{H}=v^2\ (say)$$

Therefore ,

$$H''+v^2H=0 \qquad (61)$$

$$R''+\frac{1}{r}R'+\left[\left(\lambda^2+\mu^2\right)-\frac{v^2}{r^2}\right]R=0 \qquad (62)$$

Equation $(59)-(61)$ have particular solution of the form

$$\beta=e^{-a\lambda^2 t}$$

$$H=c\ \cos c\theta+D\ \sin v\theta$$

$$Z=Ae^{\mu z}+Be^{-\mu z}$$

The differential equation (62) is called Bessel's equation of order v and its general solution is known as

$$R(r)=c_1 j_v\left(\sqrt{\lambda^2+\mu^2 r}\right)+c_2 Y_v\left(\sqrt{\lambda^2+\mu^2 r}\right)$$

where $j_v(r)\,and\,Y_v(r)$ are Bessel functions of order v of the first and second kind, respectively. Of course, $Eq.(62)$ is singular when $r=0$. The physically meaningful solutions must be twice continuously differentiable in $0\le r<a$. Hence, $Eq.(62)$ has only one bounded solution, i.e.,

$$R(r)=j_v\left(\sqrt{\lambda^2+\mu^2 r}\right)$$

Finally, the general solution of $Eq.(58)$ is given by

$$T(r,\theta,z,t)=e^{-a\lambda^2 t}\left[Ae^{\mu z}+Be^{-\mu z}\right]\left[C\cos v\theta+D\sin v\theta\right]j_v\left(\sqrt{\lambda^2+\mu^2 r}\right)$$

Example. Determine the temperature $T(r,t)$ in the infinite cylinder $0 \le r \le a$ when the initial temperature is $T(r,0)=f(r), and$ the surface $r=a$ is maintained at 0^0 temperature.

Solution. The governing PDE from the data of the problem is

$$\frac{\partial T}{\partial t}=a\nabla^2 T$$

where T is a function of r and t only. Therefore,

$$\frac{\partial^2 T}{\partial r^2}+\frac{1}{r}\frac{\partial T}{\partial r}=\frac{1}{a}\frac{\partial T}{\partial t} \tag{63}$$

The corresponding boundary and initial coordination are given

BC: $T(a,t)=0$

IC: $T(r,0)=f(r)$ (64)

The general solution of $Eq.(63)$ is

$$T(r,t)=A\exp\left(-a\lambda^2 t\right)j_0(\lambda r)$$

Using the BC (64) , we obtain

$$j_0(\lambda a)=0$$

which has an infinite number of roots, $\xi_n a(n=1,2,\dots,\infty)$. Thus , we get from the superposition principle the equation

$$T(r,t)=\sum_{n=1}^{\infty}A_n\exp\left(-a\xi_n^2 t\right)j_0(\xi_n r)$$

Now using the IC: $T(r,0)=f(r), we get$

$$f(r)=\sum_{n=1}^{\infty}A_n j_0(\xi_n r)$$

To compute A_n, we multiply both side of the above equation by $rj_0(\xi_m r)$ and integrate with respect to r to get

$$\int_0^a r f(r) j_0(\xi_m r) dr = \sum_{n=1}^{\infty} A_n \int_{0=}^{a} r j_0(\xi_m r) J_0(\xi_n r) dr$$

$$= \begin{cases} 0 & for\, n \neq m \\ A_m\left(\dfrac{a^2}{2}\right) J_1^2(\xi_m a) & for\, n = m \end{cases}$$

which gives

$$A_m = \frac{2}{a^2 j_1^2(\xi_m a)} \int_0^a u f(u) j_0(\xi_m u) du$$

Hence, the final solution of the problem is given by

$$T(r,t) = \frac{2}{a^2} \sum_{m=1}^{\infty} \frac{j_0(\xi_m r)}{j_1^2(\xi_m a)} exp\left(-\alpha \xi_m^2 t\right) \left[\int_0^a u f(u) j_0(\xi u) du\right]$$

6.5.2. Solution of Diffusion Equation In spherical Coordinates

In this section, we shall examine the solution of diffusion or heat conduction equation in the spherical coordinates system. Let us consider the three – dimensional diffusion Eq. (9) and let $T = T(r, \theta, \phi, t)$. In the spherical coordinates system Eq. (9) can be written as

$$\frac{\partial^2 T}{\partial r^2} + \frac{2}{r}\frac{\partial T}{\partial r} + \frac{1}{r^2 \sin\theta}\frac{\partial}{\partial \theta}\left(\sin\theta \frac{\partial T}{\partial \theta}\right) + \frac{1}{r^2 \sin^2\theta}\frac{\partial^2 T}{\partial \phi^2} = \frac{1}{\alpha}\frac{\partial T}{\partial t} \tag{65}$$

This equation is separated by assuming the temperature function T in the form

$$T - R(r)H(\theta)\phi(\phi)\beta(t) \tag{66}$$

Substituting $Eq.(66)\, into\, Eq.(65)$, we get

$$\frac{R''}{R} + \frac{2}{r}\frac{R'}{R} + \frac{1}{r^2 \sin\theta}\frac{1}{H}\frac{d}{d\theta}\left(\sin\theta \frac{dH}{d\theta}\right) + \frac{1}{\phi r^2 \sin^2\theta}\frac{d^2\phi}{d\phi^2} = \frac{1}{a}\frac{\beta'}{\beta} = \lambda^2 (say)$$

where λ^2 is a separation constant. Thus,

$$d\beta + \lambda^2 \alpha\beta = 0$$

whose solution is

$$\beta = c_1 e^{a\lambda^2 t} \tag{67}$$

Also,

$$r^2 \sin^2\theta \left[\frac{1}{R}\left(\frac{d^2R}{dr^2} + \frac{2}{r}\frac{dR}{dr} \right) + \frac{1}{Hr^2 \sin\theta\, d\theta}\left(\sin\theta \frac{dH}{d\theta} \right) + \lambda^2 \right] = -\frac{1}{\Phi}\frac{d^2\Phi}{d\phi^2} = m^2\,(say).$$

ich gives

$$\frac{d^2\Phi}{d\phi} + m^2\,\Phi = 0$$

whose solution is

$$\Phi(\phi) = c_1 e^{im\phi} + c_2 e^{im\phi} \tag{68}$$

Now, the other separated equation is

$$\frac{1}{R}\left(\frac{d^2R}{dr^2} + \frac{2}{r}\frac{dR}{dr} \right) + \frac{1}{Hr^2 \sin\theta}\frac{d}{d\theta}\left(\sin\theta \frac{dH}{d\theta} \right) + \lambda^2 = \frac{m^2}{r^2 \sin^2\theta}$$

or

$$\frac{r^2}{R}\left(R'' + \frac{2}{r}R' \right) + \lambda^2 r^2 = \frac{m^2}{\sin^2\theta} - \frac{1}{H \sin\theta}\frac{d}{d\theta}\left(\sin\theta \frac{dH}{d\theta} \right)$$

On re-arrangement, this equation can be written as

$$R'' + \frac{2}{r}R' + \left\{ \lambda^2 - \frac{n(n+1)}{r^2} \right\} R = 0 \tag{69}$$

and

$$-\frac{1}{H \sin\theta}\left(\sin\theta \frac{d^2H}{d\theta^2} + \cos\theta \frac{dH}{d\theta} \right) + \frac{m^2}{\sin^2\theta} = n(n+1)$$

or

$$\frac{d^2H}{d\theta} + \cot\theta \frac{dH}{d\theta} + \left\{ n(n+1) - \frac{m^2}{\sin^2\theta} \right\} H = 0 \tag{70}$$

Let $R = (\lambda r)^{-1/2}\, \psi(\mathrm{r})$ then Eq. (69) becomes

$$(\lambda r)^{-\frac{1}{2}}\left[\psi''(r)+\frac{1}{r}\psi'(r)+\left\{\lambda^2-\frac{\left(n+\frac{1}{2}\right)^2}{r^2}\right\}\psi\right]=0$$

Since $(\lambda r)\neq 0$, we have

$$\psi''(r)+\frac{1}{r}\psi'(r)+\left\{\lambda^2-\frac{\left(n+\frac{1}{2}\right)Z^2}{r^2}\right\}\psi(r)=0$$

which is Bessel's differential equation of order $\left(n+\frac{1}{2}\right)$, whose solution is $\psi(r)Aj_{n+\frac{1}{2}}(\lambda r)+BY_{n+\frac{1}{2}}(\lambda r)$

Therefore

$$R(r)=(\lambda r)^{-\frac{1}{2}}\left[Aj_{n+\frac{1}{2}}(\lambda r)+BY_{n+\frac{1}{2}}(\lambda r)\right] \tag{71}$$

where j_n *and* Y_n are Bessel functions of first and second kind, respectively . Now, $Eq.(70)$ can be put in a more convenient from by introducing a new independent variable, $\mu=\cos\theta$.

So that

$$\cot\theta=\mu/\sqrt{1-\mu^2}$$

$$\frac{dH}{d\theta}=-\sqrt{1-\mu^2}\frac{dH}{d\mu}$$

$$\frac{d^2H}{d\theta^2}=\left(1-\mu^2\right)\frac{d^2H}{d\mu^2}-\mu\frac{dH}{d\mu}$$

Thus, $Eq.(70)$ becomes

$$\left(1-\mu^2\right)\frac{d^2H}{d\mu^2}-2\mu\frac{dH}{d\mu}+\left[n\left(n+1\right)-\frac{m^2}{1-\mu^2}\right]H=0 \tag{72}$$

which is an associated Legendre differential equation whose solution is

$$H\left(\theta\right)=A'\,p_n^m\left(\mu\right)+B'Q_n^m\left(\mu\right) \tag{73}$$

$where\, p_n^m\left(\mu\right) and\, Q_n^m\left(\mu\right)$ are associated Legendre functions of degree n and of order $m,$ of the and second kind, respectively. Hence the physically meaningful general solution of the diffusion equation in spherical geometry is of the form

$$T\left(r,\theta,\phi,t\right)=\sum_{\lambda,m,n} A_{\lambda mn}\left(\lambda r\right)^{-\frac{1}{2}} j_{n+\frac{1}{2}}\left(\lambda r\right)p_n^m(\cos\theta)e^{\pm im\phi-a\lambda^2 t} \tag{74}$$

In this general solution, the functions $Q_n^m\left(\mu\right)$ and $\left(\lambda r\right)^{-\frac{1}{2}} Y_{n+\frac{1}{2}}\left(\lambda r\right)$ are exclude because these functions have poles at $\mu=\pm1 \, and\, r=0$ respectively.

Example. Find the temperature in a sphere of radius $a.$ when its surface is kept at zero temperature and its initial temperature is $f\left(r,\theta\right).$

Solution. Here, the temperature is governed by three-dimensional heat equation in spherical polar coordinates independent of $\cdot$ Therefore, the task is to find the solution of PDE

$$\frac{1}{\alpha}\frac{\partial T}{\partial t}=\frac{\partial^2 T}{\partial r^2}+\frac{2}{r}\frac{\partial T}{\partial r}+\frac{1}{r^2\sin\theta}\frac{\partial}{\partial\theta}\left(\sin\theta\frac{\partial T}{\partial\theta}\right) \tag{75}$$

Subject to

BC: $T\left(a,\theta,t\right)=0$

IC: $T\left(r,\theta,0\right)=f\left(r,\theta\right)$ (76)

The general solution of $Eq.\left(75\right)$ with the help of $Eq.\left(74\right),$ can be written as

$$T\left(r,\theta,t\right)=\sum_{\lambda,n} A_{\lambda n}\left(\lambda r\right)^{-\frac{1}{2}} j_{n+\frac{1}{2}}\left(\lambda r\right)p_n\left(cos\,\theta\right)e^{-a\lambda^2 t} \tag{78}$$

Applying the $BC(76)$, we get

$$j_{n+\frac{1}{2}}(\lambda a)=0$$

This equation has infinitely many positive roots. Denoting them by ξ_i, we have

$$T(r,\theta,t)=\sum_{n=0}^{\infty}\sum_{n=1}^{\infty}A_{ni}(\xi_i r)^{-\frac{1}{2}}J_{n+\frac{1}{2}}(\xi_i r)P_n(\cos\theta)\exp\left(-\alpha\xi_t^2 t\right) \tag{79}$$

Now applying the IC and denoting $\cos\theta\, by\, \mu,$ we get

$$f\left(r,\cos^{-1}(\mu)\right)-\sum_{n=0}^{\infty}\sum_{i=1}^{\infty}A_{ni}(\xi_1 r)^{-1/2}J_{n+\frac{1}{2}}(\xi_i r)p_n(\mu).$$

Multiplying both side by $p_n(\mu)d\mu$ and integrating between the limits, $-1 to 1,$ we obtain

$$\int_{-1}^{1}f\left(r,\cos^{-1}(\mu)\right)p_m(\mu)d\mu=\sum_{n=0}^{\infty}\sum_{i=1}^{\infty}A_{ni}(\xi_i r)^{-1/2}j_{n+1/2}(\xi_i)\int_{-1}^{1}p_m(\mu)p_n(\mu)d\mu$$

$$=\sum_{n=0}^{\infty}\sum_{i=1}^{\infty}A_{ni}(\xi_i r)^{-\frac{1}{2}}j_{n+\frac{1}{2}}(\xi_i r)\left(\frac{2}{2n+1}\right)$$

or

$$\left(\frac{2n+1}{2}\right)\int_{-1}^{1}p_n(\mu)f\left(r,\cos^{-2}(\mu)\right)d\mu=\sum_{i=1}^{\infty}A_{ni}(\xi_i r)^{\frac{1}{2}}j_{n+\frac{1}{2}}(\xi_i r)\, for\, n=0\ ,1,2,3,\ldots$$

Now, to evaluate the constants A_{ni}, we multiply both sides of the above equation by $r^{\frac{3}{2}}j_{n+\frac{1}{2}}(\xi_j r)$ and integrating with respect to r between limits 0 to a and use the orthogonality property of Bessel functions to get

$$\xi_i^{\frac{1}{2}}\left(\frac{2n+1}{2}\right)\int_0^a r^{\frac{3}{2}}j_{n+\frac{1}{2}}(\xi_j r)dr\left[\int_{-1}^{1}p_n(\mu)f\left(r,\cos^{-1}(\mu)\right)d\mu\right]$$

$$=\sum_{i=1}^{\infty} A_{ni} \int_0^a r j_{n+\frac{1}{2}}(\xi_i r) j_{n+\frac{1}{2}}(\xi_j r)\, dr$$

$$=\frac{1}{2}\sum_{i=1}^{\infty} A_{ni}\left[j'_{n+\frac{1}{2}}(\xi_i r)\right]^2, \; n=0,1,2,3,\ldots \tag{80}$$

Thus, Eq. (79) and (80) together constitute the solution for the given problem.

6.6. MAXIMUM-MINIMUM PRINCIPLE AND CONSEQUENCES

Theorem. (Maximum- minimum principle) .

Let $u(x,t)$ be a continuous function and a solution of

$$u_t = \alpha u_{xx} \tag{81}$$

for $0 \le x \le l, 0 \le t \le T$, where $T > 0$ is a fixed time. Then the maximum and minimum values u are attained either at time $t=0$ or at the end points $x=0$ and $x=l$ at some time in the interval $0 \le t \le T$.

Proof. To start with, let us assume that the assertion is false . Let the maximum value of $u(x,t)$ for $t=0$ $(0 \le x < l)$ or for $x=0$ or $x=l$ $(0 \le t \le T)$ be denoted by M. we shall assume that the function $u(x,t)$ attains its maximum at some interior point (x_0, t_0) in the rectangle defined by $0 \le x_0 \le l$, $0 \le t_0 \le T$, and them arrive at a contradiction. This mean that

$$u(x_0, t_0) = M + \varepsilon \tag{82}$$

Now, we shall compare the signs in Eq. (81) at point (x_0, t_0), it is well known from calculus that the necessary condition for the function $u(x,t)$ to posses maximum at (x_0, t_0) is.

$$\frac{\partial u}{\partial x}(x_0, t_0) = 0, \quad \frac{\partial^2 u}{\partial x^2}(x_0, t_0) \le 0 \tag{83}$$

In addition, $u(x_0, t_0)$ attains maximum for $t = t_0$, implying

$$\frac{\partial u}{\partial t}(x_0, t_0) \geq 0 \tag{84}$$

Thus, with the help of Eq. (83) and (84) we observe that the signs on the left - and right - hand sides of Eq. (81) are different. However, we cannot claim that we have reached contradiction, since the left- and right hand sides can simultaneously by zero. To complete the proof, let us consider another point (x_l, t_l) at which $\frac{\partial^2 u}{\partial x^2} \leq 0 \, and \, \frac{\partial u}{\partial t} > 0.$

$$v(x,t) = u(x,t) + \lambda(t_0 - t) \tag{85}$$

Where λ isaconstant.Obviously, $v(x_0, t_0) = M + \varepsilon \, and \, \lambda(t_0 - t) \leq \lambda T$.

Suppose we choose $\lambda > 0$, such that $\lambda < \frac{\varepsilon}{2T}$; then the maximum of $v(x,t)$ for $t = 0$ or for $x = 0, x = l$ cannot exceed the value $M + \frac{\varepsilon}{2}$. But $v(x,t)$ is a continuous function and , therefore ,a point (x_1, t_1) exists at which it assumes its maximum. It implies

$$M + \frac{\varepsilon}{2} \leq v(x_1, t_1) \geq v(x_0, t_0) = M + \varepsilon$$

This pair of inequalities is inconsistent and therefore contradicts the assumption that v takes on its maximum at (x_0, t_0). Therefore, the assertion that u attains its maximum either at $t = 0$ or at end points is true .

We can establish a similar results for minimum values of $u(x,t)$ .if u satisfies Eq. (81) , $-u$ also satisfies Eq. (81) . Hence , both maximum and minimum values are attained either initially or at the end points. Thus the proof is complete. We Shall give some of the consequences of the maximum-minimum principle in the following theorems We Shall give some of the consequences of the maximum-minimum principle in the following theorems.

Theorem. (Uniqueness theorem). Given a rectangular region defined by $0 \le x \le l, 0 \le t \le T$, and a continuous function (u, x, t) defined on the boundary of the rectangle satisfying the heat equation

$$u_t = \alpha u_{xx}$$

This equation possess one and only one solution satisfying the initial and boundary conditions

$$u(x,0) = f(x)$$

$$u(0,t) = g_1(t), u(l,t) = g_2(t)$$

where $f(x), g_1(t), g_2(t)$ are continuous on their domains of definition.

Proof. Suppose there are two solution $u_1(x,t), u_2(x,t)$ satisfying the heat equation as well as the same initial and boundary conditions. Now let us consider the difference

$$v(x,t) = u_2(x,t) - u_1(x,t)$$

Itisalsoasolutionoftheheatconductionequationfor $0 \le x \le l, 0 \le t \le T$ andis continuous in $x \, and \, t$. Also, $v(x,t) = 0, 0 \le x \le l \, and \, v(l,t) = 0, 0 \le t \le T$. Hence. $v(x,t)$ satisfies the condition required for the application of maximum –minimum principle .Thus, $v(x,t) = 0$ in the rectangular region defined by $0 \le x \le l, 0 \le t \le T$. it follows therefore that $u_1(x,t) = u_2(x,t)$.

Another important consequence of the maximum-minimum principle is the stability property which stated in the following theorem without proof.

Theorem. (stability) theorem. The Solution $u(x,t)$ of the Dirichlet problem

$$u_t = \alpha u_{xx}, \quad 0 \le x \le l, 0 \le t \le T$$

$$u(x,0) = f(x), \quad 0 \le x \le l$$

$$u(0,t) = g(t), \quad u(l,t) = h(t), 0 \le t \le T$$

depend continuously on the initial and boundary conditions.

6.7. MISCELLANEOUS EXAMPLE.

Example. A homogenous solid sphere of radius R has the initial temperature distribution $f(r), 0 \le r \le R$. where r is the distance measured form the centre. The surface temperature is maintained at 0^0. Show that the temperature $T(r,t)$ in the sphere is the solution of

$$T_1 = c^2\left(T_{rr} + \frac{2}{r}T_r \right)$$

where c^2 is a constant. Show also that the temperature in the sphere for $t > 0$ is given by

$$T(r,t) = \frac{1}{r}\sum_{n=1}^{\infty} B_n \sin\left(\frac{n\pi}{R} r \right)\exp(-\lambda_n^2 t,)\ \lambda_n = \frac{cn\pi}{R}$$

Solution. The temperature distribution in a solid sphere is governed by the parabolic heat equation

$$T_t = c^2\ \nabla^2 T$$

From the data, T is a function of r and t alone. In view of the symmetry of the sphere the above equation with the help of Eq. reduce to

$$T_t = c^2\left(T_{rr} + \frac{2}{r}T_r \right) \qquad (86)$$

Setting $v = rT$, the given BC gives

$$v(R,t) = rt(R,t) = 0$$

which the IC gives

$$v(r,0) = rT(r,0) = rf(r)$$

Since T must be bounded at $r = 0$, we require

$$v(0,t) = 0$$

Now,

$$v_t = rT_t,\ T_r = \left(\frac{v}{r} \right)_r = \frac{v_r v - v}{r^2}$$

Similarly, finding T_{rr} and substituting into Eq. (86), we obtain

$v_t = c^2 v_{rr}$

Using the variables separable method, we may write $v(r,t) = R(r)\tau(t)$ and get

$$R(r) = A\cos kr + B\sin kr$$

$$\tau(t) = \exp\left(-c^2 k^2 t\right)$$

Thus, using the principle of superposition , we get

$$v(r,t) = \sum_{n=1}^{\infty} (A_n \cos kr + B_n \sin kr)\big|_{r=0} = 0$$

Also, using $v(0,t) = 0$, we have

$$(A_n \cos kr + B_n \sin kr)\big|_{r=0} = 0$$

Implying $A_n = 0.$ Also, $v(R,t) = 0$ gives $B_n \sin kr = 0$, implying $\sin kr = 0$, as $B_n \neq 0$. Therefore,

$$kR = n\pi,\ k = \frac{n\pi}{R},\ n = 1, 2, \ldots$$

Thus, the possible solution is

$$v(r,t) = \sum_{n=1}^{\infty} B_n \sin\left(\frac{n\pi}{R} r\right) \exp\left(-\frac{c^2 n^2 \pi^2 t}{R^2}\right)$$

Finally, applying the IC: $v(r,0) = rf(r)$, we get

$$rf(r) = \sum_{n=1}^{\infty} B_n \sin\left(\frac{n\pi}{R} r\right)$$

which is a half-range Fourier series . Therefore,

$$B_n = \frac{2}{R} \int_0^R rf(r) \sin\left(\frac{n\pi}{R} r\right) dr$$

But $v(r,t) = rT(r,t)$. Hence, the temperature in the sphere is given by

$$T(r,t) = \frac{1}{r}\sum_{n=1}^{\infty} B_n \sin\left(\frac{n\pi}{R} r\right) \exp -\left(\frac{c^2 n^2 \pi^2 t}{R^2}\right)$$

Example. A circular cylinder of radius a has its surface kept at a constant temperature T_0. If the initial temperature is zero throughout the cylinder, prove that for $t > 0$.

$$T(r,t) = T_0\left\{1 - \frac{2}{a}\sum_{n=1}^{\infty} \frac{j_0(\xi_n a)}{\xi_n j_1(\xi_n a)} \exp\left(-\xi_n^2 kt\right)\right\}$$

where $\pm\xi_1, \pm\xi_2, \ldots, \pm\xi_n$ are the roots of $j_0(\xi a) = 0$, and k is the thermal conductivity which is a constant.

Solution. It is evident that T is a function of r and t alone and, therefore, the PDE to be solved is

$$\frac{\partial^2 T}{\partial r^2} + \frac{1}{r}\frac{\partial T}{\partial r} = \frac{1}{k}\frac{\partial T}{\partial t} \tag{87}$$

Subject to

$$IC: T(r,0) = 0,\ 0 \le r < a$$

$$BC: T(a,t) = T_0,\ t \ge 0$$

Let

$$T(r,t) - T_0 + T_1(r,t)$$

So that

$$T_1(r,0) = -T_0 \tag{88}$$

$$T_1(a,t) = 0 \tag{89}$$

where T_1 is the solution of Eq. (87). By the variables separable method we have,

$$T_1(r,t) = Aj_0(\lambda r)\exp\left(-\lambda^2 kt\right)$$

Using the BC: $T_1(a,t)=0$, we get

$$Aj_0(\lambda a)\exp(-\lambda^2 kt)=0$$

which gives $j_0(\lambda a)=0\, as\, A\neq 0 \,.let\, \xi_1, \xi_2, \ldots, \xi_n,$ be the roots of $j_0(\lambda a)=0$. Then the possible solution using the superposition principle is

$$T_1(r,t)=\sum_{n=1}^{\infty} A_n\, j_0(\xi_n r)\exp(-\xi_n^2 kt) \tag{90}$$

Using the $IC: T_1(r,0)=-T_0$ into Eq. (90) , we obtain

$$\sum_{n=1}^{\infty} A_n\, j_0(\xi_n r)=-T_0$$

Multiply both sides by $rj_0(\xi_m r)$ and integrating, we get

$$-T_0\int_0^a rj_0(\xi_m r)\,dr=\sum_{n=1}^{\infty} A_n\int_0^a r\, j_0(\xi_m r)\, j_0(\xi_n r)\,dr$$

$$=A_m\int_0^a rj_0^2(\xi_m r)\, if\ m=n\ ;otherwise\, 0$$

$$=A_m\frac{a^2}{2}j_1^2(\xi_m a)$$

But,

$$-T_0\int_0^a rj_0(\xi_m r)\,dr=-T_0\int_0^{\xi_m a}\frac{x}{\xi_m}j_0(x)\frac{dx}{\xi_m}(x=\xi_m r)$$

$$=-\frac{T_0}{\xi_m^2}\int_0^{\xi_m a}\frac{d}{dx}\left[xj_1(x)\right]dx$$

$$=-\frac{T_0}{\xi_m^2}\left[xj_1(x)\right]_0^{\xi_m a}=-\frac{aT_0}{\xi m}j_1(\xi_m a)$$

Therefore,

$$A_m \frac{a^2}{2} J_1^2(\xi_m a) = -\frac{aT_0}{\xi_m} j_1(\xi_m a)$$

or

$$A_n = -\frac{2T_0}{n\xi_n} \frac{1}{j_1(\xi_n a)}$$

Hence, Eq. (90) becomes

$$T_1(r,t) = -\frac{2}{a} T_0 \sum_{n=1}^{\infty} \frac{j_0(\xi_n r)}{j_1(\xi_n a)} \frac{\exp(-\xi_n^2 kt)}{\xi_n}$$

Finally, the complete solution is found to be

$$T(r,t) = T_0 \left[1 - \frac{2}{a} \sum_{n=1}^{\infty} \frac{j_0(\xi_n r)}{j_1(\xi_n a)} \frac{\exp(-\xi_n^2 kt)}{\xi_n} \right]$$

Example. Determine the temperature in a sphere of radius a, when its surface is maintained at zero temperature while its initial temperature if $f(r,\theta)$.

Solution. Here the temperature is governed by the three-dimensional heat equation in polar coordinates independent of ϕ, which is given by

$$\frac{1}{k}\frac{\partial u}{\partial t} = \frac{\partial^2 u}{\partial r^2} + \frac{2}{r}\frac{\partial u}{\partial r} + \frac{1}{r^2 \sin\theta}\frac{\partial}{\partial \theta}\left(\sin\theta \frac{\partial u}{\partial \theta} \right) \tag{91}$$

Let

$$u(r,\theta,t) - R(r)H(\theta)T(t)$$

By the variables separable method, the general solution of Eq. (91) is found to be

$$v(r,\theta,t) = \sum_{\lambda}\sum_{n} A_{\lambda n} (\lambda r)^{-\frac{1}{2}} j_{n+\frac{1}{2}}(\lambda r) p_n(\cos\theta) \exp(-k\lambda^2 t) \tag{92}$$

In the present problem, the boundary and initial conditions are

$$BC: u(a,\theta,t) = 0 \tag{93}$$

$$IC : u(r,\theta,0) = f(r,\theta) \tag{94}$$

Substituting the BC (93) into Eq. (92), we get

$$j_{n+\frac{1}{2}}(\lambda a) = 0 \tag{95}$$

Let $\xi_1 a, \xi_2 a, \ldots, \xi_1 a, \ldots$ be the roots of Eq. (95). Then the general solution can be put in the form

$$u(r,\theta,t) = \sum_{n=1}^{\infty}\sum_{i=1}^{\infty} A_{ni}(\xi_l r)^{-\frac{1}{2}} j_{n+\frac{1}{2}}(\xi_l r) p_n(\cos\theta)\exp(-k\xi_i^2 t) \tag{96}$$

Now using the IC, we obtain

$$f(r,\theta) = \sum_{n=1}^{\infty}\sum_{i=1}^{\infty} A_{ni}(\xi_i r)^{-\frac{1}{2}} J_{n+\frac{1}{2}}(\xi_i r) p_n(\cos\theta)$$

Multiplying both sides by $p_n(\cos\theta)d(\cos\theta)$ and integrating, we have

$$\int_{-1}^{1} p_n(\cos\theta) f(r,\theta) d(\cos\theta) = \sum_{n=1}^{\infty}\sum_{i=1}^{\infty} A_{ni}(\xi_i r)^{-\frac{1}{2}} j_{n+\frac{1}{2}}(\xi_i r) \int_{-1}^{1} p_n^2(\cos\theta) d(\cos\theta)$$

Using the orthogonality property of Legendre polynomials, we get

$$\int_{-1}^{1} p_n(\cos\theta) f(r,\theta) d(\cos\theta) = \sum_{n=1}^{\infty}\sum_{i=1}^{\infty} A_{ni}(\xi_i r)^{-\frac{1}{2}} j_{n+\frac{1}{2}}(\xi_i r)\left(\frac{2}{2n+1}\right)$$

Rearranging and multiplying both sides of the above equation by $r^{3/2} J_{n+\frac{1}{2}}(\xi_i r)$ and integrating between the limits 0 to a with respect to r, we get

$$\frac{2n+1}{2}\int_0^a r^{\frac{3}{2}} j_{n+\frac{1}{2}}(\xi_i r) dr \int_{-1}^{1} p_n(\cos\theta) f(r,\theta) d(\cos\theta) = A_{ni}\int_0^a r j_{n+\frac{1}{2}}^2 (\xi_i^{-\frac{1}{2}} dr$$

$$= A_{ni}\frac{a^2}{2}\left\{ j'_{n+\frac{1}{2}}(\xi_i a)\right\}^2$$

Therefore,

$$A_{ni} = \frac{(2n+1)\xi_i^{1/2}}{a^2\left\{j'_{n+\frac{1}{2}}(\xi_i a)\right\}^2}\int_0^a J_{n+1/2}(\xi_i r)\,dr\int_{-1}^{1} p_n(\cos\theta) f(r,\theta)\,d(\cos\theta) \tag{97}$$

Hence, we obtain the solution to the given problem for Eq. (96), where A_{ni} is given by Eq. (97).

Example. The heat conduction in a thin round insulated rod with heat sources present is described by the PDE

$$u_t - au_{xx} = F\frac{(x,t)}{\rho c},0 \qquad 0 < x\langle l,t\rangle 0 \tag{98}$$

Subject to '

$$BCs: u(0,t) = u(l,t) = 0$$

$$IC: u(x,0) = f(x),\ 0 \le x \le l \tag{99}$$

where ρ *and* c are constant and F is a continuous function of x *and* t. Find $u(x,t)$.

Solution. It can be noted that the boundary conditions are of homogeneous type. Let us consider the homogeneous equation

$$u_t - au_{xx} = 0 \tag{100}$$

Setting $u(x,t) = X(x)T(t), we\ get$

$$\frac{T'}{\alpha T} = \frac{X''}{X} = -\lambda^2(say) \tag{101}$$

which gives $X'' + \lambda^2 X = 0$. The corresponding BCs are

$$X(0) = X(l) = 0$$

The solution of Eq. (101) gives the desired eigen functions and eigen values, which are

$$X_n(x) = \sin\lambda_n x,\ \lambda_n^2 = \left(\frac{n\pi}{l}\right)^2,\ n \ge 1 \tag{102}$$

For the non-homogenous problem (98), let us propose a solution of the form

$$u(x,t) = \sum_{n=l}^{\infty} T_n(t) X_n(x) \tag{103}$$

It is clear that Eq. (103) satisfies the BCs (99).From the orthogonality of eigen functions, it follows that

$$T_m(t) = \frac{2}{l}\int_0^l u(x,t) X_m(x)\,dx$$

However,

$$T_m(0) = \frac{2}{l}\int_0^1 f(x)\sin\left(\frac{m\pi}{l}x\right)dx \tag{104}$$

which is an IC for $T_m(t)$. introducing Eq. (103) into the governing equation (98) . we get

$$\sum_{n=1}^{\infty} T_n' X_n - a\sum_{n=1}^{\infty} T_n X_n'' = \frac{F(x,t)}{\rho c} \tag{105}$$

Now, we shall expand $F\, x,t)/\rho c,$ so that it is represented by a convergent series on $0 < x\langle l,t\rangle 0$ in the form

$$\frac{F}{\rho c} = \sum_{n=1}^{\infty} q_n(t) X_n(x) \tag{106}$$

where

$$q_n(t) = \frac{2}{l}\int_0^1 \frac{F(x,t)}{\rho c}\sin\left(\frac{n\pi}{l}x\right)dx \tag{107}$$

Thus, $q_n(t)$ is known. Now, Eq. (105), with the help of Eq.(101) becomes

$$\sum_{n=1}^{\infty} X_n\left(T_n' + \lambda_n^2 \alpha T_n - q_n\right) = 0$$

Therefore, it follows that

$$T_n'(t) + \lambda_n^2 \alpha T_n(t) = q_n(t) \tag{108}$$

Its solution with the help of IC (104) is

$$T_n(t)=T_n(0)\exp\left(-\lambda_n^2\alpha t\right)+\int_0^t exp\left[\lambda_n^2\,\alpha(\tau-t)\right]q_n(\tau)\,d\tau \tag{109}$$

From Eq. (103) and (109) , the complete solution is found to be

$$u(x,t)=\sum_{n=1}^{\infty}\left[T_n(0)\exp\left(-\lambda_n^2\alpha t\right)+\int_0^1 \exp[\lambda_n^2\alpha(\tau-t)q_n(\tau)\,d\tau\right]X_n(x)$$

In the expanded form, it becomes

$$u(x,t)=\sum_{n=1}^{\infty}\left[\left\{\frac{2}{l}\int_0^l f(\xi)X_n(\xi)\,d\xi\right\}\exp(-\lambda_n^2\,at)\right.$$

$$\left.+\frac{2}{l}\int_0^l \exp\left\{-\lambda_n^2\,a(\tau-t)\right\}\int_0^l\frac{F(\zeta,\tau)}{\rho c}X_n(\xi)\,d\xi\,d\tau\right]X_n(x) \tag{110}$$

It can be verified that the series in Eq. (110) converges uniformly for $t>0$. By changing the order of integration and summation in Eq.(110) , we get

$$u(x,t)=\int_0^l\left[\sum_{n=1}^{\infty}\frac{\exp\left(-\lambda_n^2\,at\right)X_n(x)X_n(\xi)}{1/2}\right]f(\xi)\,d\xi$$

$$+\int_0^1\int_0^1\left[\sum_{n=1}^{\infty}\frac{\exp\left\{-\lambda_n^2\,a(l-t)\right\}X_n(x)X_n(\xi)}{l/2}\right]\frac{F(\xi,t)}{\rho c}\,d\xi\,dt$$

which can also be written in the form

$$u(x,t)=\int_0^1 G(x,\xi;t)f(\xi)\,d\xi+\int_0^1\int_0^1 G(x,\xi;t-\tau)\frac{F(\xi,\tau)}{\rho c}\,d\xi\,d\tau \tag{111}$$

where

$$G(x,\xi;t)=\sum_{n=1}^{\infty}\frac{\exp\left(-\lambda_n^2 at\right)X_n(x)X_n(\xi)}{l/2}$$

is called Green's function. More details on Green's function are given .

6.7.1. Summary and Discussion

The diffusion phenomena such as conduction of heat in solids and diffusion of vorticity in the case of viscous flow past a body are governed by a partial differential equation of parabolic type. For example, the flow of heat in a conducting medium is governed by the parabolic equation

$$\rho C \frac{\partial T}{\partial t} = div\left(K\nabla T\right) + H\left(r.T,t\right) \tag{1}$$

where ρ is the density, C is the specific heat of the solid, T is the temperature at a point with position vector r. K *is* the thermal conductivity , t is the time ,and $H\left(r,T,t\right)$ is the amount of heat generated per unit time in the element dV situated at a point $\left(x,y,z\right)$ whose position vector is r. This equation is known as diffusion equation or heat equation. We shall now derive the heat equation form the basic concepts.

Then the differential equation of heat conduction with heat source is

$$\frac{1}{\alpha}\frac{\partial T\left(r,\right)}{\partial t} = \nabla^2 T\left(r,t\right) + \frac{H\left(r,T,t\right)}{K} \tag{2}$$

In the absence of heat sources, Eq. (8) reduces to

$$\frac{\partial T\left(r,t\right)}{\partial t} = \alpha \nabla^2 T\left(r,t\right) \tag{3}$$

This is called Fourier heat conduction equation or diffusion equation. The fundamental problem of heat conduction is to obtain the solution of Eq. (2) subject to the initial and boundary conditions which are called initial boundary value problems, hereafter referred to as IBVPs.

6.8. BOUNDARY CONDITIONS

The heat conduction equation may have numerous solution unless a set of initial and boundary conditions are specified. The boundary conditions are mainly of three types which we now briefly explain.

Boundary Condition 1: The temperature is prescribed all over the boundary surface. That is, the temperature $T\left(r\ ,t\right)$ is a function of both position and time. In other words, $T = G\left(r\ ,t\right)$ which is some prescribed function on the boundary .This type of boundary condition is called the

Dirichlet condition. Specification of boundary conditions depends on the problem under investigation.

A special case includes $T(r,t)=0$ on the surface of the boundary, which is called a *homogeneous boundary condition.*

Boundary Condition II : The flux of heat , i.e., the normal derivative of the temperature $\frac{\partial T}{\partial n}$, is prescribed on the surface of the boundary . it may be a function of both position and time, i.e.,

$$\frac{\partial T}{\partial n}=f(r,t)$$

This is called the Neumann condition .Sometimes, the normal derivatives of temperature may be a function of position only or a function of time only. A special case includes

$$\frac{\partial T}{\partial n}=0 \, on\, the$$

boundary

This homogenous boundary condition is also called insulated boundary condition which sates the heat flow is zero.

Boundary condition III. A linear combination of the temperature and its normal derivatives is prescribed on the boundary , i.e.,

$$K\frac{\partial T}{\partial n}+hT=G(r,t)$$

Where $K\, and\, h$ are constants. This type of boundary condition is called Robin's condition. it means that the boundary surface dissipates heat by convection .Following Newton's law of the cooling, which states that the rate at which heat is transferred from the body to the surrounding is Proportional to the difference in temperature between the body and the surrounding , we have

$$-K\frac{\partial T}{\partial n}=h(T-T_a)$$

As a special case, we may also

$$K\frac{\partial T}{\partial n}+hT=0$$

which is homogeneous boundary . This mean that hear is converted by dissipation from the boundary surface into a surrounding maintained at zero temperature.

The other boundary conditions such as the heat transfer due to radiation obeying the fourth power temperature law and those associated with change of phase .like melting, ablation, etc., give rise to non-linear boundary conditions.

This limiting function $\delta(t)$ defined by $Eq.(33)$ and (34) is known as Dirac delta function or the unit impulse function. Its profile is depicted in . Dirac originally called it an improper function as there is no proper function with these properties . In fact, we can observed that

$$1=\int_{-\infty}^{\infty}\delta(t)dt=LT_{\varepsilon\to 0}\int_{|t|>\varepsilon}\delta_{\varepsilon}(t)dt=LT_{\varepsilon\to 0}0=0$$

Obviously , this contradiction implies that $\delta(t)$ cannot be a function in the ordinary sense. Some important properties of Dirac delta function are presented now:

Property 1: $\int_{-\infty}^{\infty}\delta(t)dt=1$

Property II: For any continuous functions $f(t)$, $\int_{-\infty}^{\infty}f(t)\delta(t)dt=f(0)$

.

Property III: let $f(t)$ be any continuous function . Then $\int_{-\infty}^{\infty}\delta(t-a)f(t)dt=f(a)$

.

Property IV: $\delta(-t)=\delta(t)$

Property V: $\delta(at)=\frac{1}{a}\delta(t),\ a>0$

Property VI: if $\delta(t)$ is a continuously differentiable . Dirac delta function Vanishing for large t, then

$$\int_{-\infty}^{\infty} f(t)\delta'(t)dt = \left[f(t)\delta(t)\right]_{-\infty}^{\infty} - \int_{-\infty}^{\infty} f'(t)\delta(t)dt$$

Since $\delta t \to 0\, as\, t \to \infty, -\infty$, we have $\int_{-\infty}^{\infty} f(t)\delta'(t)dt = -f'(0)$

operty VII: $\int_{-\infty}^{\infty} \delta'(t-a)f(t)dt = -f'(a)$

Having discussed the one-dimensional Dirac delta function , wc can extend the definition to two dimension, Thus, for every f which is continuous over the region S containing the point (ξ,η), we define $\delta(x-\xi, y-\eta)$ in such a way that

$$\iint_S \delta(x-\xi y-\eta) f(x,y)d\sigma = f(\xi,\eta)$$

Note that $\delta(x-\xi, y-\eta)$. is a forma limit of a sequence of ordinary functions, i.e.,

$\delta(x-\xi, y-\eta) = LT_{\varepsilon \to 0}\, \delta_{\varepsilon}(r)$,Where $r^2 = (x-\xi)^2 + (y-\eta)^2$.

CHAPTER 7

LAPLACE TRANSFORM METHODS

CONTENTS

7.1. INTRODUCTION

Laplace transform is essentially a mathematical tool which can be used to solve several problems in science and engineering. This transform was first introduced was first introduced by Laplace, a French mathematician, in the year 1790 in his work on probability theorem. This technique became popular when Heaviside applied to the solution of an ordinary differential equation referred hereafter as ODE. representing a problem in electrical engineering. To the basic question as to why one should learn Laplace transform technique when other technique available. The answer is very simple. Transform are used to accomplish the solution of certain problems with less effort and in a simple routine way. To illustrate, consider the problem of finding the value of x from the equation simple routine way. To illustrate, consider the problem of finding the value of x from the equation

$$x^{1.85} = 3 \qquad (1)$$

It is an extremely tedious task to solve this problem algebraically. However, taking logarithms on both side, we have the transformed equation as

$$1.85\, In\, x = In\ 3 \qquad (2)$$

In this transformed equation, the algebraic operation and exponentiation have been changed to multiplication which immediately gives

$$x = \frac{In\, 3}{1.85}$$

In

To get the required result, it is enough if we take the antilogarithm on both side of the above equation, which yields

$$x = In^{-1}\left(\frac{In\, 3}{1.85}\right)$$

with the help of any ordinary calculator , we can now compute x. Following this simple example, the Laplace transform method reduce the solution of an ODE to the solution of an algebraic equation. In fact, this method has a particular advantage in finding the solution of an ODE with appropriate. ICs, without first finding the general solution and then using ICs for evaluating the arbitrary constants. Also, when the Laplace transform technique is applied to a PDE, it reduce the number of independent variables by one.

Definition. Suppose $f(t)$ is a piecewise continuous function and if it has an additional property that there exists a real numbers $\tilde{a}_0$ and a finite positive number M such that

$$Lt_{t\to\infty}\left|f(t)\right|e^{-\gamma t}\leq M \quad for\,\gamma>\gamma_0$$

And the limit does not exist when $\gamma<\gamma_0$, then such a function is said to be of exponential order γ_0. Also written as

$$\left|f(t)\right|=0\left(e^{\gamma 0t}\right)$$

Variables such as velocity and current are always finite. Which means that $f(t)$ is bounded. Thus for any bounded function $f(t), |f((t)\,e^{-\gamma t}\to 0$ for all $\gamma>0$. The order of such a function in zero. However, variables such as electrical charge and mechanical displacement may increase without limit but of course proportional to t. Such functions are also of exponential order.

For illustration, let us consider the following examples:

$(i)\ Lt_{t\to\infty}te^{-\gamma t}=0$

The fact that t^n is of exponential order zero can be seen as follows:

$$Lt_{t\to\infty}t^n e^{-\gamma t}=Lt_{t\to\infty}\left(\frac{t^n}{e^{\gamma t}}\right)=Lt_{t\to\infty}\left(\frac{nt^{n-1}}{\gamma e^{\gamma t}}\right)$$ (using L' Hospital's rule)

Applying the L' Hospital rule repeatedly, we get

$$LT_{t\to\infty}t^n e^{-\gamma t}=0\ if\,\gamma>a$$

Thus the function e^{ut} is of exponential order a.

(iii) exp $(t^n)(n>1)$ is not of exponential order, since

$$LT_{t\to\infty}\exp\left(t^n\right)e^{-\gamma t}=LT_{t\to\infty}\exp\left[t^{n-1}-\gamma\,)\right]=\infty$$

for any finite value of γ.

Definition. Let $f(t)$ be a continuous and single-valued function of the real variable t definition for all $t, 0<t<\infty$, and is of exponential order.

Then the Laplace transform of $f(t)$ is defined as a function $F(s)$ denoted by the integral

$$L[f(t);s]=F(s)=\int_0^{\infty} e^{-st} f(t)dt \tag{3}$$

For any finite value of γ .

Over the range of values of s for which the integral exist. Here, s is a parameter, real or complex . Obviously, $L[f(t):s]$ is a function of s Thus,

Where L is the operator which transform $f(t) into F(s)$ called laplace transform operator, and L^{-1} is the inverse Laplace transform operator.

The Laplace transform belongs to the family of " integral transform". An integral transform $F(s)$ pf the function $f(t)$ is defined by an integral of the form

$$\int_c^b k(s,t) f(t)dt = F(s) \tag{4}$$

Where $k(s,t)$, a function of two variables $s and t$, is called the Kernel of the integral transform The kernel and limits of integration of various integral transform are given in Table 1 (which is not exhaustive).

Table 1. Kernel and Limits for Various Integral Transforms

Name of the transform	$k(s,t)$	a	b
Laplace transform	e^{-st}	0	∞
Fourier transform	$e^{ist} / \sqrt{2\pi}$	$-\infty$	∞
Fourier sine transform	$\sqrt{\frac{2}{\pi}} \sin st$	0	∞
Fourier cosine transform	$\sqrt{\frac{2}{\pi}} \cos st$	0	∞
Hankel transform	$tJ_n(st)$	0	∞
Mellin transform	t^{s-1}	0	∞

The integral transform defined above are applicable , either for semi-infinite or infinite domains. Similarly , finite integral transform can be defined domains. Now, we are in a position to verify the following important result.

Theorem. If $f(t)$ is piecewise continuous in the range $t \geq 0$ and is of exponential order γ. then the Laplace transform $F(s) of f(t)$ exists for all $s > \gamma$.

Proof. From the definition of Laplace transform,

$$L[f(t);s] - \int_0^\infty e^{-st} f(t)dt - \int_0^T e^{-st} f(t)dt = I_1 + I_2$$

Since $f(t)$ is piecewise continuous on every finite interval $0 < t < T$,I_1 exists, whereas

$$|I_2| \leq \int_T^\infty |e^{-st} f(t)| dt$$

But (t) is a function of exponential order' therefore,

$$\left|f(t)\right| < Me^{\lambda t} \quad for\ \lambda\ real$$

Hence,

$$\left|e^{-st} f(t)\right| < Me^{-(s-\lambda)t}$$

Thus,

$$\left|I_2\right| = \int_T^{\infty} e^{-(s-\gamma)t} M\,dt = \frac{Me^{-(s-\gamma)T}}{s-\gamma}, \; s > \gamma$$

In other words, I_2 can be made as small as we like provided T is large enough and , therefore I_2 exists. Hence $L[f(t);s]$ exist for $s > \gamma$.

7.2. TRANSFORM OF SOME ELEMENTARY FUNCTIONS

Following the definition of Laplace transform by the integral (3), we shall compute the Laplace transform of some elementary functions.

Example. Find the Laplace transform of

$(i)1 . (ii)0 ,(iii)t , (iv)\ e^{at} , \ (v)e^{-at}$

Solution. Using the definition of Laplace transform, we have

$$(i) L[l\ ;s] = \int_0^{\infty} e^{-st}(1)dt = \left(\frac{e^{-st}}{-s}\right)_0^{\infty} = \frac{1}{s} if\ s > 0$$

(ii)
$$L[0\ ;s\ )] = \int_0^{\infty} e^{-st}(0)dt = 0$$

(iii)
$$L[t\ ;s] = \int_0^{\infty} e^{-st}\ .dt = \int_0^{\infty} td\left(\frac{e^{-st}}{-s}\right) = \left(t\frac{e^{-st}}{-s}\right)_0^{\infty} - \int_0^{\infty} \frac{e^{-st}}{-s}dt = \frac{1}{s^2}$$

(iv)
$$L[t\ ;s] = \int_0^{\infty} e^{-st}\ e^{at}\,dt = \int_0^{\infty} e^{-(s-a)t}\,dt = \left[\frac{e^{-(s-a)t}}{-(s-a)}\right]_0^{\infty} = \frac{1}{s-a} s > a$$

(v) $L\left[e^{-at};s\right] = \int_0^\infty e^{-st}\, e^{-at}\, dt = \int_0^\infty e^{-(s+a)}\, dt = \frac{1}{s+a}$

Example. Find the Laplace transform of

(i) $\cos at$, $(ii) \sin at$.

Solution Following the definition of Laplace transform, we have

$$(i) L[\cos at\,;s] = \int_0^\infty e^{-st}\cos at\;dt = Re\int_0^\infty e^{-st}\,e^{iat}dt = Re\,L\left[e^{iat}\;;s\right]$$

$$Re\frac{1}{s-ia} = Re\frac{s+ia}{s^2+a^2} = \frac{s}{s^2+a^2}$$

(ii) $L[\sin at\,;s] = Im\,L\left[e^{iat}\;;s\right] = Im\frac{s+ia}{s^2+a^2} = \frac{a}{s^2+a^2}$

Example. Find the Laplace transform of

(i) $\cosh at$, $(ii) at$.

Solution. Using the result established in , we have

$$(i) L[\cosh at\,;s] = L\left[\frac{e^{at}+e^{-at}}{2}\right] = \frac{1}{2}\left\{L\left[e^{at}\;;s\right] + L\left[e^{-at}\;;s\right]\right\}$$

$$= \frac{1}{2}\left(\frac{1}{s-a}+\frac{1}{s+a}\right) = \frac{s}{s^2-a^2}$$ c

(ii) $L[\sinh at\,;s] = L\left[\frac{e^{at}-e^{-at}}{2}\;;s\right] - \frac{1}{2}\left\{L\left[e^{at}\;;s\right] - L\left[e^{-at}\;;s\right]\right\}$

$$= \frac{1}{2}\left(\frac{1}{s-a}-\frac{1}{s+a}\right) = \frac{a}{s^2-a^2}$$

Example. Find the Laplace transform of t^n, where n is a positive integer

Solution . Using the definition of Laplace transform , we have

$$L\left[t^n ;s\right]=\int_0^\infty e^{-st}\ t^n\ dt=\int_0^\infty t^n\ d\left(\frac{e^{-st}}{-s}\right)=\left(t^n\ \frac{e^{-st}}{-s}\right)_0^\infty+\frac{n}{s}\int_0^\infty t^{n-1}\ e^{-st}\ dt=\frac{n}{s}\int_0^\infty t^{n-1}e^{-st}\,dt$$

Hence ,

$$L\left[t^n ;s\right]=\frac{n}{s}L\left[t^{n-1} ;s\right]$$

Similarly, we can prove the following

$$L\left[t^{n-1} ;s\right]=\frac{n-1}{s}L\left[t^{n-2} ;s\right]$$

$$L\left[t^{n-2} ;s\right]=\frac{n-2}{s}L\left[t^{n-2} ;s\right]$$

$\vdots$

$$L\left[t^2 ;s\right]=\frac{2}{s}L\left[t ;s\right]$$

$$L\left[t ;s\right]=\frac{1}{s^2}$$

Therefore, $L\left[t^n ;s\right]=\frac{n}{s}.\frac{n-1}{s}.\frac{n-2}{2}\ldots\frac{2}{s}.\frac{1}{s^2}=\frac{n!}{s^{n+1}}$ which can be expressed in Gamma functions as

$$L\left[t^n ;s\right]=\frac{\sqrt{n+1}}{s^{n+1}}$$

7.3. PROPERTIES OF LAPLACE TRANSFORM

We present a few important properties of the Laplace transform in the following theorems which will be enable us to find the Laplace transform of a combination of functions whose transform are known.

Theorem. (Linearity property) . if $c_1\ and\ c_2$ are any two constants and if $F_1(s)\ and\ F_2(s)$ are the Laplace transforms, respectively of $f_1(t)\ and\ f_2(t)$, then

$$L\left[\{c_1 f_1(t)+c_2 f_2(t)\};s\right]=c_1 L\left[f_1(t);s\right]+c_2 L\left[f_2(t);s\right]=c_1 F_1(s)+c_2 F_2(s)$$

Proof. Following the definition of Laplace transform , we have

$$L\left[\{c_1 f_1(t)+c_2 f_2(t)\};s\right]=\int_0^\infty e^{-st}\{c_1 f_1(t)+c_2 f_2(t)\}\,dt$$

$$=c_1\int_0^\infty e^{-st} f_1(t)\,dt+c_2\int_0^\infty e^{-st} f_2(t)\,dt$$

$$=c_1 L\left[f_1(t);s\right]+c_2 L\left[f_2(t);s\right]$$

$$=c_1 F_1(s)+c_2 F_2(s)$$

Theorem. (shifting property) . if a function is multiplied by e^{at}, the transform of the resultant is obtained by replacing s by $(s-a)$ in the transform of the original function. That is , if

$$L\left[f(t);s\right]=F(s)$$

Then

$$L\left[e^{at} f(t);s\right]=F(s-a)$$

Proof . From the definition of Laplace transform,

$$L\left[e^{at} f(t);s\right]=\int_0^\infty e^{-st}e^{at} f(t)\,dt=\int_0^\infty e^{-(s-a)t} f(t)\,dt=F(s-a)$$

Similarly,

$$L\left[e^{-at} f(t);s=F(s+a)\right]$$

Theorem . (Multiplication by power of) . if

$$L\left[f(t);s\right]=F(s)$$

then

$$L\left[t^n f(t);s\right]=(-1)^n\frac{d^n}{ds^n}F(s)=(-1)^n F^{(n)}(s)$$

where $n=1,2,3,\ldots$

Proof. Form the definition of Laplace transform

$$F(s)=L[f(t);s]=\int_0^\infty e^{-st} f(t)dt$$

Hence,

$$\frac{d}{dt}[F(s)]=\frac{d}{ds}\left[\int_0^\infty e^{-st} f(t)dt\right]$$

Interchanging the operations of differentiation and integration for which we assume that the necessary conditions are satisfied and since there are two variables $s \, and \, t,$ we use the notation of partial differentiation and obtain

$$\frac{d}{dt}\{F(s)\}=\int_0^\infty \frac{\partial}{\partial s}\{e^{-st} f(t)\}dt=-\int_0^\infty e^{-st} tf(t)\}dt=-L[tf(t);s]$$

Therefore ,

$$L[tf(t);s]=-\frac{}{ds}F(s)$$

By repeated application of the above result, it can be shown that

$$L[t^n f(t);s]=(-1)^n\frac{d^n}{ds^n}F(s)=(-1)^n F^n(s)$$

Theorem. (Differentiate property) , if

$$L[f(t);s]=F(s)$$

Then

$$L[f^n t;s]=s^n F(s)-s^{n-1}f(0)-s^{n-2}f'(0)-\ldots-s f^{n-2}(0)-f^{n-1}(0)$$

Proof Form the definition of Laplace transform , we have

$$L[f'(t);s]=\int_0^\infty e^{-st} f(t)dt=[e^{-st} f(t)]_0^\infty+s\int_0^\infty e^{-st} f(t)dt$$

Similarly , it can be shown that

$$L[f''(t);s]=sL[f'(t);s]-f'(0)=s\{sF(s)-f(0)\}-f'(0)=s^2 F(s)-sf(0)-f'(0)$$

$$L\left[f'''(t)\,;s\right]=s^{3}F(s)-s^{2}f(0)-sf'(0)-f''(0)$$

Thus, in general,

$$L\left[f^{n}\,t;s\right]=s^{n}F(s)-s^{n-1}(0)-s^{n-2}f(0)-\ldots-sf^{n-2}(0)-f^{n-1}(0)$$

This property is very useful for solving differential equations.

Example. Find the Laplace transform of

$(i)e^{at}\cos bt$, $(ii)e^{at}\sin bt$, $(iii)e^{at}\cosh bt$, $(iv)e^{at}t^{n}$, *and* $(v)\cos at\cosh bt$.

Solution. Using the shifting property

$$(i)\ L\left[e^{at}\sin bt\,;s\right]=\left.\frac{s}{s^{2}+b^{2}}\right|_{s\to(s-a)}=\frac{s-a}{(s-a)^{2}+b^{2}}$$

$$(ii)L\left[e^{at}\ \sin bt\,;s\right]=\left.\frac{b}{s^{2}-b^{2}}\right|_{s\to(s-a)}=\frac{s-a}{(s-a)^{2}+b^{2}}$$

$$(iii)\ L\left[e^{at}\ \cosh bt\,;s\right]=\left.\frac{s}{s^{2}-b^{2}}\right|_{s\to(s-a)}=\frac{s-a}{(s-a)^{2}-b^{2}}$$

$$(iv)\,L\left[e^{at}t^{n}\,;s\right]=\left.\frac{n!}{s^{n+1}}\right|_{s\to(s-a)}=\frac{n!}{(s-a)^{n+1}}$$

$$(v)L[\cos at\ \cosh bt\,;s]=L\left[\cos at\left(\frac{e^{bt}-e^{-bt}}{2}\right);s\right]$$

$$-\frac{1}{2}\left\{L\left[e^{bt}\cos at\,;s\right]-L\left[e^{-bt}\cos at\,;s\right]\right\}$$

$$=\frac{1}{2}\left\{\frac{s-a}{(s-b)^{2}+a^{2}}-\frac{s+b}{(s+b)^{2}+a^{2}}\right\}$$

Example. Find the Laplace transform of the following :

$(i)t^{2}e^{at}$, $(ii)t\sin at$, $(iii)t^{2}\cos at$, $(iv)\ t^{n}e^{-at}$

Solution. Using the result established in the Theorem , we have

$$(i) L\left[t^2\ e^{at}\ ;s\right]=(-1)^2\frac{d^2}{ds^2}L\left[e^{at}\ ;s\right]=\frac{d^2}{ds^2}\left(\frac{1}{s-a}\right)=\frac{d}{ds}\left(\frac{-1}{(s-a)^2}\right)=\frac{2}{(s-a)^3}$$

Alternatively,

$$L\left[\ e^{at}t^2\ ;s\right]=\left.\frac{2!}{s^3}\right|_{s\to(s-a)}=\frac{2}{(s-a)^3}$$ (using the shifting property)

$$(ii) L[t\ \sin at\ ;s]=(-1)^1\frac{d}{ds}L[\sin at\ ;s]=-\frac{d}{ds}\left(\frac{a}{s^2+a^2}\right)=\frac{2as}{\left(s^2+a^2\right)^2}$$

(iii) $$L\left[t^2\cos at\ ;s\right]=(-1)^2\frac{d^2}{ds^2}L[\ \cos at\ ;s]=\frac{d^2}{ds^2}\left(\frac{s}{s^2+a^2}\right)$$

$$=\frac{d}{ds}\left[\frac{a^2-s^2}{\left(s^2+a^2\right)^2}\right]=\frac{2s^3-6sa^2}{\left(s^2+a^2\right)^3}$$

$$(iv)\ L\left[e^{-at}t^n\ ;s\right]=\left.\frac{n!}{s^{n+1}}\right|_{s\to(s+a)}=\frac{n!}{(s+a)^{n+1}}$$ (using the shifting property)

Example. Verify the initial value theorem for the function $f(t)=1+e^{-t}\ (\sin t+\cos t\ )$

Solution . Given $f(t)=1+\ e^{-t}\ (\sin t+\cos t)$, *we have*

$$F(s)=L[\ f(t);s\ ]=L[1\ ;s]+L\left[\left(e^{-t}\ \sin t+e^{-t}\ \cos t\right);s\right]$$

$$=\frac{1}{s}+\left(\frac{1}{s^{2+1}}+\frac{s}{s^2+1}\right)_{s\to(s+1)}$$

$$=\frac{1}{s}+\left(\frac{2+s}{s^2+2s+2}\right)$$

Hence,

$$sF(s) = 1 + \frac{2s + s^2}{s^2 + 2s + 2}$$

Therefore,

$$Lt_{s \to \infty} sF(s) = Lt_{s \to \infty} 1 + \frac{\frac{2}{s+1}}{1 + \frac{2}{s} + \frac{2}{s^2}} = 1 + 1 = 2$$

But $f(0) = 1 + 1 = 2$. Thus,

$$Lt_{s \to \infty} sF(s) = f(0)$$

Hence the result

Theorem . (Division by t) . if

$$L[f(t); s = F(s)]$$

Then

$$L\left[\frac{f(t)}{t}; s\right] = \int_s^\infty F(s)\, ds$$

Proof. From the definition of Laplace transform

$$L[f(t); s] = F(s) = \int_0^\infty e^{-st} f(t)\, dt$$

Integrating the above with respect to s between the limits s to ∞, we get

$$\int_s^\infty F(s)\, ds = \left[\int_s^\infty e^{-st} f(t)\, dt\right] ds = \int_0^\infty f(t) \left(\int_0^\infty e^{-st} ds\right) dt$$

(by changing the order of integration)

$$= \int_0^\infty f(t) \left(\frac{e^{-st}}{-t}\right)_s^\infty dt = \int_0^\infty \frac{f(t)}{t} e^{-st}\, dt = L\left[\frac{f(t)}{t}\right]; s$$

Hence the result .

Note: In applying this rule , one should be careful . Since $\frac{f(t)}{t}$ may have an infinite discontinuity at $t = 0.$ it may not be integrable. If $\frac{f(t)}{t}$ is not integrable, then its Laplace transform does not exist. For example , at $t = 0$, the function $\sin t t$ doses not have an infinite discontinuity , while the function $\cos \frac{t}{t}$ *has an* infinite discontinuity .

7.4. TRANSFORM OF A PERIODIC FUNCTION

A function $f(t) is$ called periodic with period $T, if\ (t+T) = f(t)$ for all values of t and $T > 0.$ For example, the trigonometric functions $\sin t$ *and* $\cos t$ are periodic functions of periods $2\pi.$ Periodic functions occur very often in a variety of engineering problems.

Theorem. *if* $f(t)$ is a periodic function with periods T, then

$$L[f(t);s] = \int_0^T e^{-st} f(t)\,dt / (1 - e^{-st})$$

Proof. From the definition of Laplace transform , we have

$$L[f(t);s] = \int_0^\infty e^{-st} f(t)\,dt = \int_0^T e^{-st} f(t)\,dt + \int_T^\infty e^{-st} f(t)\,dt$$

If we substitute $t = u + T$ in the second integral on the right-hand side and write $dt = du.$ we obtain

$$L[f(t);s] = \int_0^T e^{-st} f(t)\,dt + \int_0^\infty e^{-s(u+T)} f(u+T)\,du$$

$$= \int_0^T e^{-st} f(t)\,dt + e^{-st} \int_0^\infty e^{-st} f(u)\,du$$

$$= \int_0^T e^{-st} f(t)\,dt + e^{-sT} L[f(t);s]$$

Rearranging , we get

$$\left(1-e^{sT}\right)L\left[f(t);s\right]=\int_0^T e^{-st} f(t)dt$$

Thus,

$$L\left[f(t);s\right]=\int_0^T e^{-st} f(t)dt \,/\left(1-e^{-st}\right)$$

Hence the result.

Example. Obtain the Laplace transform of the periodic saw-tooth wave function given by

$f(t)=\frac{t}{T}$ of period $T, 0<t<T$

Solution . The graph of the periodic saw-tooth function is described in . since $f(t)$ is periodic with $period\, T,$ we have

$$L f(t);s\,]=\frac{1}{1-e^{-st}}\int_0^T e^{-st}\frac{t}{T}dt=\frac{1}{T\left(1-e^{-sT}\right)}\int_0^T e^{-st}\,tdt=\frac{1}{T\left(1-e^{-st}\right)}\int_0^T td\left(\frac{e^{-st}}{-s}\right)$$

$$=\frac{1}{T\left(1-e^{-st}\right)}\left[\left(\frac{te^{-st}}{-s}\right)_0^T+\frac{1}{s}\int_0^T e^{-st}\,dt\right]$$

$$=\frac{1}{T\left(1-e^{-st}\right)}\left[\frac{Te^{-st}}{-s}-\frac{1}{s^2}\left(e^{-st}-1\right)\right]$$

Therefore,

$$L\left[f(t);s\right]=\frac{1}{s^2T}-\frac{e^{-st}}{s\left(1-e^{-st}\right)}$$

Example. Find the Laplace transform of

$$f(t)=\begin{cases}1, & 0\le t<2\\ -1, & 2\le t<4\end{cases}$$

$$f(t+4)=f(t)$$

Solution. In this problem, $f(t)$ is a periodic function of period 4; we, therefore, have \

$$L[f(t);s] = \frac{1}{1-e^{-4s}} \int_0^4 e^{-st} f(t)\,dt$$

$$= \frac{1}{1-e^{-4s}} \left(\int_0^2 e^{-st}(1)\,dt + \int_2^4 e^{-st}(-1)\,dt \right)$$

$$= \frac{1}{1-e^{-4s}} \left(\frac{-2e^{-2s}}{s} + \frac{e^{-4s}}{s} + \frac{1}{s} \right)$$

7.5. TRANSFORM OF ERROR FUNCTION

The error function denoted by erf(t) is defined as

$$\operatorname{erf}(t) = \frac{2}{\sqrt{\pi}} \int_0^t e^{-u^2}\,du$$

This function occurs in many branches of science and engineering ; for example , in probability theory, the theory of heat condition, and so on . In terms of the power series , we have

$$\operatorname{erf}(t) = \frac{2}{\sqrt{\pi}} \int_0^t \sum_{n=0}^{\infty} \frac{(-1)^n t^{2n+1}}{n!(2n+1)}$$

We can easily verify that these series converge everywhere and, therefore , $\operatorname{erf}(t) is$ an entire function .From the definition , it can be verified at once that

$$\operatorname{erf}(0) = 0$$

$$\operatorname{erf}(\infty) = \frac{2}{\sqrt{\pi}} \int_0^{\infty} e^{-u^2}\,du = \frac{\sqrt{1/2}}{\sqrt{\pi}} = 1$$

The graph of error function is shown in

In solving heat conduction equation, it has been found useful to introduce the complementary error Function defined as

$$erfc(t) = \frac{2}{\sqrt{\pi}} \int_0^{\infty} e^{-u^2}\, du = \frac{2}{\sqrt{\pi}} \left(\int_0^{\infty} e^{-u^2}\, du - \int_0^{t} e^{-u^2}\, du \right)$$

Therefore,

$$erfc(t) = 1 - \mathrm{erf}(t)$$

Now we shall the Laplace transform of erf (t): From the definition Of Laplace Transform

$$L[\mathrm{ert}(t)\,;s] = \int_0^{\infty} e^{-st} \frac{2}{\sqrt{\pi}} \int_0^{t} e^{-u^2}\, du\, dt$$

Changing the order on integration , we obtain

$$L[\mathrm{erf}(t);s] = \frac{2}{\sqrt{\pi}} \int_0^{\infty} e^{-u^2} \int_0^{\infty} e^{-st}\, dt\, du$$

$$= \frac{2}{s\sqrt{\pi}} \int_0^{\infty} e^{-(u^2+su)}\, du$$

$$= \frac{2}{s\sqrt{\pi}} e^{\frac{s^2}{4}} \int_0^{8} e^{-\left(u+\frac{s}{2}\right)^2}\, du$$

Setting $x = u + \frac{s}{2}$, we get

$$L[\mathrm{erf}(t);s] = \frac{2}{s\sqrt{\pi}} e^{\frac{s^2}{4}} \int_{\frac{s}{2}}^{\infty} e^{-x^2}\, dx$$

$$= \frac{1}{s} e^{s^2/4}\, erfc\left(\frac{s}{2}\right)$$

Example. Find the Laplace transform of erf $\left(t^{1/2}\right)$.

Solution. From the definition of Laplace transform , we have

$$L\left[\operatorname{erf}(t^{\frac{1}{2}}\ ;s\right]=\int_0^\infty e^{-st}\frac{2}{\sqrt{\pi}}\int_0^{r^{1/2}} e^{-u^2}\,du\,dt$$

Changing the order of integration, we get

$$L\left[\operatorname{erf}\left(t^{1/2}\right);s\right]=\frac{2}{\sqrt{\pi}}\int_0^\infty e^{-st}\frac{2}{\sqrt{\pi}}\int_0^{r^{1/2}} e^{-n^2}\,du\,dt$$

$$=\frac{2}{s\sqrt{\pi}}\int_0^\infty e^{-u^2}\,du\int_{u^2}^\infty e^{-st}\,dt$$

$$\frac{2}{s\sqrt{\pi}}\int_0^\infty e^{-u^2-su^2}\,du$$

$$\frac{2}{s\sqrt{\pi}}\int_0^\infty e^{-(1+s)u2}\,du$$

Setting $(1+s)u^2=x^2\ or\ \sqrt{1+su}=x,$ we have

$du=dx\,/\sqrt{1+s}$

Then

$$L\left[\operatorname{erf}\left(t^{\frac{1}{2}}\right);s\right]=\frac{2}{s\sqrt{\pi}\sqrt{1+S}}\int_0^\infty e^{-x^2}\,dx=\frac{1}{s\sqrt{1+s}}\left(\frac{2}{\sqrt{\pi}}\int_0^\infty e^{-x^2}\,dx\right)$$

or

$$L\left[\operatorname{erf}\left(t^{1/2}\right);s\right]=\frac{1}{s\sqrt{1+s}}$$

7.6. TRANSFORM OF BESSEL'S FUNCTION

Bessel function arise in several problem involving circular or cylindrical geometry, it is therefore useful to find the Laplace transform of Bessel functions if the first kind .

Example. Find the Laplace transform of

$(i)\ j_0(t),\quad (ii)tj_0(t),(iii)e^{-at}j_0(t).$

Solution. (i) From the definition of the Bessel function, we have

$$j_n(t) = \sum_{r=0}^{\infty} \frac{(-1)^r}{r!\sqrt{n+r+1}} \left(\frac{t}{2}\right)^{n+2r}$$

For $n = 0$, we have

$$j_0(t) = \sum_{r=0}^{\infty} \frac{(-1)^r}{r!\sqrt{r+1}} \left(\frac{t}{2}\right)^{2r} = \sum_{r=0}^{\infty} \frac{(-1)^r}{(r!)^2} \left(\frac{t}{2}\right)^{2r}$$

$$= 1 - \frac{t^2}{2^2} + \frac{t^4}{2^2 \times 4^2} - \frac{t^6}{2^2 \times 4^2 \times 6^2} + \ldots$$

Thus,

$$L[j_0(t);s] = L[1;s] - \frac{1}{2^2} L[t^2;s] + \frac{1}{2^2 \times 4^2} L[t^4;s] - \ldots$$

$$= \frac{1}{s} - \frac{1}{2^2}\frac{2!}{s^3} + \frac{1}{2^2 \times 4^2}\frac{4!}{s^5} - \frac{1}{2^2 \times 4^2 \times 6^2}\frac{6!}{s^7} + \ldots$$

$$\frac{1}{s}\left[1 - \frac{1}{2}\left(\frac{1}{s^2}\right) + \frac{1\times 3}{2\times 4}\left(\frac{1}{s^2}\right)^2 - \frac{1\times 3\times 5}{2\times 4\times 6}\left(\frac{1}{s^2}\right)^3 + \ldots\right]$$

$$= \frac{1}{s}\left(1 + \frac{1}{s^2}\right)^{-1/2} = \frac{1}{\sqrt{1+s^2}}$$

Hence,

$$L[j_0(t);s] = \frac{1}{\sqrt{1+s^2}}$$

(ii) From the properties of Laplace transform, we have

$$L[t^n f(t);s] = (-1)^n \frac{d^n}{ds^n} F(s)$$

Therefore,

$$L[tj_0(t);s] = (-1)\frac{d}{ds}\left[L[j_0(t);s]\right] = -\frac{d}{ds}\left(\frac{1}{\sqrt{1+s^2}}\right)$$

Thus,

$$L\left[tj_0(t);s\right] = \frac{s}{\left(1+s^2\right)^{\frac{3}{2}}}$$

(iii) From the shifting property of Laplace transform , we have

$$L\left[e^{-at} f(t) ;s\right] = F(s+a)$$

Therefore,

$$L\left[e^{-at} j_0(t);s\right] = \left.\frac{1}{\sqrt{1+s^2}}\right|_{s\to(s+a)} = \frac{1}{\sqrt{1+(s+a)^2}}$$

7.7. TRANSFORM OF DIRAC DELTA FUNCTION

The concept of impulse function or Dirac delta function has been introduced in itself. In certain applications involving a sudden excitation of a system or a large voltage over a short interval of time, the Laplace transform of Dirac Delta function is useful. Form the property of Dirac Delta function, we have

$$\int_0^\infty \delta(t-a) f(t)\, dt = f(a)$$

In particular , if $f(t)=e^{-st}$, then

$$L\left[\delta(t-a);s\right] = \int_0^\infty e^{-st}\,\delta(t-a)\, dt = e^{-as},\ a>0$$

7.7.1. Inverse Transform

So far we have discussed various properties of the Laplace transform and studies the Laplace transform of some simple functions. However, if the Laplace transform technique is to be useful in applications, we have to consider the reverse problem too, i.e., we have to find the original function x $f(t)$ when we know its Laplace transform $F(s)$. Thus, if

$$L\left[f(t);s\right] = F(s)$$

Then

$f(t)=L^{-1}\left[F(s);t\right]$

In other words , the inverse Laplace transform of a given function $F(s)$ is that function $f(t)$ whose Laplace transform is $F(s)$. It can be established that $f(t)$ is unique . Here, L^{-1} is known as inverse Laplace transform operator. Form the elementary definition (24) and form the results obtained thus far in finding the Laplace transform of some elementary functions , we can immediately generate the following table of transforms:

Table of Laplace Transform

$f(t)$	$L\left[f(t);s\right]-F(s)$	$F(s)$	$L^{-1}\left[F(s);t\right]=f(t)$
0	0	0	0
1	$\frac{1}{s}$	$\frac{1}{s}$	1
e^{at}	$\frac{1}{(s-a)}$	$\frac{1}{(s-a)}$	e^{at}
e^{-at}	$\frac{1}{(s+a)}$	$\frac{1}{(s+a)}$	e^{-at}
t	$1/s^2$	$1/s^2$	t
t^n	$\frac{n!}{s^{n+1}}$	$1/s^{n+1}$	$t^n/n!$
$\sin at$	$\frac{a}{s^2+a^2}$	$\frac{a}{s^2+a^2}$	$\sin at$
$\cos at$	$\frac{s}{s^2+a^2}$	$\frac{s}{s^2+a^2}$	$\sinh at$
$\sinh at$	$\frac{a}{s^2-a^2}$	$\frac{a}{s^2-a^2}$	$\sinh at$
$\cosh at$	$\frac{s}{s^2-a^2}$	$\frac{s}{s^2-a^2}$	$\cosh at$
$t\sin at$	$\frac{2as}{\left(s^2+a^2\right)^2}$	$\frac{2as}{\left(s^2+a^2\right)^2}$	$t\sin at$
$t\cos at$	$\frac{s^2\ a^2}{\left(s^2+a^2\right)^2}$	$\frac{s^{2\ 2}}{\left(s^2+a^2\right)^2}$	$t\cos at$

In most of the problems we have considered earlier, $L[f(t);s]$ is a simple rational function. The linearity property holds true even in the case of inverse transform. That is, if $F_1(s)$ and $F_2(s)$

Are the Laplace transform of $f_1(t) and\, f_2(t)$, and if $c_1\, and\, c_2$ are any two constants, then

$$L^{-1}\left[\{c_1F_1(s)\pm c_2F_2(s)\};t\right]=c_1\,L^{-1}\left[F_1(s);t\right]\pm c_2\,L^{-1}\left[F_2(s):t\right]$$

By expressing $L[f(t);s]$ as partial fractions, we should be able to recognize them as the transform of some known function , with the helps of which we can write down the inverse transform . Similarly shifting property is also useful in constructing the inverse transform of some functions, which is stated in the following theorem:

Theorem.

If $L[f(t);s]=F(s),$ then

$$L^{-1}\left[F(s+a);t\right]=e^{-at}\,L^{-1}\left[F(s);t\right]$$

Proof.

Since $L[f(t);s]=F(s),$ we have

$$L^{-1}\left[F(s);t\right]=f(t)$$

Recalling the shifting property of Laplace transform, we find that

$$L\left[e^{-at}(t)\;;s\right]=F(s+a)$$

$$L^{-1}\left[F(s+a);t\right]=e^{-at}\,f(t)$$

Thus,

$$L^{-1}\left[F(s+a);t\right]=e^{at}\,L^{-1}\left[F(s);t\right]$$

Example. Obtain the inverse Laplace transform of

$$\frac{4s^2-3s+5}{(s+1)(s^2-2s+2)}$$

Solution Using partial fraction expansion , we can write

$$\frac{4s^2-3s+5}{(s+1)(s^2-2s+2)}=\frac{A}{s+1}+\frac{Bs+C}{s^2-2s+2}$$

Therefore,

$$\frac{4s^2-3s+5}{(s+1)(s^2-2s+2)}=\frac{A}{s+1}+\frac{Bs+C}{s^2-2s+2}$$

Therefore,

$$4s^2-3s+5=A(s^2-2s+2)+(Bs+C)(s+1)$$

Let $s=-1$; then $A=\frac{12}{5}.$ Equating the coefficient of s on both sides, we have

$$B+C=\frac{9}{5}$$

Equating the coefficient of constant on both sides, we get $2A+C=5$ which gives $C=\frac{1}{5},$ and hence $B=\frac{8}{5}.$ The given expression can now be written as

$$\frac{4s^2-3s+5}{(s+1)(s^2-2s+2)}=\frac{12}{5}\frac{1}{s+1}+\frac{8}{5}\frac{s}{(s-1)^2+1^2}+\frac{1}{5}\frac{1}{(s-1)^2+1^2}$$

Thus, we find that

$$L^{-1}\left[\frac{4s^2-3s+5}{(s+1)(s^2-2s+2)};t\right]=\frac{12}{5}L^{-1}\left[\frac{1}{s+1};t\right]+\frac{8}{5}L^{-1}\left[\frac{(s-1)+1}{(s-1)^2+1^2};t\right]+\frac{1}{5}L^{-1}\left[\frac{1}{(s-1)^2+1^2};t\right]$$

$$+\frac{1}{5}e^tL^{-1}\left[\frac{1}{s^2+1};t\right]$$ (by using the shifting property)

$$=\frac{12}{5}e^{-t}+\frac{8}{5}e^t(\cos t\sin t)+\frac{1}{5}e^t\sin t$$

$$=\frac{12}{5}e^{-t}+\frac{8}{5}e^t\cos t+\frac{9}{5}e^t\sin t$$

Theorem. (change of scale property). If

$$L^{-1}\left[F(s),t\right]=f(t)$$

then

$$L^{-1}\left[F(\alpha s);t\right]=\frac{1}{\alpha}f\left(\frac{t}{\alpha}\right)$$

Proof. Form the definition of Laplace transform, we have

$$F(s)=\int_0^\infty e^{-st}f(t)\,dt$$

Therefore,

$$f(\alpha s)=\int_0^\infty e^{-\alpha st}f(t)\,dt$$

Let $\alpha t = x$, so that $dt=\frac{dx}{a}$.*Then* we get

$$F(\alpha s)=\frac{1}{a}\int_0^\infty e^{-sx}f\left(\frac{x}{a}\right)dx=\frac{1}{a}L\left[f\left(\frac{x}{a}\right);s\right]=\frac{1}{a}L\left[f\left(\frac{t}{a}\right);\right]$$

Thus,

$$L^{-1}\left[F(as);t\right]=\frac{1}{a}f\left(\frac{t}{a}\right)$$

Example. Find the inverse Laplace transform of

(i) In $\frac{s^2+1}{s(s+2)}$, $(ii)\cot^{-1}\left(\frac{s}{k}\right)$

Solution. (i) From Theorem , we have

$$L\left[t^n f(t);s\right]=(-1)^n F^n(s)$$

In particular $n=1$ gives

$$L\left[tf(t);s\right]=-F'(s)$$

i.e.,

$$\frac{d}{ds}F(s) = -L\left[tf(t);s\right]$$

Let

$$L\left[f(t);s\right] = In\frac{s^2+1}{s(s+1)}$$

Then,

$$\frac{d}{ds}F(s) = \frac{d}{ds}\left[In\left(s^2+1\right) - Ins - In(s+1)\right]$$

$$= \frac{2s}{s^2+1} - \frac{1}{s} - \frac{1}{s-1} = L\left[tf(t);s\right]$$

We get

$$L\left[ft(t);s\right] = \frac{1}{s} + \frac{1}{s+1} - \frac{2s}{s^2+1}$$

Hence,

$$tf(t) = L^{-1}\left(\frac{1}{s};t\right) + L^{-1}\left(\frac{1}{s+1};t\right)\left(\frac{1}{s+1};t\right) - 2L^{-1}\left(\frac{s}{s^2+1};t\right)$$

$$= 1 + e^{-t} - 2\cos t$$

Therefore,

$$f(t) = L^{-1}\left[In\frac{s^2+1}{s(s+1)};t\right] = \frac{1+e^{-t}-2\cos t}{t}$$

(ii) Let $L\left[f(t);s\right] = \cot^{-1}\left(\frac{s}{k}\right) = F(s)$

Then ,

$$\frac{d}{ds}F(s) = \frac{d}{ds}\left(\cot^{-1}\frac{s}{k}\right) = -\frac{k}{k^2+}$$

Now we obtain

$$\frac{d}{ds}F(s) = \frac{k}{k^2+s^2} = L\left[tf(t);s\right]$$

Therefore,

$$tf(t)=L^{-1}\left[\frac{k}{k^2+s^2};t\right]=\sin kt$$

Hence,

$$f(t)=L^{-1}\left[\cos^{-1}\left(\frac{s}{k}\right);t\right]=\frac{\sin kt}{t}$$

7.8. CONVOLUTION THEOREM (FALTUNG THEOREM)

We often across function which are not the transform of some known function, but then, they can possibly be expressed as a product of two functions, each of which is the transform of a known function. Thus we may be able to write the given function as $F(s)G(s).$ where of a known to be transform of the functions $f(t)$ and $g(t),$ respectively.

Theorem. If $F(s) and\, G(s)$ are the Laplace transform of $f(t) and\, g(t)$ respectively, then $F(s)G(s)$ is the Laplace transform of

$$\int_0^t f(t-u)g(u)du$$

i.e.,

$$L^{-1}\left[F(s)G(s)\right]=\int_0^t f(t-u)g(u)du$$

This integral is called the convolution of $f\ and\ g$ and is denoted by the symbol $f^*g.$

Proof. Form the definition of Laplace transform ,we have

$$F(s)G(s)=\left[\int_0^\infty e^{-sv}f(v)dv\right]\left[\int_0^\infty e^{-su}g(u)du\right]$$

$$=\int_0^\infty\int_0^\infty e^{-s(v+u)}f(v)g(u)dv\,du$$

$$=\int_0^\infty g(u)\left\{\int_0^\infty e^{(v+u)} f(v)dv\right\}du$$

Let $u+v=t$ in the inner integral. Then,

$$F(s)G(s) \quad \int_0^\infty g(u)\left\{\int_0^\infty e^{-st} f(t-u)dt\right\}du$$

Change the order of integration

Then, we get

$$F(s)G(s)=\int_0^\infty\left\{\int_0^t e^{-st} f(t-u)g(u)du\right\}dt$$

$$=\int_0^\infty e^{-st}\left\{\int_0^t f(t-u)g(u)du\right\}dt$$

$$=L\left[\int_0^t f(t-u)g(u)du\ ;t\right]$$

Hence the result.

Definition. We define the Laplace convolution of $f(t) and\ g(t)$ by the integral

$$f^*g=\int_0^t f(t-u)g(u)du$$

It can be verified that f and g can be interchanged in the convolution, i.e., $f\ and\ g$ are commutative Let $t-u-v$ so that $-du=dv$. Then ,

Therefore,

$$f^*g=g^*f$$

Example. Prove that

$$\int_0^t j_0(t)\,j_0\,(t-u)du=\sin t$$

Solution. Form table ,we not that

$$L^{-1}\left[\frac{1}{s^2+1};t\right]=\sin t$$

We shall also write

$$\frac{1}{s^2+1}=\frac{1}{\sqrt{s^2+1}}\frac{1}{\sqrt{s^2+1}}$$

Now taking

$$F(s)=\frac{1}{\sqrt{s^2+1}},G(s)=\frac{1}{\sqrt{s^2+1}}$$

Their inverse transform give

$$L^{-1}\left[\frac{1}{\sqrt{s^2+1}};t\right]=j_0(t)\quad=f(t)=g(t)$$

Hence, using the convolution theorem, we have

$$L^{-1}\left[\frac{1}{s^2+1};t\right]=\int_0^t j_0(t)j_0(t-u)du=\sin t$$

7.8.1. Summary and Discussion

Laplace transform is essentially a mathematical tool which can be used to solve several problems in science and engineering . This transform was first introduced was first introduced by Laplace, a French mathematician, in the year 1790 in his work on probability theorem. This technique became popular when Heaviside applied to the solution of an ordinary differential equation referred hereafter as ODE. representing a problem in electrical engineering . To the basic question as to why one should learn Laplace transform technique when other technique available. The answer is very simple. Transform are used to accomplish the solution of certain problems with less effort and in a simple routine way.

Definition. Let $f(t)$ be a continuous and single-valued function of the real variable t definition for all $t, 0<t<\infty,$ and is of exponential order. Then the Laplace transform of $f(t)$ is defined as a function $F(s)$ denoted by the integral

$$L\left[f(t);s\right]=F(s)=\int_0^\infty e^{-st}f(t)\,dt \tag{3}$$

For any finite value of γ .

Over the range of values of s for which the integral exist. Here, s is a parameter, real or complex . Obviously, $L\left[f(t):s\right]$ is a function of s Thus,

Where L is the operator which transform $f(t)\, into\, F(s)$ called laplace transform operator, and L^{-1} is the inverse Laplace transform operator.

The Laplace transform belongs to the family of " integral transform". An integral transform $F(s)$ pf the function $f(t)$ is defined by an integral of the form

$$\int_c^b k(s,t)f(t)\,dt=F(s) \tag{4}$$

where $k(s,t)$, a function of two variables $s\, and\, t$, is called the Kernel of the integral transform The kernel and limits of integration of various integral transform are given in Table 1 (which is not exhaustive) .

Theorem. If $f(t)$ is piecewise continuous in the range $t\geq 0$ and is of exponential order γ. then the Laplace transform $F(s) of\, f(t)$ exists for all $s>\gamma$.

$$(i)\,L[l\,;s]=\int_0^\infty e^{-st}(1)\,dt=\left(\frac{e^{-st}}{s}\right)_0^\infty=\frac{1}{s}\,if\;\; s>0$$

(ii) $$L[0\,;s\,)]=\int_0^\infty e^{-st}(0)\,dt=0$$

(iii) $$L[t\,,s]=\int_0^\infty e^{-st}\,.dt-\int_0^\infty td\left(\frac{e^{-st}}{-s}\right)-\left(t\frac{e^{-st}}{-s}\right)_0^\infty\int_0^\infty\frac{e^{-st}}{-s}\,dt=\frac{1}{s^2}$$

(iv) $$L[t\ ;s]=\int_0^\infty e^{-st}\ e^{at}\,dt=\int_0^\infty e^{-(s-a)t}\,dt=\left[\frac{e^{-(s-a)t}}{-(s-a)}\right]_0^\infty=\frac{1}{s-a}\ s>a$$

(v) $$L\left[e^{-at}\ ;s\right]=\int_0^\infty e^{-st}\ e^{-at}\,dt=\int_0^\infty e^{-(s+a)}\,dt=\frac{1}{s+a}$$

$$(i)L[\cos at\ ;s]=\int_0^\infty e^{-st}\cos at\ dt=Re\int_0^\infty e^{-st}\ e^{iat}dt=Re\,L\left[e^{iat}\ ;s\right]$$

$$Re\frac{1}{s-ia}=Re\frac{s+ia}{s^2+a^2}=\frac{s}{s^2+a^2}$$

(ii) $$L[\ \sin at\ ;s]=Im\,L\left[e^{iat}\ ;s\right]=Im\frac{s+ia}{s^2+a^2}=\frac{a}{s^2+a^2}$$

$$(i)L[\ \cosh at\ ;s]=L\left[\frac{e^{at}+e^{-at}}{2}\right]=\frac{1}{2}\left\{L\left[e^{at}\ ;s\right]+L\left[e^{-at}\ ;s\right]\right\}$$

$$=\frac{1}{2}\left(\frac{1}{s-a}+\frac{1}{s+a}\right)=\frac{s}{s^2-a^2}$$
c

(ii) $$L[\sinh at\ ;s]=L\left[\frac{e^{at}-e^{-at}}{2}\ ;s\right]=\frac{1}{2}\left\{L\left[e^{at}\ ;s\right]-L\left[e^{-at}\ ;s\right]\right\}$$

$$=\frac{1}{2}\left(\frac{1}{s-a}-\frac{1}{s+a}\right)=\frac{a}{s^2-a^2}$$

Table Of Laplace Transform

$f(t)$	$L[f(t);s]=F(s)$	$F(s)$	$L^{-1}[F(s);t]=f(t)$
0	0	0	0
1	$\frac{1}{s}$	$\frac{1}{s}$	1
e^{at}	$\frac{1}{(s-a)}$	$\frac{1}{(s-a)}$	e^{at}
e^{-at}	$\frac{1}{(s+a)}$	$\frac{1}{(s+a)}$	e^{-at}
t	$1/s^2$	$1/s^2$	t
t^n	$\frac{n!}{s^{n+1}}$	$1/s^{n+1}$	$t^n/n!$
$\sin at$	$\frac{a}{s^2+a^2}$	$\frac{a}{s^2+a^2}$	$\sin at$
$\cos at$	$\frac{s}{s^2+a^2}$	$\frac{s}{s^2+a^2}$	$\sinh at$
$\sinh at$	$\frac{a}{s^2-a^2}$	$\frac{a}{s^2-a^2}$	$\sinh at$
$\cosh at$	$\frac{s}{s^2-a^2}$	$\frac{s}{s^2-a^2}$	$\cosh at$
$t\sin at$	$\frac{2as}{(s^2+a^2)^2}$	$\frac{2as}{(s^2+a^2)^2}$	$t\sin at$
$t\cos at$	$\frac{s^2\ a^2}{(s^2+a^2)^2}$	$\frac{s^2\ 2}{(s^2+a^2)^2}$	$t\cos at$

7

CHAPTER 8

GREEN'S FUNCTION

CONTENTS

8.1. INTRODUCTION

Consider the differential equation

$$Lu(x) = f(x)L^{-1} \tag{1}$$

Where L is an ordinary linear differential operator, $f(x)$ is a known function, while is a unknown function. To solve the above differential equation, one method is to find the operator in the form of an integral operator with a kernel $G(x,\xi)$ such that

$$u(x) = L^{-1}f(x) = \int G(x,\xi)f(\xi)d\xi \tag{2}$$

The kernel of this integral operator is called Green's function for the differential operator. Thus the solution of the non-homogenous differential equation can be written down, once the Green's function for the problem is known. Applying the differential operator to both sides of , we get

$$f(x) = LL^{-1} f(x) = \int LG(x,\xi)f(\xi)d\xi \tag{3}$$

This equation is satisfied if we choose $G(x,\xi)$ such that

$$LG(x,\xi) = \delta(x-\xi) \tag{4}$$

Where $\delta(x-\xi)$ is a Dirac δ - function. The solution of Eq. is called a singularity solution of Eq. In previous section , we have already introduced the concept of Dirac $\delta - function$ and studied its various properties. Now, they become handy to understand more about Green's function

Definition . Let us consider an auxiliary function $\phi(x)of$ a real variable x which possess derivative of all orders and vanishes outside a finite interval. Such function are called test functions.

Now we shall introduced the concept of the derivative of a $\delta -$ function in terms of the derivative of a test function . we say that $\delta'(x)$ is the derivative of $\delta(x)$ if

$$\int_{-\infty}^{\infty} \delta'(x)\phi(x)dx = -\phi'(0) \tag{5}$$

For every test function $\phi(x)$. similarly , we define $\phi''(x)$ by

$$\int_{-\infty}^{\infty} \delta''(x)\phi(x)dx = -\phi''(0) \tag{6}$$

With this definition of a derivative, we can show that the $\delta - function$ is the derivative of a Heavside unit step function $H(x)$ defined by

$$H(x) = \begin{cases} 1 & for\, x \geq 0 \\ 0 & for\, x < 0 \end{cases} \tag{7}$$

To see this result, we consider

$$\int_{-\infty}^{\infty} H'(x)\phi(x)dx = -\int_{-\infty}^{\infty} H(x)\phi'(x)dx = -\int_{0}^{\infty} \phi'(x)dx = \phi(0)$$

By comparing the above result with property III of i.e., with

$$\int_{-\infty}^{\infty} \delta(x)\phi(x)dx = \phi(0)$$

we obtain

$$H'(x) = \delta(x) \tag{8}$$

Similarly, the notion of δ $function$ and its derivative enables us to give a meaning to the derivative of a function that has a jump discontinuity at $x = \xi$ of magnitude unity. Let

$$H(x \quad \xi) = \begin{cases} 1, & x \geq \xi \\ 0, & x < \xi \end{cases}$$

Then, for any test function $\phi(x)$. we have

$$\int_{-\infty}^{\infty} H'(x-\xi)\phi(x)dx = \int_{-\infty}^{\infty} \delta(x-\xi)\phi(x)dx = \phi(\xi) \tag{9}$$

It can also be noted that

$$\frac{d}{dx}\left[(x-\xi)H(x-\xi)\right]=(x-\xi)H'(x-\xi)+H(x-\xi)$$

Integrating both sides with respect to x between the limits $-\infty to \infty$,we get

$$(x-\xi)H(x-\xi)=\int_{-\infty}^{\infty}(x-\xi)H'(x-\xi)dx+\int_{-\infty}^{\infty}H(x-\xi)dx$$

$$=\int_{-\infty}^{\infty}H(x-\xi)dx \tag{10}$$

We now consider an example to illustrate the inversion of a differential operator by considering the BVP:

$$\frac{d^2u}{dx^2}=f(x),\ u(0)=u(1)=0$$

In this Eq. (4) becomes

$$LG=\frac{d^2G}{dx^2}=\delta(x-\xi) \tag{11}$$

Noting that the $\delta-$function is the derivative of the Heaviside unit step function and integrating

$$\frac{d}{dx}\left[G(x,\xi)\right]=H(x-\xi)+C_1(\xi)$$

where $C_1(\xi)$ is an arbitrary function. Integrating the above result once with respect to $x.$ we get

$$G(x,\xi)=\int H(x-\xi)dx+C_1(\xi)x+C_2(\xi)$$

$$=(x-\xi)H(x-\xi)+C_1(\xi)x+C_2(\xi)$$

where $C_1(\xi) and\ C_2(\xi)$ can be determined form the boundary conditions. Thus , from Eq. (2) we have

$$u(x)=\int_0^x(x-\xi)H(x-\xi)f(\xi)d\xi+x\int_{-\infty}^{\infty}C_1(\xi)f(\xi)d\xi+\int_{-\infty}^{\infty}C_2(\xi)f(\xi)d\xi$$

Now. Using the boundary condition $:u(0)=0,$ we get

$$0 = 0 + 0 + \int_{-\infty}^{\infty} C_2(\xi) f(\xi) d\xi$$

Implying that $C_2(\xi) = 0.$ Using the second boundary condition : $u(1) = 0.$ we have

$$0 = \int_0^1 (1-\xi) f(\xi) d\xi + \int_{-\infty}^{\infty} C_1(\xi) f(\xi) d\xi$$

Implying that $C_1(\xi) = -(1-\xi), 0 \le \xi \le 1,$ and zero for all other values of $\xi.$ Hence

$$u(x) = \int_0^x (x-\xi) H(x-\xi) f(\xi) d\xi - x\int_0^1 (1-\xi) f(\xi) d\xi \tag{12}$$

Comparing this result with Eq. (2) , we have the Kernel of the integral operator, which is known as Green's function or source function given by

$$G(x,\xi) = (x-\xi) H(x-\xi) - x(1-\xi), \quad 0 \le \xi \le 1 \tag{13}$$

Satisfying the boundary condition: $G(0,\xi) = G(1,\xi) = 0$

This concept can be extended to partial differential equation also. To make the ideas clearer, let us consider

$$L[u(X)] = f(X) \tag{14}$$

where L is some linear v partial differential operator in three independent variables x, y, z and X is a vector in three-dimensional space. Then the Green' $G(X;X)$ which satisfies the equation.

$$L[G(X;X')] = \delta(X - X') \tag{15}$$

On expansion , this equation becomes

$$L[G(x,y,z;x',y',z')] = \delta(x-x')\delta(y-y')\delta(z-z') \tag{16}$$

Here the expansion $\delta(X - X')$ is the generiliazation of the concept of delta function in three-dimesnional space IR and $G(X,X)$ represents the effect at that point X due to a source function or delta function input applied at $'$.Equation has the following interpretation in heat conduction or electrostatic: $G(X;X)$ can be viewed as the temperature (the electrostatic potential) at any point X in IR due to a unit source (due to a unit charge) located at X, we get Eq. (15) on both sides by $f(X)$ and integrating over the volume V with respect to X'. we get

$$L\left[\int_{V_x} G(X;X)f(X)dV_X\right] = \int_{V_X} f(X)\delta(X - X')dV_x' = f(X)$$

Comparing with Eq. (14) ,we arrive at

$$u(X) = \int_{V_X} G(X;X)f(X)dV_x \tag{17}$$

Which is the solution of Eq. (14) . This leads to the simple definition that a function $u(x, y, z ; x', y', z')$ is a fundamental solution of the equation , for example, $\nabla^2 u = 0$ if is a solution of the non-homogenous equation $\nabla^2 u = \delta(x, y, z; x', y', z')$

This idea can be easily extended to higher dimension . Thus the Green's function technique can be applied, In principle , to find the solution of any linear non-homogenous partial differential equation . Although a neat formula (17) has been for solution of non-homogenous PDE, in practice , it is not any easy task to construct the Green's function . We shall now present a few singularity solutions. Also called fundamental solutions to the well-known operators. These will guide us to construct Green's function for the solutions of partial differential equations which occurs most frequently in mathematical physics.

To start with, the fundamental solution for a three-dimensional potential problem satisfies

$$\nabla^2 u = \delta(X) \qquad u \tag{18}$$

or

$$div\, grad\, u = \delta(X)$$

where u can be interpreted, for example, as the electrostatic potential. We seek a solution which depends only on the source distance $r = |\mathbf{X}|$; thus, for $r > 0, u(r)$ satisfies

$$\nabla^2 u = \frac{1}{r^2}\frac{\partial}{\partial r}\left(r^2 \frac{\partial u}{\partial r}\right) = 0$$

On integration, we get

$$u = \frac{A}{r} + B$$

Using the fact that the potential vanishes at infinity ,we have

$$u = \frac{A}{r}$$

Integrating Eq. (18) the potential vanishes at infinity ,we have

$$\int_{R_\varepsilon} [div\, grad\, u] dV = 1$$

Using the divergence theorem, we obtain

$$\int_{\sigma_\varepsilon} \left.\frac{\partial u}{\partial r}\right|_{r=\varepsilon} dS = 1$$

Where σ_ε is the surface of R_ε .Hence ,

$$-A\int \left.\frac{1}{r^2}\right|_{\epsilon} dS = -\frac{A}{\epsilon^2} \times 4\pi\epsilon^2 = -4\pi A = 1$$

Therefore,

$$A = -\frac{1}{4\pi}$$

Thus, the singularity solution or the fundamental solution of $\nabla^2 u = 0$ is

$$u - -\frac{1}{4\pi r} \tag{19}$$

The two-dimensional case of Eq. (18) is

$$\frac{1}{r}\frac{\partial}{\partial r}(r\left(\frac{\partial u}{\partial r}\right)=0, \quad r>0$$

On integration, we get

$$u = A\,In\,r + B$$

Integrating Eq. (18) over a disc $R_{ò}$ of small radius $ò$, whose surface is $\sigma_{ò}$, we get

$$\int_{R_\varepsilon} div\,grad\,u = 1$$

Hence,

$$\int_{\sigma_\varepsilon} \left.\frac{\partial u}{\partial r}\right|_{r=ò} dS = \int \left.\frac{A}{r}\right|_{ò} dS = \frac{A}{ò}\times 2\pi ò = 1$$

Therefore, $A=\frac{1}{2\pi}.$ The constant B remains arbitrary and can be set equal to zero for convenience Thus, the fundamental solution is

$$u(r)=\frac{1}{2\pi}In\,r \tag{20}$$

If $if\ r(x,y\ ,z)\,and\ r'(x',y',z')$ are two distinct points in three-dimensional space IR, then the singularity solution of Laplace equation is

$$u=\frac{1}{4\pi|r-r'|} \tag{21}$$

Similarly, the singularity solution for the diffusion equation

$$u_t - k\nabla^2 u = 0$$

In three-space variables assumes the form

$$\frac{1}{8\left[\pi k(t-\tau)\right]^{3/2}}\exp\left[\frac{-|r-r'|^2}{4k(t-\tau)}\right] \tag{22}$$

For Helmholtz equation in three space variables, viz.

$\nabla^2 u + k^2 u = 0$

The singularity solution is

$$\frac{e^{ik|r-r|}}{|r-r'|} \tag{23}$$

Before we attempt to solve Eq.(17) we shall examine the form of three-dimensional $\delta - function$ in genral curvilinear coordinates as a preparation to study the solution of partial differential equations in polar coordinates, sphereical polar coordinates etc. suppose we are looking for a transformation form Cartesian coordinates , x, y , to curvilinear coordinates ξ, η through the relations

$$x - f(\xi,\eta),\quad y = g(\xi,\eta) \tag{24}$$

where $f\ and\ g$ are single-valued, continuously differentiable functions of their arguments. Suppose that under this transformation, $\xi = \beta_1\ and\ \eta = \beta_2$ correspond to $x = \alpha_1$ and $y = \alpha_2$ respectively Also,

$$\begin{pmatrix} dx \\ dy \end{pmatrix} = \begin{pmatrix} f_\xi & f_\eta \\ g_\xi & g_\eta \end{pmatrix}\begin{pmatrix} d\xi \\ d\eta \end{pmatrix} = \frac{\partial(f,g)}{\partial(\xi,\eta)} = j\begin{pmatrix} d\xi \\ d\eta \end{pmatrix}$$

If we transform the coordinates following Eq. (24) , the relation

$$\iint \phi(x,y)\delta(x-\alpha_1)\delta(y-\alpha_2)\,dx\,dy = \phi(\alpha_1,\alpha_2)$$

Becomes

$$\iint \phi(f,g)\delta\left[(\xi,\eta)-a_1\right]\delta\left[g(\xi,\eta)-a^2\right]|J| = d\eta = \phi(\alpha_1,\alpha_2) \tag{25}$$

where j is the Jacobian of the transformation Eq. (25) states that the Dirac δ -function $\delta\left[f(\xi,\eta)-a_1\right]\partial\left[g(\xi,\eta)\ \ a_2\right]|j|$ assign to any test function $\phi(f,g)$ the value of that test function at the point where $f' = a_1\ , g = a_2, i.e.,$ at the points $\zeta = \beta_1\ , \eta = \beta.$ Thus we may write

$$\delta\left[f(\xi,\eta)-a_1\right]\delta\left[g(\xi,\eta)-a_2\right]|j| = \delta(x-a_1)\delta(y-a_2)|j| = \delta(\xi-\beta_1)\delta(\eta-\beta_2)$$

Hence

$$\delta(x-a_1)\delta(y-a_2)=\frac{\delta(\xi-\beta_1)\delta(\eta-\beta_2)}{|j|} \tag{26}$$

In the next few sections, we shall discuss the Green's function method for solving partial differential equations with particular emphasis on elliptic equations. The discussion on wave equation and heat equation is also included in respectively .

8.1.1. Green's Function for Laplace Equation

To find analytic solution of the boundary value problems , the Green' function method of the convenient technique. In this section, we shall give the definition of Green's function for the Laplace equation and study its basic properties . To begin with, we shall define Green's function for the Dirichlet problem, i.e., we shall find u such that $\nabla^2 u=0$ is valid inside some finitely bounded region IR, enclosed by $\partial IR,$ as sufficiently smooth surface , when $u=f$ is prescribed on the boundary ∂IR.

Suppose that u is known at every poibt of the boundary ∂IR and that it satisfies the relation

$$\nabla^2 u=0 \; in\, IR$$

The task is to find $u(p)\, when\, P\in IR$. Let $=\boldsymbol{r}$, and let c be a sphere with centre at P and radius . ọ. Also , let Σbe denoted by $\partial\Sigma$, and

$$u'=\frac{1}{|r-r'|} \tag{27}$$

where r' is another point Q either in Σ or on the boundary $\partial\Sigma$. if u and $\frac{\partial}{\partial n}$ are twice continuously differentiable functions in IR and have first order derivatives on $\partial IR,$ then by Green 's theorem the region IR we have , form Eq. (19) , the relation

$$\int\!\!\int_{IR}\!\!\int \left(u\nabla^2 u'-u'\nabla^2 u\right)dV=\int\!\!\int_{\partial IR}\left(u\frac{\partial u'}{\partial n}-u'\frac{\partial u}{\partial n}\right)dS \tag{28}$$

Here, n I the unit vector normal to $dSIR$ drawn outwards from and $\frac{\partial}{\partial n}$ denotes differentiation in that direction. Since $\nabla^2 u = \nabla^2 u' = 0$ within $\partial\Sigma$, *we have* , in the region Σ, the relation

$$\int_C \left(u \frac{\partial u'}{\partial n} - u' \frac{\partial u}{\partial n} \right) dS' + \iint_{\partial IR} \left(u \frac{\partial u'}{\partial n} - u' \frac{\partial u}{\partial n} \right) dS = 0$$

Inserting u' form relation (27), the above equation reduces to

$$\iint_C \left[u(r') \frac{\partial}{\partial n} \left(\frac{1}{|r-r'|} - \frac{1}{|r-r'|} \frac{\partial u}{\partial n} \right)(r') \right] dS'$$

$$+ \iint_{\partial IR} \left[u(r) \frac{\partial}{\partial n} \left(\frac{1}{|r-r|} \right) - \frac{1}{|r-r|} \frac{\partial u}{\partial n}(r) \right] dS = 0 \qquad (29)$$

When Q is on C , we have

$$\frac{1}{|r-r'|} = \frac{1}{\epsilon}, \frac{\partial}{\partial n} \left(\frac{1}{|r-r'|} \right) = \frac{1}{\epsilon^2}$$

Also, dS', the surface element on $C, is ò^2 \sin\theta \ d\theta \, d\phi.$ d on C,

$$u(r') = u(r) + du$$

or

Theretore,

$$u(r') = u(r) + O(\epsilon) \ on C$$

$$\frac{\partial u}{\partial n}(r) = \frac{\partial u}{\partial n}(r) + C(\epsilon)$$

Now,

$$\iint_C u(r') \frac{\partial}{\partial n} \left(\frac{1}{|r-r'|} \right) dS' = \iint_C [u(r) + O(\epsilon)] \frac{1}{\epsilon^2} \times \epsilon^2 \ sit \theta \, d\theta \, d\phi$$

$$u(p) \iint_C \sin\theta \, d\theta \ d\phi + O(\epsilon)$$

$$u(p)\int_{\phi=0}^{2\pi}\int_{\theta=0}^{\pi}\sin\theta\,d\theta\;d\phi+O(\epsilon)$$

$$=4\pi u(r)+O(\epsilon)$$

$$\iint_C \frac{1}{r-r'|}\frac{\partial}{\partial n}(r')dS'=\frac{1}{\epsilon}\iint\left[\frac{\partial u}{\partial n}(r)+0(\epsilon)\right]\epsilon^2\sin\theta\,d\theta\,d\phi+0(\epsilon)$$

Employing these results and taking the limit as $\epsilon\to 0$; Eq. (29) becomes

$$4\pi u(r)+\iint_{\partial IR}\left[u(r)\frac{\partial}{\partial n}\left(\frac{1}{|r-r'|}\right)-\frac{1}{|r-r'|}\frac{\partial}{\partial n}u(r')\right]dS=0$$

Implying thereby

$$u(r)=\frac{1}{4\pi}\iint_{\partial IR}\left[\frac{1}{|r-r'|}\frac{\partial u}{\partial n}(r)-u(r)\frac{\partial}{\partial n}\left(\frac{1}{|r-r'|}\right)\right]dS \qquad (30)$$

Therefore , the value of at an point of the region IR is determined , if u and $\frac{\partial u}{\partial n}$ are known on the boundary , ∂IR . This lead to the conclusion that both the values of u and $\frac{\partial u}{\partial n}$ are required solution to obtain the solution of Dirichlet' s problem. But this is not so, as can be seen from the concept of the Green's function defined as follows : Let $H(r,r)$ be a function hu armonic in IR. Then the Green's function for the Dirichlet problem involving the Laplace operator is defined by are the G, the two point function for the Dirichlet problem involving the Laplace operator is defined by point function of position, as

$$G(r,r')=\frac{1}{|r-r'|}+H(r,r') \qquad (31)$$

where $H(r,r')$ satisfies the following

$$(i)\quad \left(\frac{\partial^2}{\partial x^2}+\frac{\partial^2}{\partial y^2}+\frac{\partial^2}{\partial z^2}\right)H(r,r')=0$$

$$(ii)\quad G=\frac{1}{|r-r'|}+H(r,r)=0 \quad on\, \partial IR \tag{33}$$

Thus the Green's function for the Dirichlet problem involving the Laplace operator is a function

$G(r,r)$ which satisfies the following properties:

$$(i)\quad \nabla^2 G(r,r)=\delta(r-r')\ in\ IR \tag{34}$$

$$(ii)\quad G(r,r)=0\ on\, \partial IR \tag{35}$$

(iii) G is symmetric, i.e.,

(36)

(iv) G is continuous , but $\partial G/\partial n$. has . dcontinuity at the point r. which is given by the equation

$$Lt_{\grave{o}\to 0}\iint_C \frac{\partial G}{\partial n}dS=1 \tag{37}$$

Following the procedure adopted in the derivation of Eq. (30), replacing $u'\ by\, G(r,r)$, we can show that

$$u(r)=-\frac{1}{4\pi}\iint_{\partial IR}\left[G(r,r)\frac{\partial u}{\partial n}(r)-u(r)\frac{\partial G}{\partial n}(r,r)\right]dS \tag{38}$$

From Eq. (31) and (33) we can see that $G=0\, on\, \partial IR$.. Thus the solution u. at an interior point is given by the relation

$$u(r)=-\frac{1}{4\pi}\iint_{\partial IR}u(r)\frac{\partial G}{\partial n}(r,r')dS \tag{39}$$

And therefore, the solution of the interior Dirichlet ' s problem is reduced to the determination of Green's function $G(r,r')$.

Green's function can be interpreted physically as follows: Let ∂IR be a grounded electrical conductor (boundary potential zero) and if a unit charge is located athe source point r, then G. is the sum of the potential at the point r'. due to the charge at the source point r in the free space and the potential due to the charges induced on ∂IR. Thus

$$G(r,r) = \frac{1}{|r,r'|} + H(r,r') \tag{40}$$

Hence, property $(i), viz,$. Eq. (34) , essentially mns that $\nabla^2 G = 0$ everywhere except at the source point (r) ..

Example. Consider a sphere with centre at the origin and radius 'a'. Apply the divergence theorem to the sphere and show that

$$\nabla^2\left(\frac{1}{r}\right) = -4\pi\delta(r)$$

..

$\delta(r)$ is a Dirac delta function.

Solution . Applying the divergence theorem to

$$grad\left(\frac{1}{r}\right) = \nabla\left(\frac{1}{r}\right).$$

t

$$\int\int\int_V \nabla.\nabla\left(\frac{1}{r}\right) dV = \int\int_S \nabla\left(\frac{1}{r}\right).\hat{n}\, dS$$

where $\hat{n}$ is an outward drawn normal . if $u = u(r,\theta,\phi)$, then

$$grad\, u = \hat{e}_r \frac{\partial u}{\partial r} + \hat{e}_\theta \frac{1}{r}\frac{\partial u}{\partial \theta} + e_\theta \frac{1}{r}\sin\theta \frac{\partial u}{\partial \phi}$$

$$\int\int_S grad\left(\frac{1}{r}\right).\hat{e}_r\, dS = \int\int_S \frac{1}{r^2} dS = -\frac{1}{a^2} \times 4\pi a^2 = -4\pi$$

Thus, we observe that $\nabla^2\left(\frac{1}{r}\right)$ has the following properties :

(i) it is undefined at the origin .

(ii) it vanishes if $r \neq 0$.

(iii) its integral over any sphere with centre at the origin is -4π. Hence, we conclude that

$$\iiint_V \nabla^2 \left(\frac{1}{r} \right) dV = -4\pi\delta(r)$$

Now, we shall prove the symmetric property through the following theorem.

8.2. THE EIGENFUNCTION METHOD

Let us consider the Dirichlet boundary value problem described by

$$\nabla^2 u = f \tag{41}$$

Valid in certain region IR, subject to the boundary condition

$$u = g \tag{42}$$

On ∂IR, the boundary of IR.

From the definition of Green's function which has been introduced in The Green's function must satisfy the relations

$$\nabla^2 G = \partial(x-\xi,\ y-\eta)\ in\, IR \tag{43}$$

$$G = 0\ ,\ on\, \partial IR \tag{44}$$

Now, consider the eigenvalue problem associated with the operator ∇^2 in the domain IR,

$$\nabla^2 \phi + \lambda\phi = 0\ i\, nIR \tag{45}$$

$$\phi = 0\ on\, \partial IR \tag{46}$$

Let λ_{mn} be the eigenvalue and ϕ_{mn} be the corresponding eigen functions. Suppose we give Fourier series expansion to $G\, and\, \delta$ in terms of the eigen functions ϕ_{mn} in the following form:

$$G(x,y;\xi,\eta)=\sum_m\sum_n a_{mn}(\xi,\eta)\phi_{mn}(x,y) \tag{47}$$

$$\delta(x-\xi,y-\eta)=\sum_m\sum_n b_{mn}(\xi,\eta)\phi_{mn}(x,y) \tag{48}$$

where

$$b_{mn}=\frac{1}{\|\phi_{mn}\|^2}\iint_{IR}\delta(x-\xi,y-\eta)\phi_{mn}(x,y)\,dx\,dy=\frac{\phi_{mn}(\xi,\eta)}{\phi_{mn}{}^2} \tag{49}$$

$$\phi_{mn}{}^2=\iint_{IR}\phi_{mn}^2\,dx\,dy$$

Now, substituting Eq. (47) and ((48) into Eq. (43) and (44) and noting that Eq. (45) has the form

$$\nabla^2\phi_{mn}+\lambda_{mn}\phi_{mn}=0 \tag{50}$$

We obtain

$$\nabla^2\sum_m\sum_n a_{mn}(\xi,\eta)\phi_{mn}(x,y)=\sum_m\sum_n b_{mn}(\xi,\eta)\phi_{mn}(x,y)$$

Using Eq. (49) and (50) , the above equation reduce to

$$-\sum_m\sum_n\lambda_{mn}a_{mn}(\xi,\eta)\phi_{mn}(x,y)=\sum_m\sum_n\frac{\phi_{mn}(\xi,\eta)\phi_{mn}(x,y)}{\|\phi_{mn}\|^2}.$$

From which

$$a_{mn}(\xi,\eta)=-\frac{\phi_{mn}(\xi,\eta)}{\lambda_{mn}\|\phi_{mn}\|^2} \tag{51}$$

Hence, Eq.(47) and (51) give the required Green's function for the Dirichlet problem, in the form

$$G(x,y;\xi,\eta)=-\frac{\sum_m\sum_n\phi_{mn}(\xi,\eta)\phi_{mn}(x,y)}{\lambda_{mn}\|\phi_{mn}\|^2} \tag{52}$$

Example. find the Green's function f $0\le y\le b,\left(\nabla^2+\lambda\right)u=0\ \ in\, IR$ or theirichlet problem on the rectangle IR $:0 \le x \le a,$ described by the PDE

$$(\nabla^2 + \lambda)u = 0 \quad in\ IR \tag{53}$$

And the $BC\ u = 0\, on\, \partial\, IR.$

Solution. The eigen functios of the given PDE can be obtained easily by using the variables separable method. Let us assume the solution o the given PDE in the form $u(x, y) = X(x)Y(y).$ Substituting into the given PDE, we obtain

$$\frac{X''}{X} = -\left(\frac{Y''}{Y} + \lambda\right) = -v \quad \text{(a separation constant)} \tag{54}$$

Since u is zero on the boundary $\partial\, IR\ , X$ satisfies

$$X'' + vX = 0, \quad X(0) = X(a) = 0 \tag{55}$$

Its solution, in general , is

$$X(x) = A\cos\sqrt{v}\,x + B\sin\sqrt{vx}$$

$X(0) = 0$ implies $A = 0. X(a) = 0$ gives $\sin\sqrt{va} = 0, implying\ \sqrt{v} = \frac{n\pi}{a}.$ The corresponding real valued eigen functions are $X_n = \sin\left(\frac{n\pi x}{a}\right), n = 1\ ,2\ ,\ldots while$ the eigen values are $v_n = n^2\frac{\pi^2}{a}\ , n = 1\ ,2\ ,\ldots$ Now, the factor Y satisfies

$$Y'' + (\lambda - v_n)Y = 0, \ Y(0) = Y(b) = 0$$

Following the above procedure, we can show at once that the eigen functions are given by

$$Y_m = \sin\frac{m\pi y}{b}, \quad m = 1, 2\ldots,$$

And the corresponding eigen values are

$$\lambda - v_n = \frac{m^2\pi^2}{b^2}, \quad m = 1, 2\ ,\ldots$$

Thus, we obtain the eigen functions to the given problem in the form

$$\phi_{mn}(x,y) = \sin\frac{m\pi x}{a}\sin\frac{n\pi y}{b}, m = 1,2,\ldots; n = 1,2,.. \qquad (56)$$

while the eigen values are given by

$$\lambda_{mn} - \frac{m^2\pi^2}{a^2} + \frac{n^2\pi^2}{b^2} + \pi^2\left(\frac{m^2}{a^2} + \frac{n^2}{b^2}\right) \qquad (57)$$

Computation of $\|\phi_{mn}\|$ gives

$$\|\phi_{mn}\|^2 = \int_0^a\int_0^b \sin^2\frac{m\pi x}{a}\sin^2\frac{n\pi x}{b}\,dx\,dy = \frac{ab}{4} \qquad (58)$$

Hence , the Green's function for the given Dirichlet problem can be obtained with the help of Eq. (56) and (57) as

$$G(x,y,\xi,\eta) = -\frac{4ab}{\pi^2}\sum_{m=1}^{\infty}\sum_{n=1}^{\infty}\frac{\sin\left(\frac{m\pi x}{a}\right)\sin\frac{n\pi x}{b}\sin\frac{m\pi\xi}{a}\sin\frac{n\pi\eta}{b}}{m^2b^2 + n^2a^2} \qquad (59)$$

8.3. SUMMARY AND DISCUSSION

Consider the differential equation

$$Lu(x) = f(x) \qquad (1)$$

Where L is an ordinary linear differential operator, $f(x)$ is a known function, while $u(x)$ is a unknown function. To solve the above differential equation, one method is to find the operator L^{-1} in the form of an integral operator with a kernel $G(x,\xi)$ such that

$$u(x) = L^{-1}f(x) = \int G(x,\xi)f(\xi)d\xi \qquad (2)$$

The kernel of this integral operator is called Green's function for the differential operator .

Definition . Let us consider an auxiliary function $\phi(x) of$ a real variable x which possess derivative of all orders and vanishes outside a finite interval. Such function are called test functions.

The idea of Green's function can be easily extended to higher dimension . Thus the Green's function technique can be applied, In principle , to find the solution of any linear non-homogenous partial differential equation . Green's function can be applied for the solutions of partial differential equations which occurs most frequently in mathematical physics.

Green's function for the Dirichlet problem involving the Laplace operator is a function

$G(r,r)$ which satisfies the following properties:

$(i)\quad \nabla^2 G(r,r) = \delta(r-r')\ in\ IR$

$(ii)\quad G(r,r) = 0\ on\, \partial IR$

(iii) G is symmetric, i.e., $G(r,r') = G(r',r)$

(iv) G is continuous , but $\partial G / \partial n$ has a discontinuity at the point r.

which is given by the equation $$Lt_{\grave{o} \to 0} \int_C\int \frac{\partial G}{\partial n} dS = 1$$.

REFERENCES

1. C.W. Gear Numerical Initial Value Problems in ordinary Differential Equation, Prentice-Hall, 1971.
2. E Hairer , S. P. Norsett and G Wanner , Solving Ordinary Differential Equations I, Springer-Verlag, 1987.
3. Differential equation with boundary value Problem, D . G Zill and M. R. Cullen, Brooks/Cole, USA
4. Paul Blanchard, Robert L. Devaney, Glenn R. Hall, *Differential Equations (Preliminary Edition)*,PWS Publishing, Boston, 1996.
5. William H. Boyce and Richard C. Diprima, *Elementary Differential Equations and Boundary Value Problems (6th Edition),* Wiley, New York, 1996.
6. C.H. Edwards, Jr., David E. Penney, *Elementary Dfifferential Equations with Applications (Third Edition),* Prentice-Hall, Englewood Cliffs, NJ, 1996.
7. R. Kent Nagle and Edward B. Saff, *Fundamentals of Differential Equations (Third Edition),* Addison Wesley, Reading, MA, 1993.

INDEX